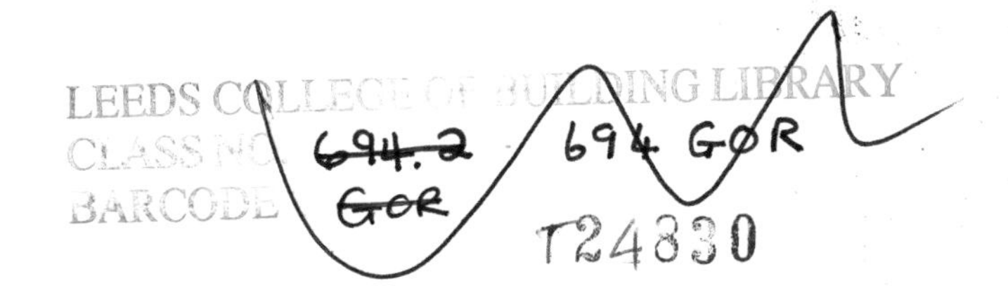

Manual of First- and Second-Fixing Carpentry

Manual of First- and Second-Fixing Carpentry

Les Goring FIOC, LCG, FTC

Associate of the Chartered Institute of Building
Former Senior Lecturer in Wood Trades at Hastings College of Arts and Technology

Drawings by the author

OXFORD AUCKLAND BOSTON JOHANNESBURG MELBOURNE NEW DELHI

Butterworth-Heinemann
Linacre House, Jordan Hill, Oxford OX2 8DP
225 Wildwood Avenue, Woburn, MA 01801-2041
A division of Reed Educational and Professional Publishing Ltd

A member of the Reed Elsevier plc group

First published in Great Britain by Arnold 1998
Reprinted by Butterworth-Heinemann 2000

British Library Cataloguing in Publication Data
A catalogue record for this book is available from the British Library

ISBN 0 340 67773 2

Cover Design: Terry Griffiths (artwork drawn by the author)

Typeset by J&L Composition Ltd, Filey, North Yorkshire
Printed and bound in Great Britain by The Bath Press, Bath

To Mary, Penny, Jon and Jenny, and other Gorings
of whom there are many

Contents

Preface

This book was written because there is a need for trade books with a strong practical bias, using a DIY step-by-step approach – and not because there was any desire to add yet another book to the long list of carpentry books already on the market. Although many of these do their authors credit, the bias is mainly from a technical viewpoint with wide general coverage and I believe there is a potential market for books (manuals) that deal with the sequence and techniques of performing the various, unmixed specialisms of the trade. Such is the aim of this book, to present a practical guide through the first two of these subjects, namely first-fixing and second-fixing carpentry.

The book, hopefully, will be of interest to many people, but it was written primarily for craft apprentices (a rare breed in this present-day economy), trainees and building students, established tradespeople, seeking to reinforce certain weak or sketchy areas in their knowledge and, as works of reference, the book may also be of value to vocational teachers, lecturers and instructors. Finally, the sequential, detailed treatment of the work should appeal to the keen DIY enthusiast.

Les Goring

Acknowledgements

The author would like to thank the following people and companies for their co-operation in supplying technical literature and other assistance:

CSC Forest Products Limited (OSB flooring panels); Gang-Nail Systems Ltd, a member company of the International Truss Plate Association; Kevin Hodger of Hastings College (inventor of the Roofmaster); Kieran Damani of House of Hastings (Thorcraft) Ltd, Hastings; Laybond Products Ltd; Mike Willard of Bexhill Locksmiths & Alarms; Peter Oldfield and Peter Shaw of Hastings College; Schauman (UK) Ltd (plywood flooring panels); South Coast Roofing Supplies Ltd, St Leonards-on-Sea; Steve Pearce of South Coastal Windows and Doors (uPVC) Ltd; Tony Fleming, Head of Construction Studies at Hastings College; and last, but not least, Tony Moon of A & M Design Consultants, St Leonards-on-Sea.

Abbreviations

bdg	boarding
bldg	building
BMA	bronze metal antique
BS	British Standards (Institution)
c/c	centre to centre (measurement)
℄	centre line
cpd	cupboard
DPC	damp-proof course
DPM	damp-proof membrane
dia, ø	diameter
EML	expanded metal lathing
ex.	prefix to material size before being worked
ffl	finished floor level
GL	ground level
hdb	hardboard
hwd	hardwood
ms	mild steel
O/A	over all (measurement)
par	planed all round
ppd	prepared (timber planed all round)
PVA	polyvinyl acetate (adhesive)
swd	softwood
T&G	tongued and grooved
TRADA	Timber Research and Development Association
vh	vertical height

Technical Data

STANDARD TIMBER-SIZES

Table 1 shows the basic sectional sizes for sawn softwood recommended by the British Standards to be available to the industry – it should be borne in mind that any non-standard requirement represents a special order and is likely to cost more.

Standard metric lengths are based on a 300 mm module, starting at 1.8 m and increasing by 0.3 m to 2.1 m, 2.4 m, 2.7 m and so on, up to 6.3 m. Non-standard lengths above this, usually from North American species, may be obtained up to about 7.2 m.

Table 1 Sawn structural timber sizes (From BS 336: 1995)

mm		75	100	125	150	175	200	225	250	275	300
	inches	3	4	5	6	7	8	9	10	11	12
22	$\frac{7}{8}$		✓	✓	✓	✓	✓	✓			
25	1	✓	✓	✓	✓	✓	✓	✓			
38	$1\frac{1}{2}$	✓	✓	✓	✓	✓	✓	✓			
47	$1\frac{7}{8}$	✓	✓	✓	✓	✓	✓	✓	✓		✓
63	$2\frac{1}{2}$		✓	✓	✓	✓	✓	✓			
75	3		✓	✓	✓	✓	✓	✓	✓	✓	✓
100	4		✓		✓		✓	✓	✓		✓
150	6				✓		✓				✓
250	10								✓		
300	12										✓

STANDARD DOOR-SIZES

Door frames and linings may vary in their opening sizes, but are normally made to accommodate standard doors. Again, it must be realized that special doors, made to fit non-standard frames or linings, would considerably increase the cost of the job. The locations given to the groups of standard door-sizes in Table 2, below, are only a guide, not a fixed rule.

Table 2 Standard doors and their usual location

	Metric			Imperial		
	Height (m)	Width (mm)	Thickness (mm)	Height (ft/in)	Width (ft/in)	Thickness (in)
Main entrance doors	2.134	914	45	7′0″	3′0″	$1\frac{3}{4}$
	2.083	864	45	6′10″	2′10″	$1\frac{3}{4}$
	2.032	813	45	6′8″	2′8″	$1\frac{3}{4}$
	1.981	838	45	6′6″	2′9″	$1\frac{3}{4}$
	1.981	762	45	6′6″	2′6″	$1\frac{3}{4}$
Room doors	2.032	813	45 or 35	6′8″	2′8″	$1\frac{3}{4}$ or $1\frac{3}{8}$
	1.981	762	45 or 35	6′6″	2′6″	$1\frac{3}{4}$ or $1\frac{3}{8}$
Bathroom/toilet doors	1.981	762 or 711	35	6′6″	2′6″ or 2′4″	$1\frac{3}{8}$
	1.981	686	35	6′6″	2′3″	$1\frac{3}{8}$
Cupboard doors	1.981	610	35	6′6″	2′0″	$1\frac{3}{8}$
	1.981	533	35	6′6″	1′9″	$1\frac{3}{8}$
	1.981	457	35	6′6″	1′6″	$1\frac{3}{8}$

Note: When door frames and doors are required to be fire-resisting, special criteria laid down by the British Standards Institution and The Building Regulations must be adhered to – a detailed reference to this is given in Chapter 6.

1

Reading Construction Drawings

1.1 INTRODUCTION

Construction drawings are necessary in most spheres of the building industry, as being the best means of conveying detailed and often complex information from the designer to all those concerned with the job. Building tradespeople, especially carpenters and joiners, should be familiar with the basic principles involved in understanding and reading drawings correctly. Mistakes on either side – in design or interpretation of the design – can be costly, as drawings form a legal part of the contract between architect/ client and builder. This applies even on small jobs, where only goodwill may suffer; for this reason, if a non-contractual drawing or sketch is supplied, it should be kept for a period of time after completion of the job, in case any queries should arise.

1.1.1 Retention of Drawings or Sketches

A simple sketch supplied by a client in good faith to a builder or joinery shop for the production of a replacement casement-type window, is shown in Figure 1.1(a). The client's mistake in measuring between plastered reveals is illustrated in Figure 1.1(b). Retention of the sketch protects the firm from the possibility of the client's wrongful accusation.

Another important rule is to study the whole drawing carefully and be reasonably familiar with the details before starting work.

The details given in this chapter are based on the recommendations laid down by the British Standards Institution, in their latest available publications entitled *Construction drawing practice*, BS 1192: Part 1: 1984, and BS 1192: Part 3: 1987. BS 1192: Part 5: 1990, which is not referred to here, is a guide for the structuring of computer graphic information.

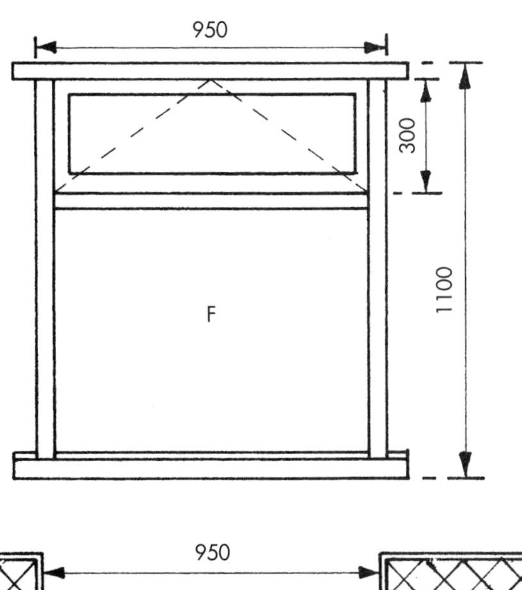

(a)

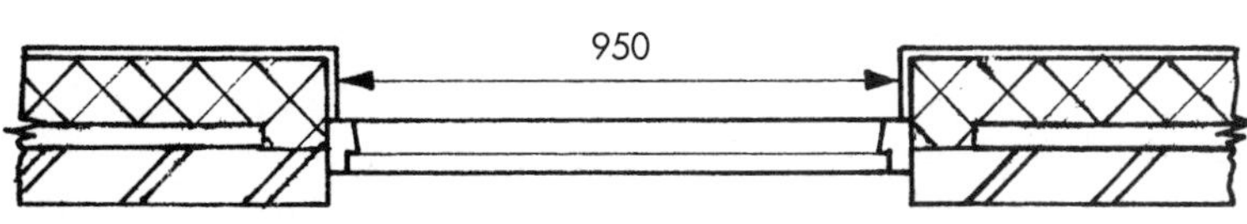

(b)

Figure 1.1(a) Client's sketch drawing.
(b) Horizontal section showing client's mistake

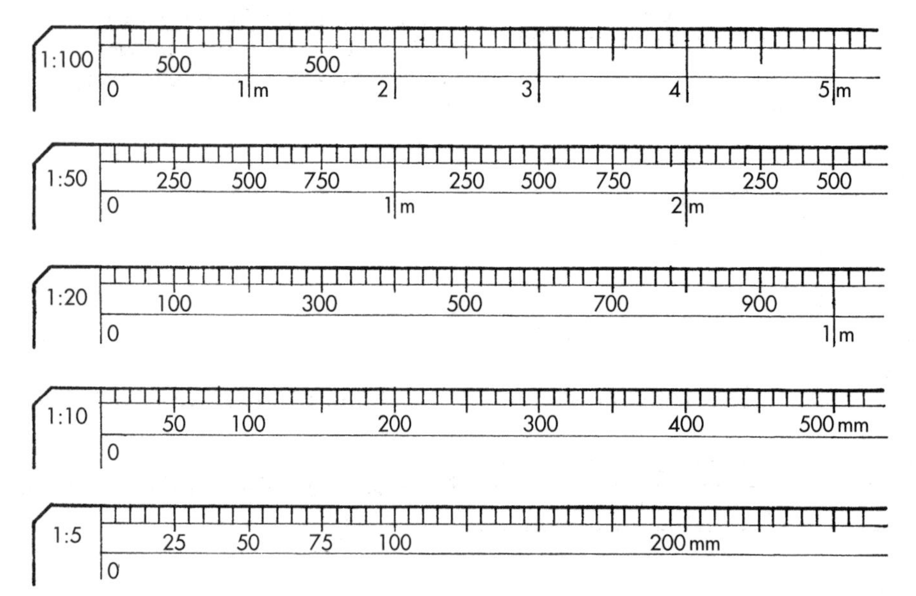

Figure 1.2 Common metric scales

1.1.2 Scales Used on Drawings

Parts of metric scale rules, graduated in millimetres, are illustrated in Figure 1.2. Each scale represents a ratio of given units (millimetres) to one unit (one millimetre). Common scales are 1:100, 1:50, 1:20, 1:10, 1:5 and 1:1 (full size). For example, scale 1:5 = one-fifth ($\frac{1}{5}$) full size, or 1 mm on the drawing equals 5 mm in reality.

Although a scale rule is useful when reading drawings, because of the dimensional instability of paper, preference should always be given to written dimensions found on the drawing.

1.1.3 Correct Expressions of Dimensions

The abbreviated expression, or unit symbol, for metres is a small letter m, and letters mm for millimetres. Symbols are not finalized by a full stop and do not use a letter 's' for the plural. Confusion occurs when, for example, $3\frac{1}{2}$ metres is written as 3.500 mm – which means, by virtue of the decimal point in relation to the unit symbol, $3\frac{1}{2}$ millimetres! To express $3\frac{1}{2}$ metres, it should have been written as 3500 mm, 3.5 m, 3.50 m, or 3.500 m. Either one symbol or the other should be used throughout on drawings; they should not be mixed. Normally, whole numbers should indicate millimetres, and decimalized numbers, to three places of decimals, should indicate metres.

1.1.4 Sequence of Dimensioning

The recommended dimensioning sequence is illustrated in Figure 1.3. Length should always be given first, width second and thickness third, for example 900 × 200 × 25 mm. However, if a different sequence is used, it should be consistent throughout.

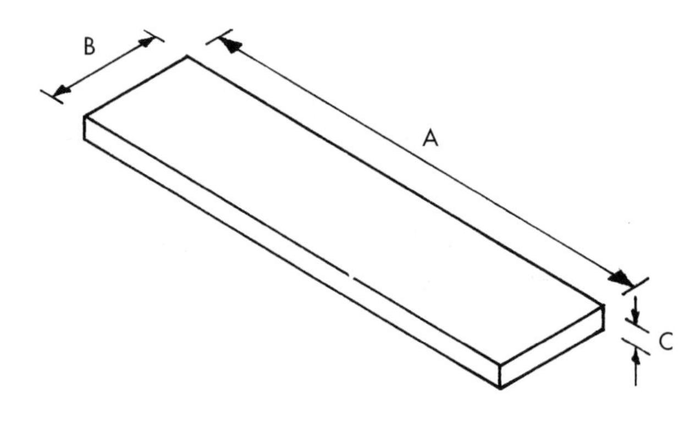

Figure 1.3 Dimensioning sequence = *A* × *B* × *C*

1.1.5 Dimension Lines and Figures

A dimension line with open arrowheads for basic/modular (unfinished) distances, spaces or components is indicated in Figure 1.4(a). Figure 1.4(b) indicates the more common, preferred dimension lines, with solid arrowheads, for general use in finished work sizes.

All dimension figures should be written above and along the line; figures on vertical lines should be written, as shown, to be read from the right-hand side.

1.1.6 Special-purpose Lines

Figure 1.5: Section lines seen on drawings indicate imaginary cutting planes, at a particular point through the drawn object, to be exposed to view. The view is called the section and is lettered A–A, B–B and so on, according to the number of sections to be exposed. It is important to bear in mind that the arrows indicate the direction of view to be seen on a separate section drawing.

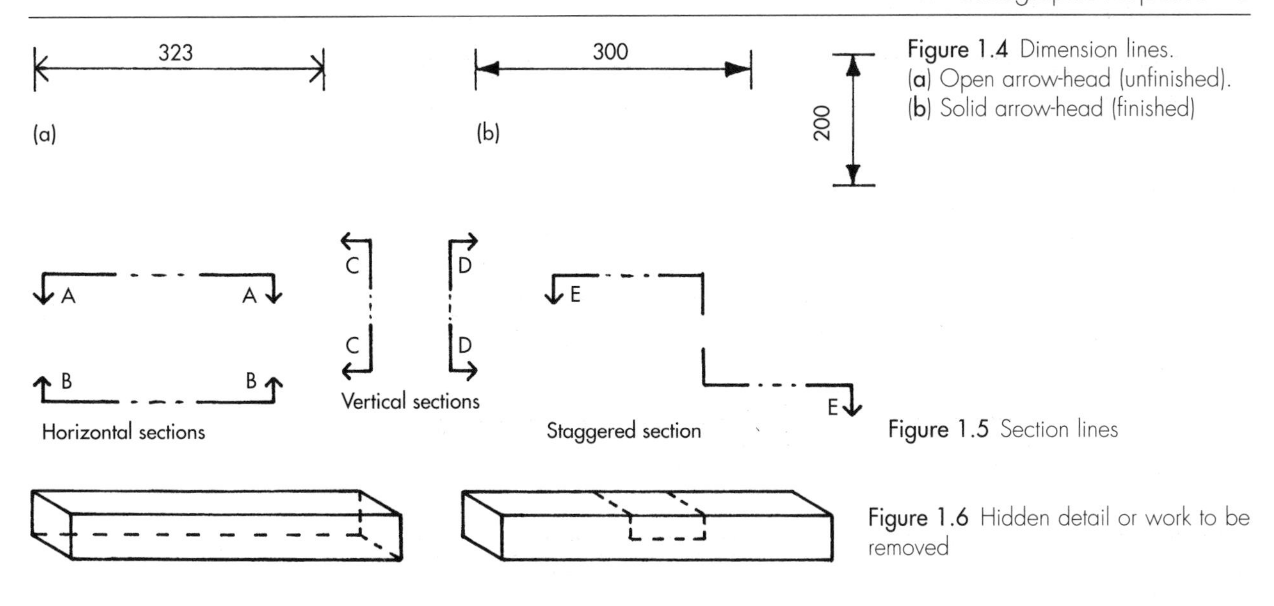

Figure 1.4 Dimension lines.
(**a**) Open arrow-head (unfinished).
(**b**) Solid arrow-head (finished)

Figure 1.5 Section lines

Figure 1.6 Hidden detail or work to be removed

Figure 1.6: Hidden detail or work to be removed, is indicated by a broken line.

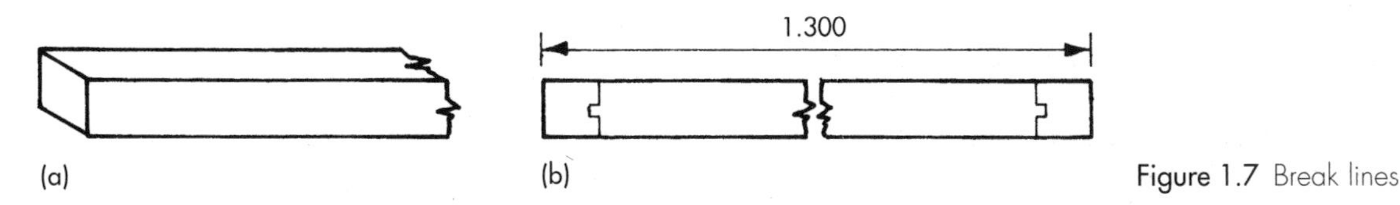

Figure 1.7 Break lines

Figure 1.7(a): End break-lines (zig-zag pattern) indicate that the object is not fully drawn.

Figure 1.7(b): Central break-lines (zig-zag pattern) indicate that the object is not drawn to scale in length.

Figure 1.8 Centre or axial line

Figure 1.8: Centre or axial lines are indicated by a thin dot-dash chain.

1.2 ORTHOGRAPHIC PROJECTION

1.2.1 Introduction

Orthography is a Latin/Greek-derived word meaning 'correct spelling' or 'writing'. In technical drawing it is used to mean 'correct drawing'; orthographic projection, therefore, refers to a conventional drawing method used to display the three-dimensional views (length, width and height) of objects or arrangements as they will be seen on one plane – namely the drawing surface.

The recommended methods are known as *first-angle* (or European) *projection* for construction drawings, and *third-angle* (or American) *projection* for engineering drawings.

1.2.2 First-angle Projection

The box in Figure 1.9(a) is used here as a means of explaining first-angle projection (F.A.P.). If you can imagine the object shown in Figure 1.9(b) to be suspended in the box, with enough room left for you to walk around it, then by looking squarely at the object from all sides and from above, the views seen would be the ones shown on the surfaces in the background.

1.2.3 Opening the Topless Box

In Figure 1.9(c) the topless box is opened out to give the views as you saw them in the box and as they should be laid out on a drawing. Figure 1.9(d) shows the BS symbol recommended for display on drawings to indicate that first-angle projection (F.A.P.) has been used.

Note that when views are separated onto different drawings, becoming unrelated orthographically, descriptive captions should be used such as 'plan', 'front elevation', 'side elevation', etc.

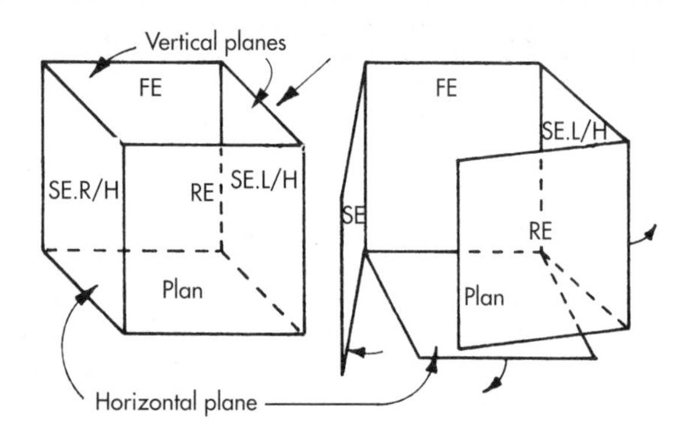

Figure 1.9(a) Theory of first-angle orthographic projection (SE = side elevation, FE = front elevation, RE = rear elevation, R/H = right-hand side, L/H = left-hand side)

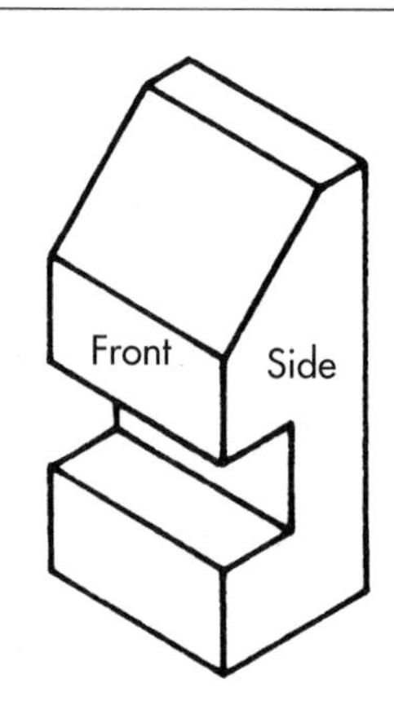

Figure 1.9(b) Example object

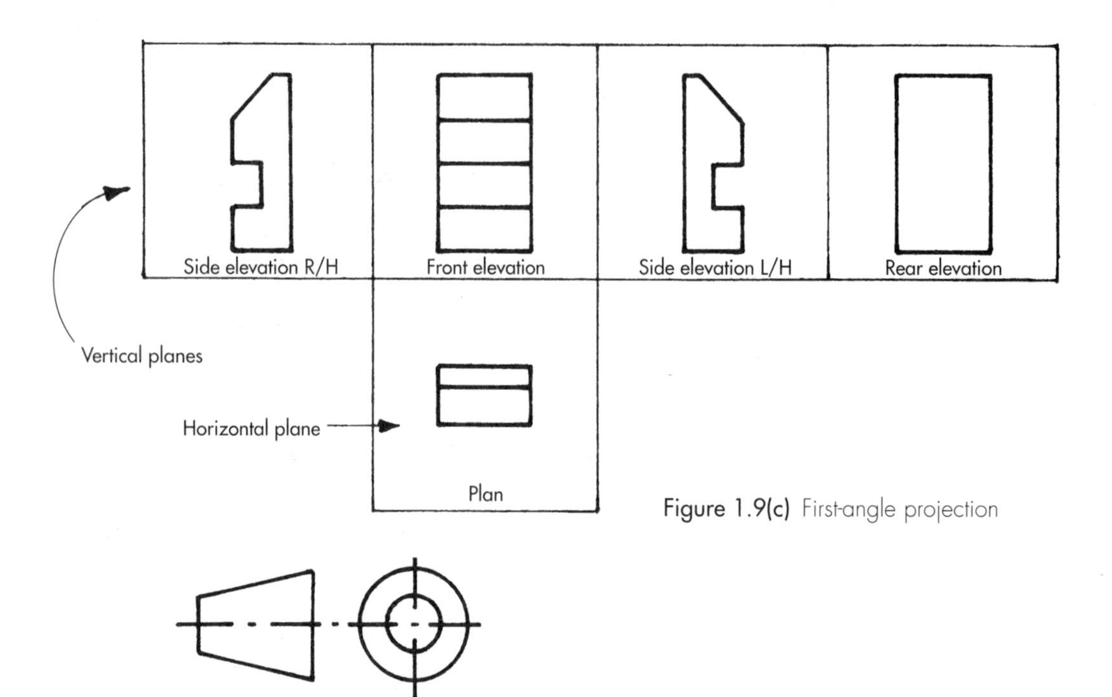

Figure 1.9(c) First-angle projection

Figure 1.9(d) F.A.P. symbol

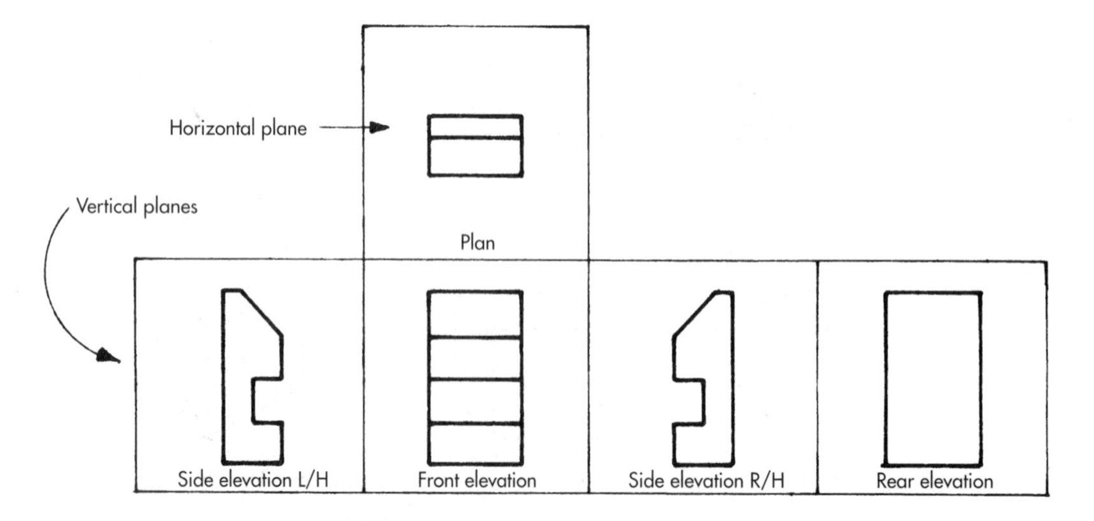

Figure 1.9(e) Third-angle projection

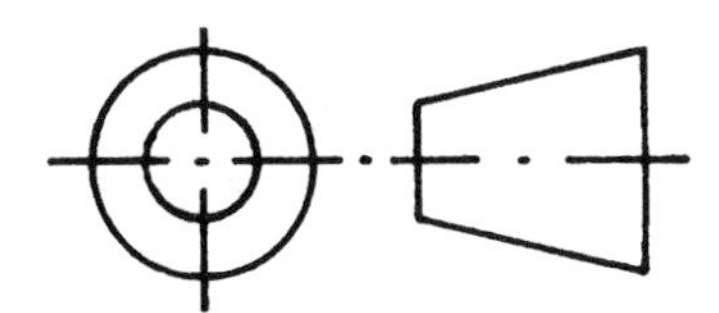

Figure 1.9(f) T.A.P. symbol

1.2.4 Third-angle Projection

This is shown in Figure 1.9(e) for comparison only. This time the box has a top instead of a bottom; the views from the front and rear would be shown on the surface in the background, as before, but the views seen on the sides would be turned around and seen on the surfaces in the foreground; the view from above (plan) would be turned and seen on the surface above. Figure 1.9(f) shows the BS symbol for third-angle projection (T.A.P.).

1.2.5 Pictorial Projections

Figure 1.10: Another form of orthographic projection produces what is known as pictorial projections, which preserve the three-dimensional view of the object. Such views have a limited value in the make-up of actual working drawings, but serve well graphically to illustrate technical notes and explanations.

1.2.6 Isometric Projection

This is probably the most popular pictorial projection used, because of the balanced, three-dimensional effect. Isometric projections consist of vertical lines and base lines drawn at 30°, as shown in Figure 1.10(a). The length, width and height of an object thus drawn are to true scale, expressed as the ratio 1:1:1.

1.3 OBLIQUE PROJECTIONS

There are three variations of oblique projections.

1.3.1 Cavalier Projection

Shown in Figure 1.10(b) with front (F) drawn true to shape, and side (S) elevations and plan (P) drawn at 45°, to a ratio of 1: 1: 1. Drawn true to scale by this method, the object tends to look mis-shapen.

1.3.2 Cabinet Projection

Shown in Figure 1.10(c), this is similar to cavalier except that the side and plan projections are only drawn to half scale, i.e. to a ratio of 1:1:$\frac{1}{2}$, making the object look more natural.

1.3.3 Planometric Projection

Shown in Figure 1.10(d), this has the plan drawn true to shape, instead of the front view. This comprises verticals, lines on the front at 30° and lines on the side elevation at 60°. It is often wrongly referred to as axonometric.

1.3.4 Perspective Projections

Figure 1.11: Parallel perspective, shown in Figure 1.11(a) refers to objects drawn to diminish in depth to a vanishing point.

Angular perspective, shown in Figure 1.11(b) refers to an object whose elevations are drawn to diminish to two vanishing points. This is of no value in pure technical drawing.

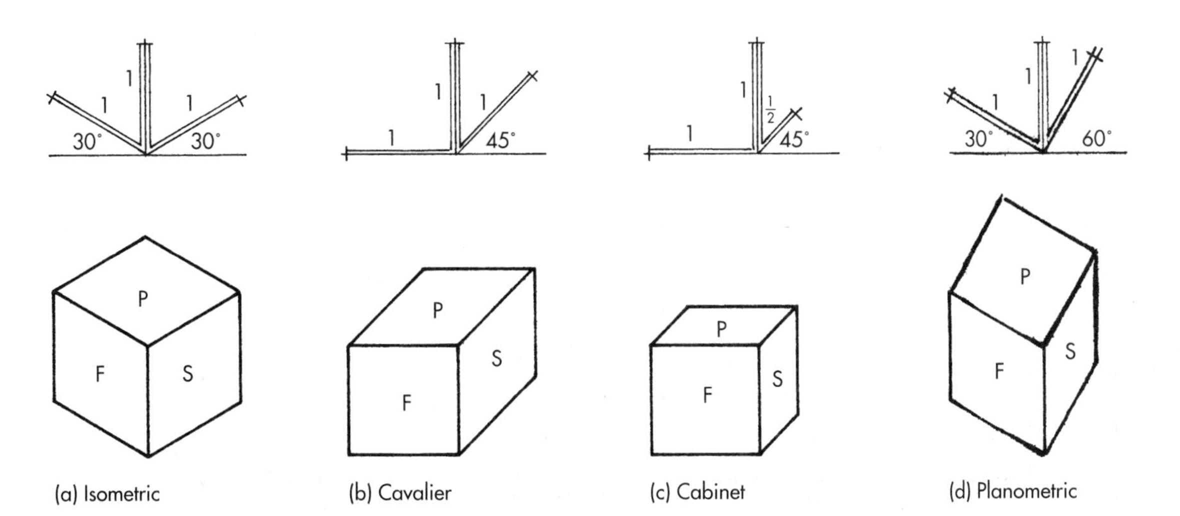

Figure 1.10 Pictorial projections (F = front, P = plan, S = side elevation)

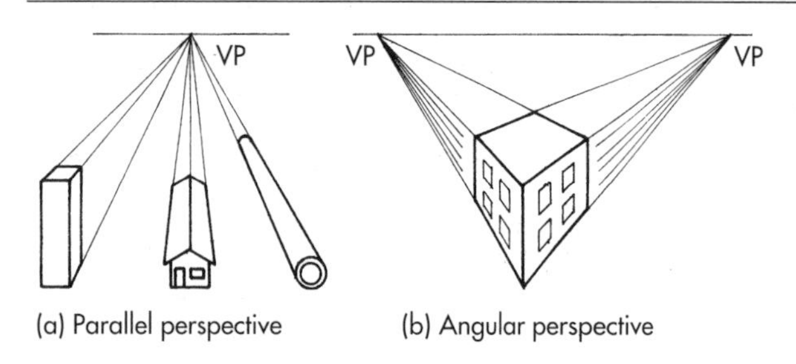

Figure 1.11 Perspective projections (VP = vanishing point)

1.3.5 Graphical Symbols and Representation

Figure 1.12: Illustrated here are a selection of graphical symbols and representations used on building drawings. *Figure 1.13*: On more detailed drawings, various materials and elements are identified by such sectional representation as shown here.

To help reduce the amount of written information on working drawings, abbreviations are often used. A selection are shown here:

BMA = bronze metal antique
DPC = damp-proof course
DPM = damp-proof membrane

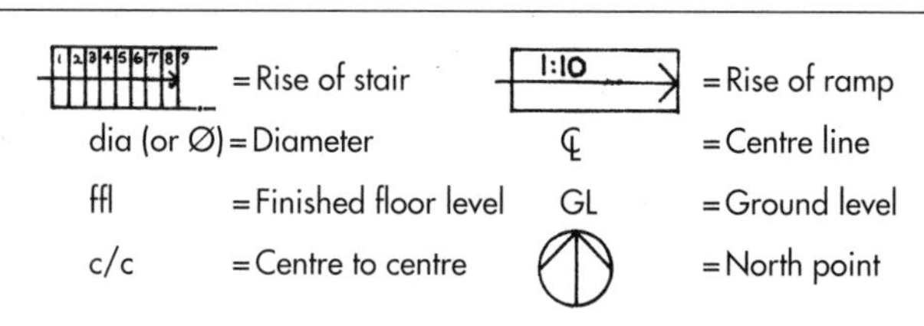

Figure 1.12 Graphical symbols and representations

EML = expanded metal lathing
par = planed all round
PVA = polyvinyl acetate
T&G = tongue and groove
bdg = boarding
bldg = building
cpd = cupboard
hdb = hardboard
hwd = hardwood
ms = mild steel
swd = softwood

1.3.6 Window Indication

Figure 1.14: Windows shown on elevational drawings usually display indications as to whether a window is fixed (meaning without any opening window or vent) or

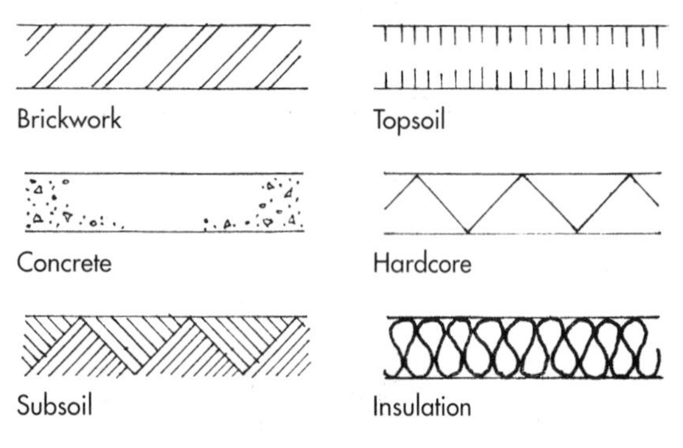

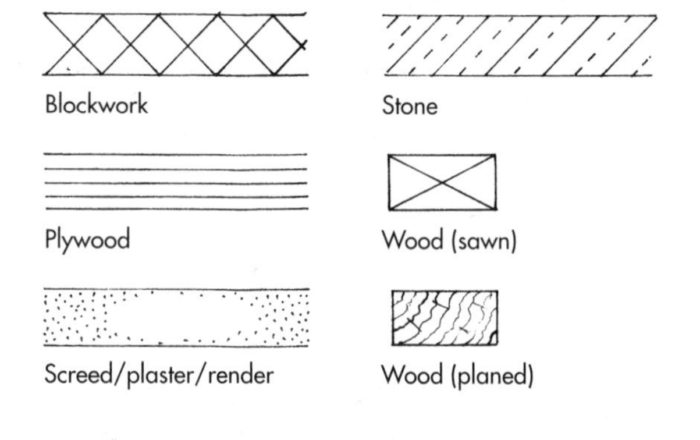

Figure 1.13 Sectional representation of materials

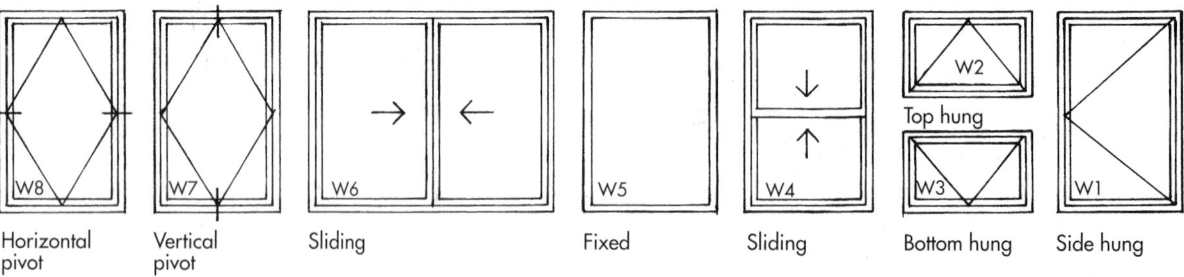
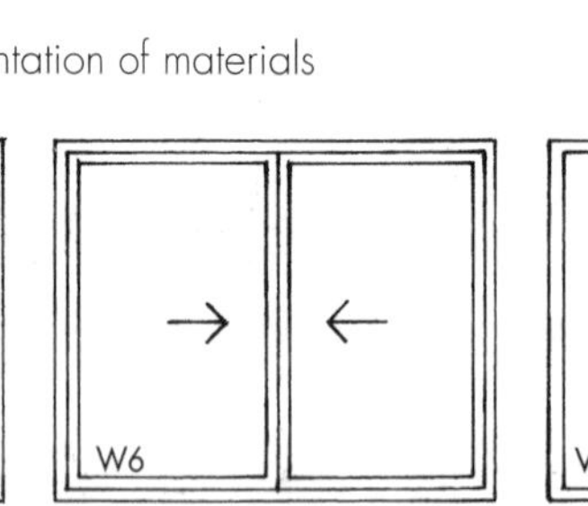
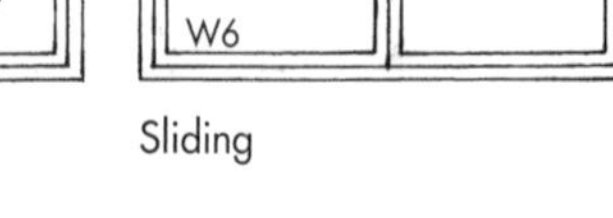

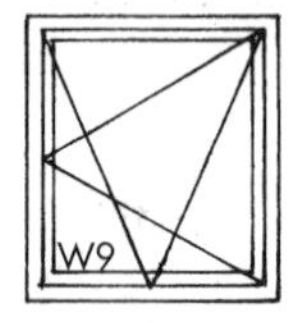

Figure 1.14 Opening/fixed window indication – numbered clockwise round the exterior of the building

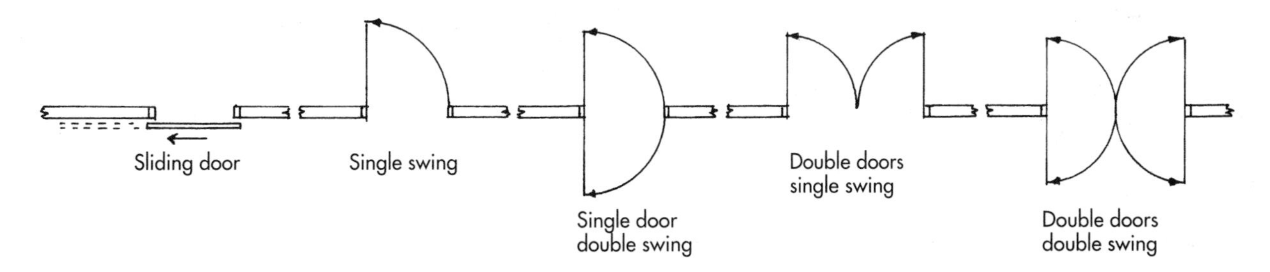

Figure 1.15 Plan view of door indication

opening (meaning that the window is to open in a particular way, according to the BS indication drawn on the glass area).

1.3.7 Door Indication

Figures 1.15 and 1.16: Doors shown on plan-view drawings are usually shown as a single line with an arrowed arc indicating their opening-direction, as illustrated. Alternatively, the 90° arrowed arc may be replaced by a 45° diagonal line, from the door-jamb's edge to the door's leading edge. Figure 1.16 is the indication for revolving doors.

1.3.8 Block Plans

Figure 1.17: Block plans shown on construction drawings, usually taken from Ordnance Survey maps, are to identify the site (e.g. No. 1 Woodman Road, as illustrated) and to locate the outline of the building in relation to its surroundings.

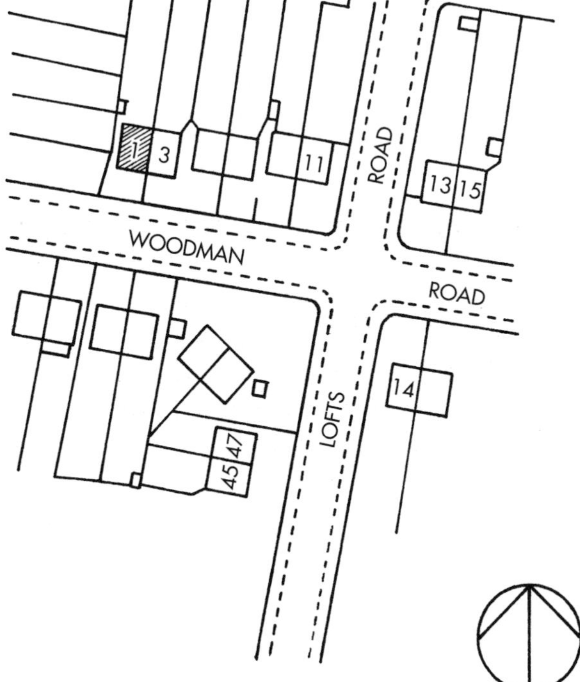

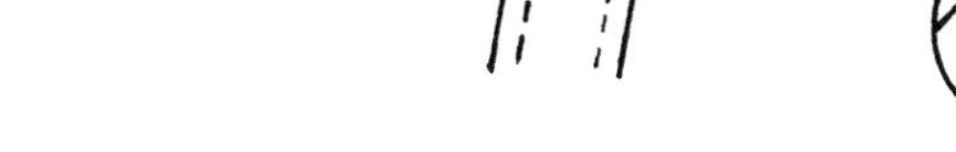

Figure 1.17 Block plan (scale 1:1250)

Figure 1.16 Revolving doors

1.3.9 Site Plans

Figure 1.18: Site plans locate the position of buildings in relation to setting-out points, means of access, and the general layout of the site; they also give information on services and drainage, etc.

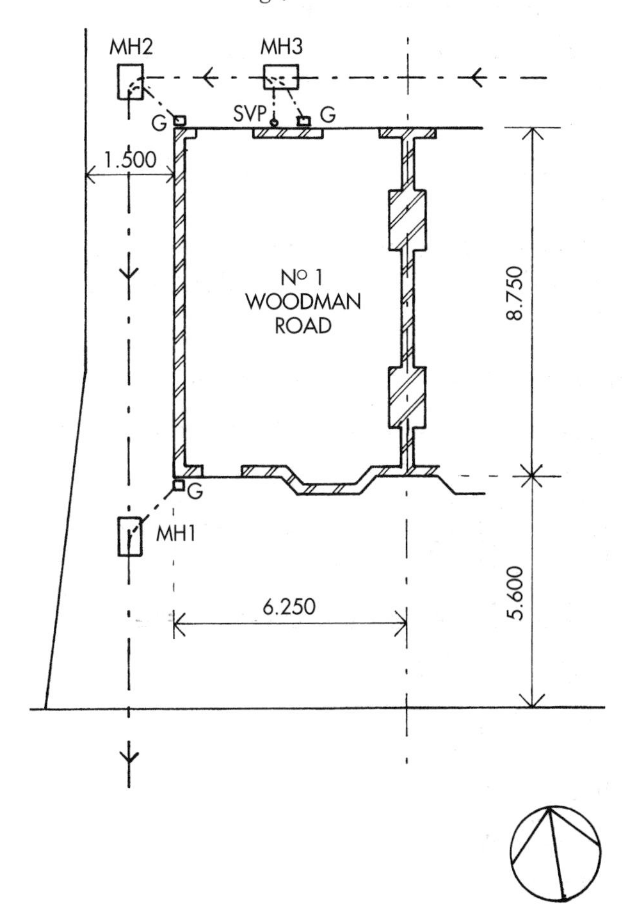

Figure 1.18 Site plan (scale 1:200)

1.3.10 Location Drawings

These are usually drawn to a scale of 1:50 and are used to portray the basic, general construction of buildings. Other, more detailed, drawings cover all other aspects.

2

Tools Required: their Care and Proper Use

2.1 INTRODUCTION

The whole range of tools for first- and second-fixing carpentry is quite extensive and includes power and battery-operated (cordless) tools in the essential list. The following details, therefore, do not cover all the tools that you *could* have, rather all the tools that you *should* have.

2.2 MARKING AND MEASURING

2.2.1 Pencils

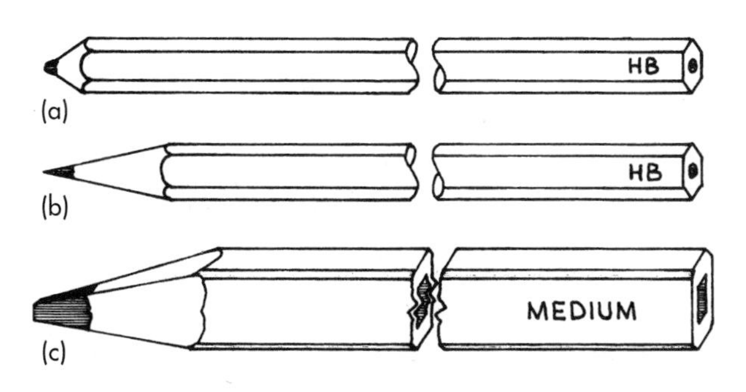

Figure 2.1 Pencils

Figure 2.1: These must be kept sharp for accurate marking. Although sharpening to a pin-point is quite common, for more accurate marking and a longer-lasting point, they can easily be sharpened to a chisel-point, similar to the sharpening illustrated in Figure 2.1(c). Stumpy sharpening (Figure 2.1(a)) should be avoided; sharpen at an angle of about 10° (Figure 2.1(b)). Use grade HB for soft, black lines on unplaned timber and – if you prefer – grade 2H on planed timber. Choose a hexagon shape for better grip and anti-roll action, and a bright colour to detect easily when left lying amongst shavings. Oval or rectangular-shaped carpenters' pencils (Figure 2.1(c)), of a soft or medium grade lead, are better for heavy work such as roofing, joisting, marking unplaned timber, etc. – although one disadvantage is that they cannot be put behind the ear for quick availability, as is the usual practice with ordinary pencils.

2.2.2 Tape Rule

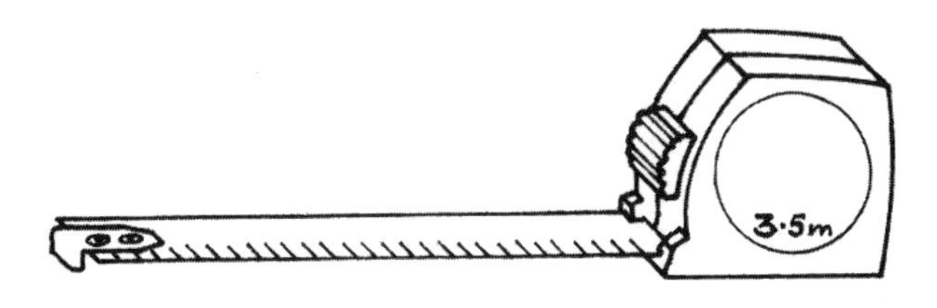

Figure 2.2 Tape rule

Figure 2.2: This is essential for fast, efficient measuring on site work. For this type of carrying-rule, sizes vary between 2 m and 10 m. Models with lockable, power-return blades and belt clips, one of 3.5 m and one of 8 m length are recommended. When retracting these power-return rules, slow down the last part of the blade with the sliding lock to avoid damaging the riveted metal hook at the end or nipping your fingers. To reduce the risk of kinking the sprung-steel blade, do not leave extended after use.

2.2.3 Folding Rule

Figure 2.3: This rule is optional, having been superseded by the tape rule. However, it is sometimes preferred for measuring/marking small sizes. Its unfolded length is 1 m and it is 250 mm folded. It is marked in single millimetre, 5 mm, 10 mm (centimetre) and 100 mm (decimetre) graduations. It is still available in boxwood or – better still – in virtually unbreakable white or grey engineering plastic with tipped ends, permanently tensioned joints and bevelled edges for easier, more accurate reading/marking when the rule is laid out flat. These rules, although tough,

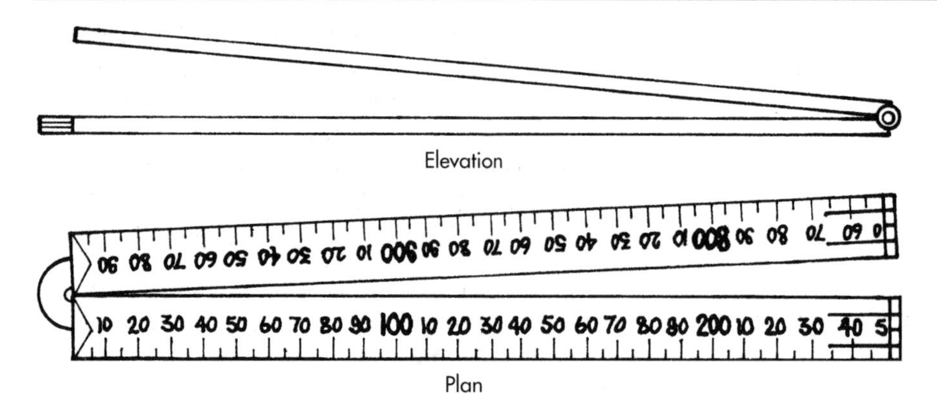

Figure 2.3 Folding rule

should always be folded after use to avoid possible hinge damage, especially from underfoot if left on the floor.

2.2.4 Chalk Line Reel

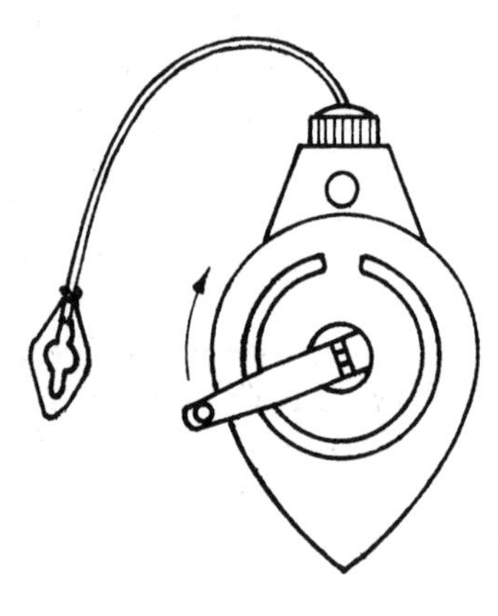

Figure 2.4 Chalk line reel

Figure 2.4: This tool is very useful for marking straight lines by holding the line taut between two extremes, lifting at any mid point with finger and thumb and flicking onto the surface to leave a straight chalk line. The line is retractable by winding a hinged handle housed in the die-cast aluminium case, that folds back after use. Powdered chalk is available in colours of red, white, blue, orange, green and yellow. The reel has a subsidiary use as a plumb bob – but it is not ideal for this purpose.

2.2.5 Spirit Level

Figure 2.5: This is an essential tool for plumbing and levelling operations. Sizes vary between 250 mm and 2 m long, but a level of 800 mm length is recommended for general usage and easy accommodation in the tool kit. Heavy-duty levels of aluminium alloy die-cast, or lightweight models of extruded aluminium, with clear, tough plexiglass vials (containing spirit and trapped bubble), epoxy-bonded into their housings to give lasting accuracy, are the most popular levels nowadays.

Figure 2.5 Spirit level

Even though these levels are shockproof, they should not be treated roughly, as body damage can affect accuracy. After use, avoid leaving levels lying on the floor or ground to be trodden on, especially when partly suspended, resting on other objects such as scrap timber. When checking or setting up a level or plumb position, be sure that the bubble is equally settled between the lines on the vial for accurate readings.

2.2.6 Straightedges

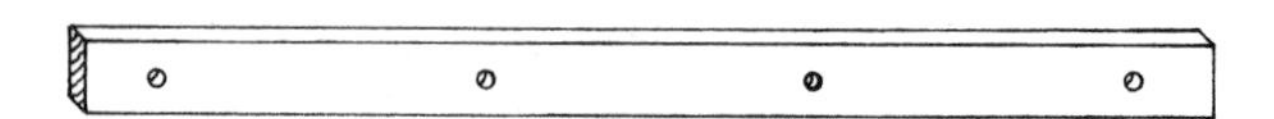

Figure 2.6 Straightedge

Figure 2.6: In the absence of very long spirit levels, straightedges may be used. These are parallel, straight softwood boards of various lengths, for setting out or (with a smaller spirit level held against one edge) for plumbing and levelling. If transferring a datum point in excess of the straightedge length, the risk of a cumulative error is reduced by reversing the straightedge end-for-end at each move. Traditionally, large holes were drilled along the centre axis to prevent the board from being claimed for other building uses.

2.2.7 Plumb Bob

Figure 2.7: There is still a use, however limited, for these traditional plumbing devices. The short one in the illustration is made of steel, blacked to inhibit rust; the other, which is a heavier type of $4\frac{1}{2}$ounces, has a red plastic body filled with steel shot and a 3 m length of nylon line. They should, as illustrated, always be suspended away from the surface being checked and measured for equal readings at top and bottom. If in possession of a tarnishable steel plumb bob, wipe with an oily rage occasionally. Although commonly called plumb bobs, if they are pointed on the underside, they are really *centre bobs*. The point is very useful for plumbing to a mark on the floor.

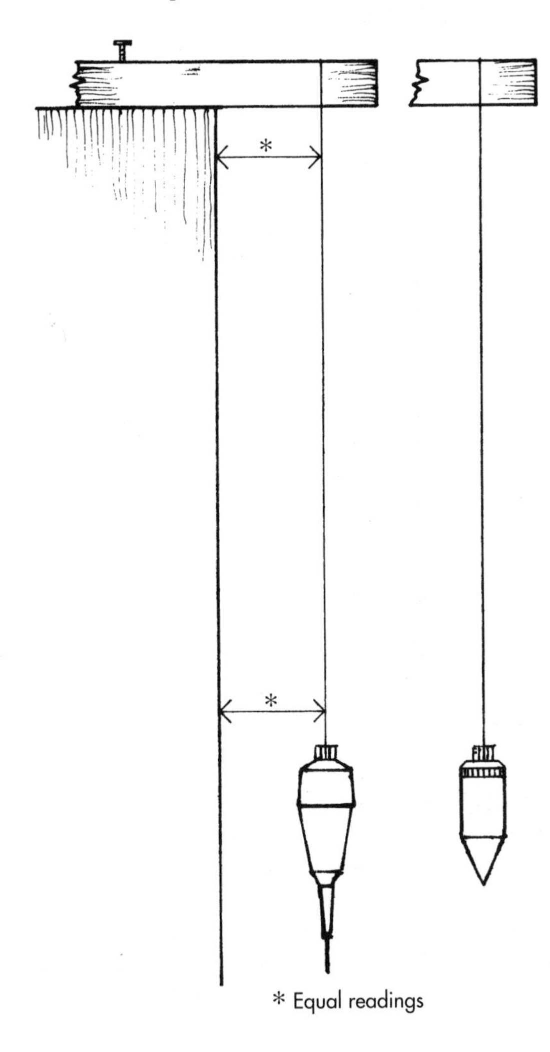

Figure 2.7 Plumb bobs

2.2.8 Combination Mitre Square

Figure 2.8: This tool was adopted from the engineering trades and is now widely favoured on site work for the following reasons: it is robust (the better, more expensive type) and withstands normal site abuse; it can be used for testing or marking narrow rebated edges, as shown, or for testing or marking angles of 90°, 45°, and 135°; the blade can be adjusted from the stock to a set measurement and, with the aid of a pencil, used as a pencil gauge. This facility is useful for marking sawn boards, for example, as opposed to using a marking gauge that may not be clearly visible on a rough sawn surface. The square's stock has an inset spirit vial and can be used for plumbing and levelling – although this is not the tool's best feature. The blade locking-nut should always be tightened after each adjustment, otherwise inaccuracies in the angle between the stock and blade will readily occur, causing errors in marking or testing. Finally, a scribing pin is usually located in the end of the stock. This is a feature carried over from the square's originally intended use as an engineering tool and is used for marking lines on metal.

2.2.9 Sliding Bevel

Figure 2.9: This is basically a slotted blued-and-hardened steel blade, sliding and rotating from a hardwood (rosewood) or plastic stock. The plastic is impact-resistant. The blade is tightened by a screw or

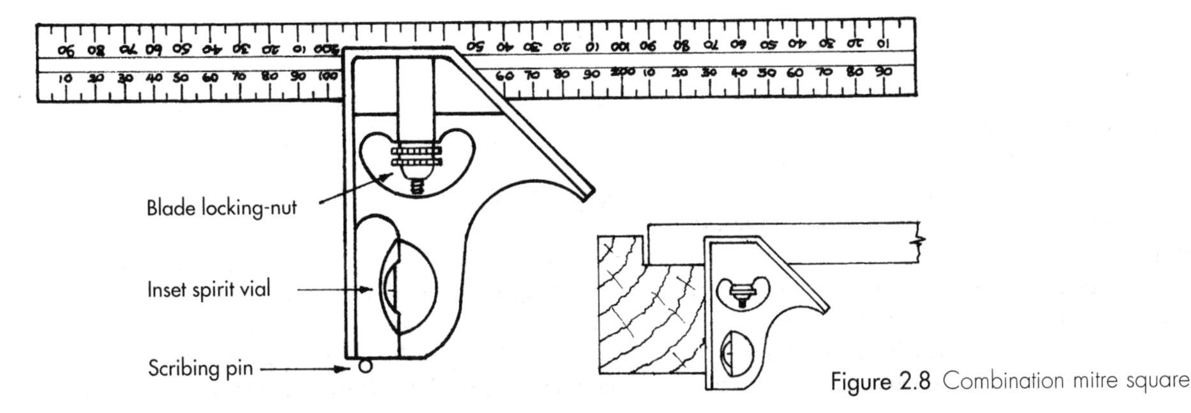

Figure 2.8 Combination mitre square

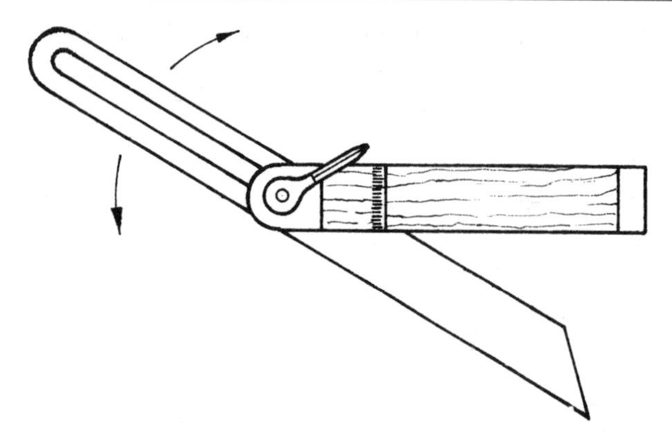

Figure 2.9 Sliding bevel

a half wing nut. The latter is best for ease and speed, being manually operated. This is an essential tool for angular work, especially roofing if using the *Roofing Ready Reckoner* method. For protection against damage, always return the blade to the stock-housing after use.

2.2.10 Steel Roofing Square/Metric Rafter Square

Figure 2.10: This tool, originally called a *steel square* or *steel roofing square*, is now metricated and referred to as a *metric rafter square*. Its size is 610 × 450 mm. The long side is called the *blade*, the short side the *tongue*. This traditional tool, primarily for developing roofing bevels and lengths (covered in the chapter on roofing), has a good subsidiary use as a try square, for marking and testing certain right angles with greater opposite sides than the combination square or normal try square can deal with effectively.

2.2.11 Roofmaster Square

Figure 2.11: This revolutionary roofing square, as mentioned in the chapter on roofing, is a recently marketed tool, well worth considering as an alternative

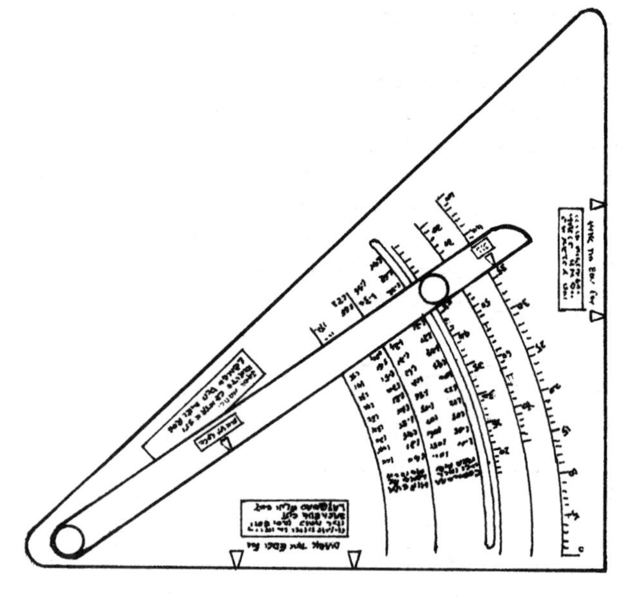

Figure 2.11 The Roofmaster

to a traditional type roofing square. It is a compact, precision instrument, measuring 335 mm on each right-angled side and is of anodized aluminium construction with easy-to-read laser-etched markings. It gives angle cuts for all roof members and the lengths of rafters without the need for separate tables. It is designed for easy use, whereby only the roof pitch angle is required to obtain all other angles and lengths. (Readers wishing to obtain a Roofmaster should contact OMI Cowley, Precision Tool & Instrument Company, Brett Drive, Bexhill-on-Sea, TN40 2JP, UK, Tel: (01424) 732674, Fax: (01424) 732966 for further information.)

2.3 HANDSAWS

2.3.1 Introduction

There is still a use, however limited, for conventional saws and they are still being sold in tool shops. This

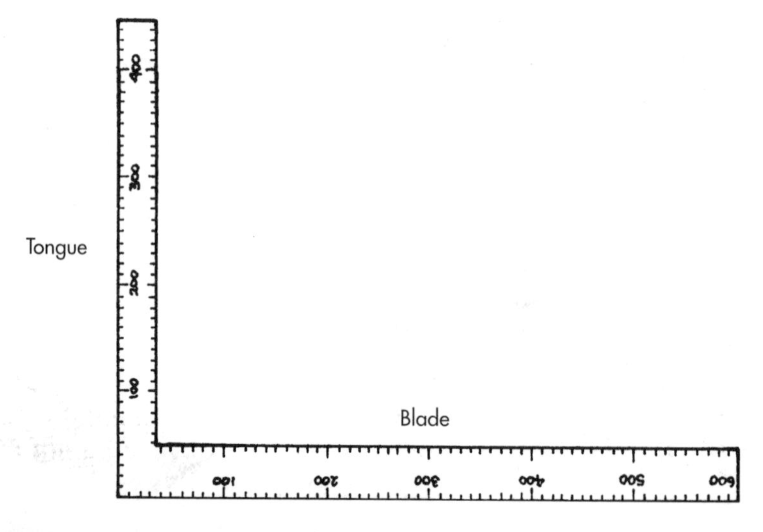

Figure 2.10 Steel roofing square

demands their inclusion here, but it must be said that they have been mostly superseded by modern hardpoint, throwaway saws. This is no doubt because they are cheap to buy, they retain their sharpness for a long time and, when blunt, they can affordably be replaced without the inconvenience – assuming a person has the skill – of resharpening. Starting with conventional saws in order of priority, you should have a *crosscut saw*, a *panel saw*, a *tenon saw* and, in the absence of a portable powered circular saw, a *rip-saw*.

2.3.2 Crosscut Saw

Figure 2.12: As the name implies, this is for cutting timber across the grain. Blade lengths and points per 25 mm (pp25) or ppi (points per inch) vary, but 660 mm (26 in) length and 7 or 8 pp25 are recommended. All handsaw teeth on conventional type saws contain 60° angular shapes leaning, by varying degrees, towards the toe of the saw. The angle of lean, relative to the front cutting edge of the teeth, is called the *pitch*. When sharpening saws (covered in a separate chapter), it helps to know the required pitch. For crosscut saws the pitch should be 80°. When crosscutting, the saw, as illustrated, should be at an approximate angle of 45° to the timber.

2.3.3 Panel Saw

Figure 2.13: This is a saw for fine crosscutting, which is particularly useful for cutting sheet material such as plywood or hardboard. A blade length of 560 mm (22 in), 10 pp25 and 75° pitch is recommended. When cutting thin manufactured boards (plywood, hardboard, etc.), the saw should be used at a low angle of about 15–25°.

2.3.4 Tenon Saw

Figure 2.14: This saw, because of its brass or steel back, is sometimes referred to as a *back saw*. Technically thought of as a general purpose bench saw for fine cutting, it is however widely used on site for certain second-fixing operations involving fine crosscutting of small sections. The brass-back type, as well as keeping the thin blade rigid, adds additional weight to the saw for easy use. The two most popular blade lengths, professionally, are 300 mm (12 in) and 350 mm (14 in). The 250 mm (10 in) saw is less efficient because of its short stroke. On different makes of saw, the teeth size varies between 13 and 15 pp25. For resharpening purposes, although dependent upon your skill and eyesight, 13 pp25 is recommended, with a pitch of 75°.

2.3.5 Rip Saw

Figure 2.15: This saw is used for cutting along or with the grain and is not in great demand nowadays because of the common use of machinery and portable powered circular saws on site. However, it is not obsolete and can be very useful in the absence of power. A blade length of 660 mm (26 in), 5 or 6 pp25 and a pitch of 87° is recommended. When ripping (cutting along or with the grain), the saw should be used at a steep angle of about 60–70° to the timber. Because of the square-edged teeth

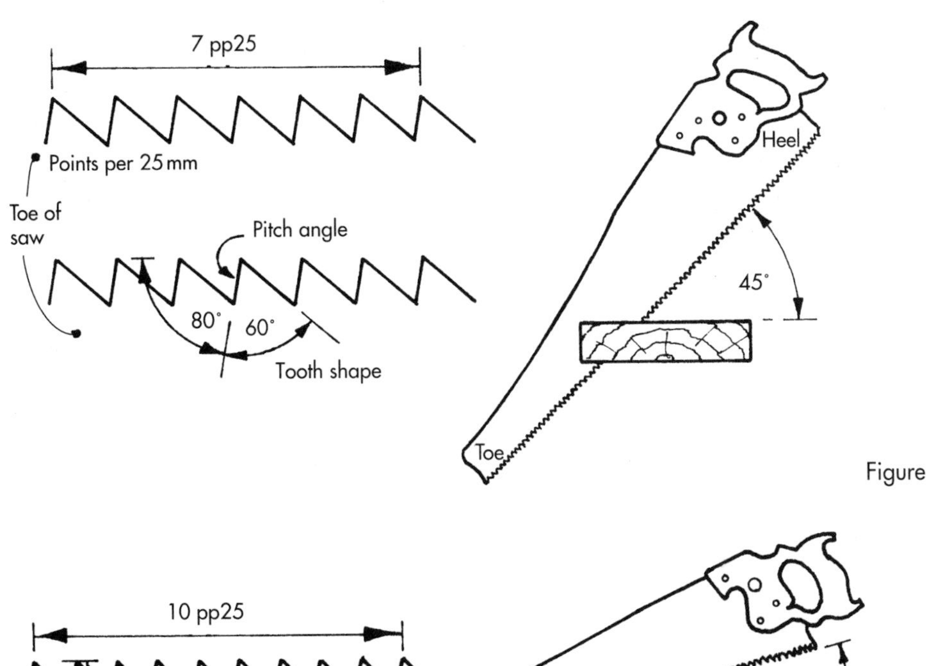

Figure 2.12 Crosscut saw

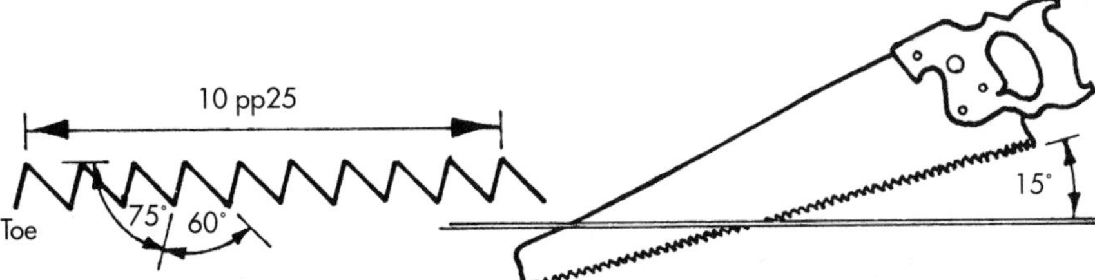

Figure 2.13 Panel saw

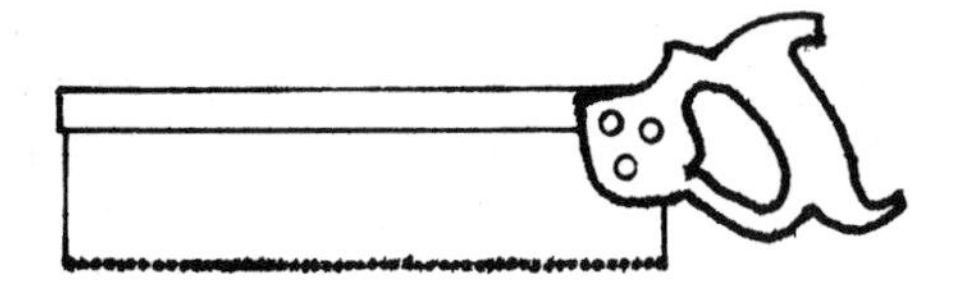

Figure 2.14 Tenon saw

and pitch angle, this saw cannot be used for crosscutting.

Conventional saws should be kept dry if possible and lightly oiled, but if rusting does occur, soak liberally with oil and rub well with fine emery cloth.

2.3.6 Hardpoint Handsaws and Tenon Saws

Figures 2.16: These modern throwaway saws have high-frequency hardened tooth-points which stay sharper for at least five times longer than conventional saw teeth. Two shapes of tooth predominate; the first, referred to as *universal*, conforms to the conventional 60° tooth-shape and 75° pitch; the second, known as the *fleam* tooth, resembles a *flame* in shape (hence its name), with a conventional front-pitch of 75°, an unconventional back-pitch of 80°, giving the fleam-tooth shape of 25°. Most of the handsaws are claimed to give a superior cutting performance across and along the grain. A recent addition to the range has a Teflon-like black, friction-reducing coating on the blade, to eliminate binding and produce a faster cut with less effort.

Range of Sizes

The saws usually have plastic handles – some with improved grip – with a 45° and 90° facility for marking mitres or right angles. Handsaw sizes available are 610 mm (24 in) × 8 pp25, 560 mm (22 in) × 8 pp25, 508 mm (20 in) × 8 pp25, 560 mm (22 in) × 10 pp25, 508 mm (20 in) × 10 pp25, 480 mm (19 in) × 10 pp25,

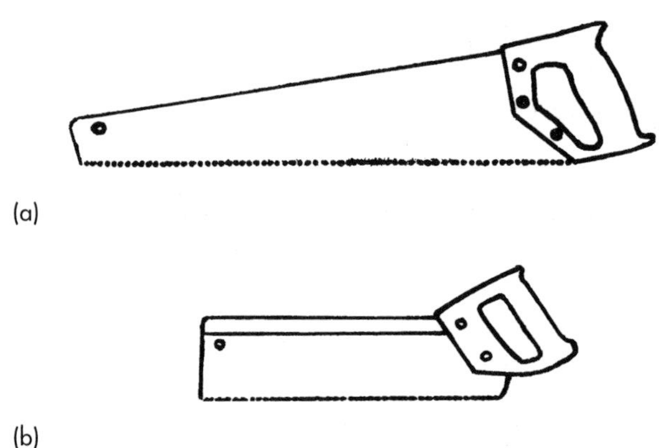

Figure 2.16(a) Hardpoint handsaw; (b) hardpoint tenon saw

455 mm (18 in) × 10 pp25 and 405 mm (16 in) × 10 pp25. Tenon saw sizes available are 300 mm (12 in) × 13 pp25, 250 mm (10 in) × 13 pp25, 300 mm (12 in) × 15 pp25 and finally 250 mm (10 in) × 15 pp25. Three recommended saws from this range would be the 610 mm (24 in) × 8 pp25 and the 560 mm (22 in) × 10 pp25 black-coated handsaws and a 300 mm (12 in) × 13 pp25 tenon saw.

2.3.7 Pullsaws

These lightweight, unconventional saws of oriental origin, cut on the pull-stroke, which eliminates buckling. They can be used for ripping or crosscutting. The unconventional precision-cut teeth, with three cutting edges, are claimed to cut up to five times faster, leaving a smooth finish without breakout or splintering. The sprung steel blade is ultra-hardened to give up to ten times longer life and can easily be replaced at the push of a button. Replacement blades cost about two-thirds the cost of the complete saw, but a complete saw is relatively inexpensive.

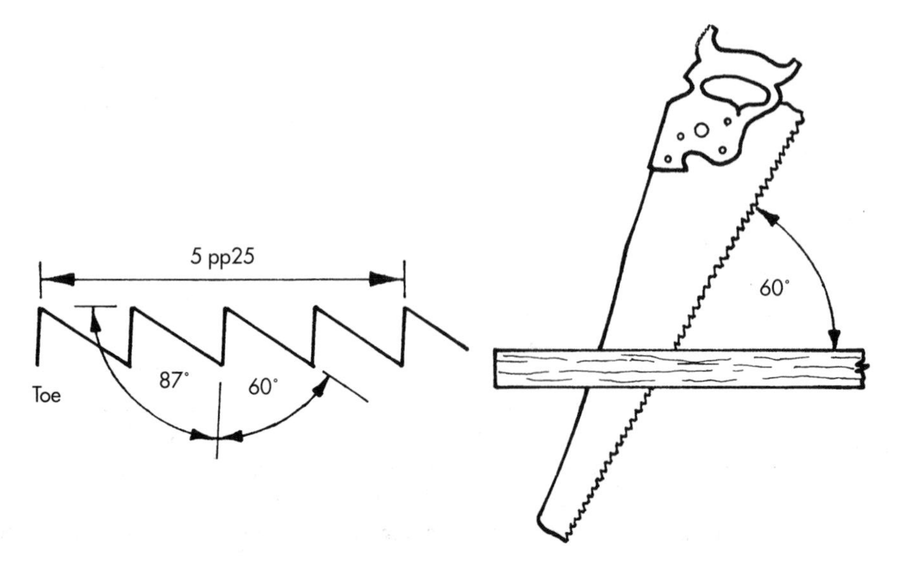

Figure 2.15 Rip saw

General Saw and Fine Saw

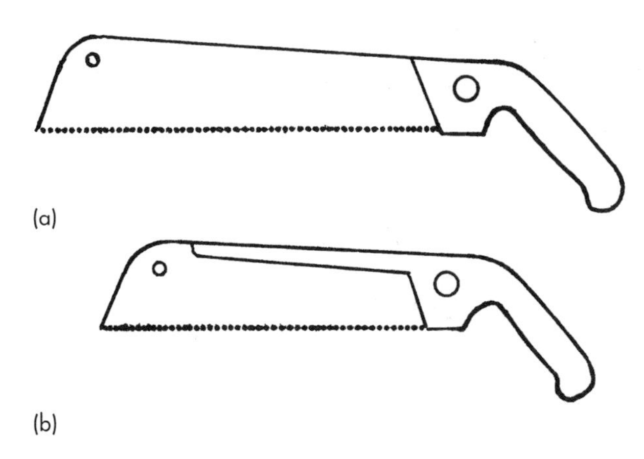

Figure 2.17(a) General carpentry saw; **(b)** fine cut saw

Figure 2.17: Only two saws from the range are illustrated here. The first is called a *general carpentry saw* and has a 455 mm (18 in) blade × 8 pp25. This model comes in two other sizes, 380 mm (15 in) × 10 pp25 and 300 mm (12 in) × 14 pp25. The latter is recommended for cutting worktops and laminates without chipping. The second model is called a *fine cut saw* and has a half-length back or full-length back support and is said to surpass conventional tenon saws. This model comes in two variations, one with a fine-cut blade of 270 mm × 15 pp25, the other with an ultra-fine blade of 270 mm × 17 pp25.

2.3.8 Coping Saw

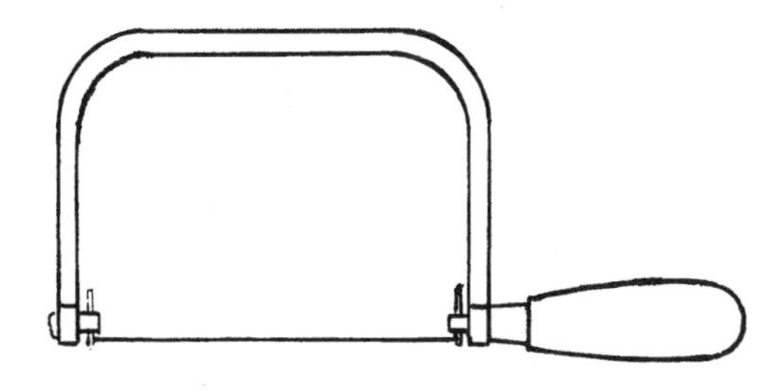

Figure 2.18 Coping saw

Figure 2.18: This traditional tool has not changed or lost its popularity and usefulness over many years. In second-fixing carpentry, it is mainly used for scribing (cutting the profile shape) of moulded skirting boards where they meet in the corners of a room (covered in another chapter), but occasionally comes in useful for other curved cuts in wood or plastic. The saw blades are very narrow with projecting pinned-ends and teeth set at 14 pp25. The blades, although easily broken with rough or unskilled sawing, have been heat-treated to the required degree of hardness and toughness and are obtainable in packs of 10. Although the narrowness of the blade demands that it be set in the frame with the front pitch of the teeth set to face the handle and working on the pull stroke, it can, if preferred, be set to cut on the forward action, providing that a degree of skill has been developed. The blade can be swivelled to cut at any angle to the frame, after unscrewing the handle slightly; the handle should be fully tightened after each adjustment of the blade.

2.3.9 Mitre Saw

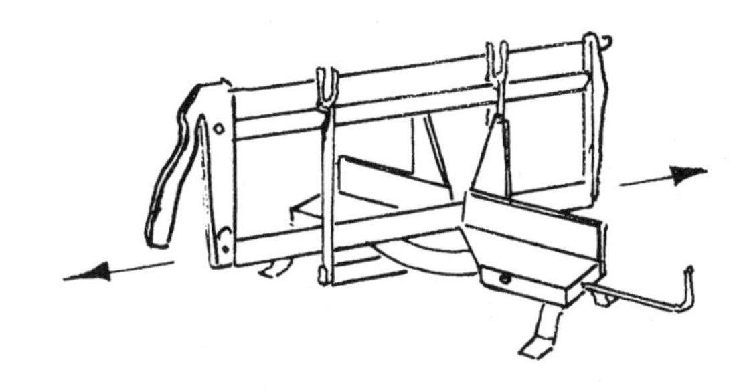

Figure 2.19 Mitre saw

Figure 2.19: Nowadays, as well as the traditional wooden mitre box or block (covered elsewhere), mitre saws are quite commonly used. They can be power-driven or hand models and can cut vertical angles within a 90° arc, starting at 45° from the fence, then moving to any intermediate position until reaching 135° from the fence. Different models have varying depths of cut and some can be tilted from the vertical cut to produce compound angles. Apart from mitring, a common use is with the saw at 90° to the fence and bed, cutting timber squarely.

2.4 HAMMERS

2.4.1 Claw Hammer

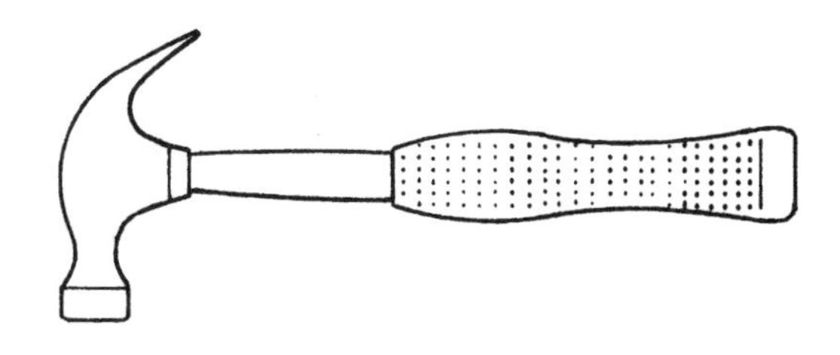

Figure 2.20 Metal-shafted claw hammer

Figure 2.20: Although this tool is basically for nailing and extracting nails, it has also been widely used over the years, using the side of the head, as an alternative to the wooden mallet. This is an acceptable practice on impact-resistant plastic chisel handles – especially as this type of handle is really too hard for the wooden mallet – but it is bad practice to use a hammer on wooden chisel handles, as they quickly deteriorate under such treatment. However, in certain awkward site situations, the mallet is too bulky and only the side of the claw hammer is effective.

Other Uses

The claw is also used for a limited amount of leverage work, such as separating nailed boards, etc. To preserve the surface shape of the head, the hammer should not be used to chip or break concrete, brick or mortar. When hammering normally, hold the lower end of the shaft and develop a swinging wrist action – avoid *throttling* the hammer (holding the neck of the shaft, just below the head). Choice of weights is between 450 g (1 lb), 565 g ($1\frac{1}{4}$ lb) and 675 g ($1\frac{1}{2}$ lb); choice of type is between steel shaft with nylon cushion grip, steel shaft with leather binding, fibreglass shaft in moulded polycarbonate jacket and the conventional wooden shaft. The latter has a limited life-span on site work. The choice is yours, but the steel-shafted type with nylon cushion grip, 675 g in weight, is recommended.

2.4.2 Mallet

Figure 2.21: The conventional wedge-shaped pattern, made of beech, is rather bulky and not generally

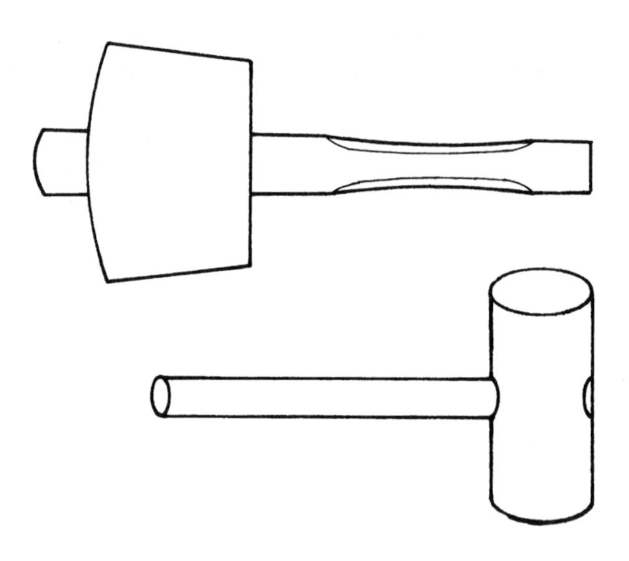

Figure 2.21 Wedge-shape and round-head mallet

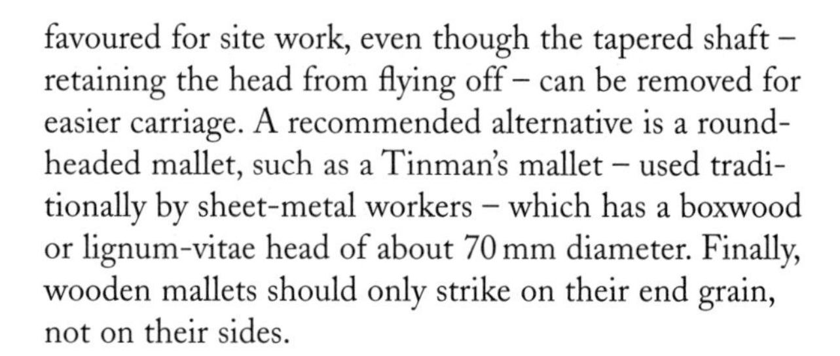

favoured for site work, even though the tapered shaft – retaining the head from flying off – can be removed for easier carriage. A recommended alternative is a round-headed mallet, such as a Tinman's mallet – used traditionally by sheet-metal workers – which has a boxwood or lignum-vitae head of about 70 mm diameter. Finally, wooden mallets should only strike on their end grain, not on their sides.

2.5 SCREWDRIVERS

Figure 2.22: Although power screwdrivers, especially the cordless type, are very popular nowadays, hand screwdrivers are still widely used and even preferred for certain jobs. Research has proved that the following selection are still in demand in the trade.

2.5.1 Ratchet Screwdriver

The *ratchet* screwdriver is available with flared slotted tip in four blade-lengths of 75 mm, 100 mm, 150 mm and 200 mm. They are also available with a No 2 Supadriv/Pozidriv tip and a No 2 Phillips' tip in blade-lengths of 100 mm only.

2.5.2 Spiral Pump Screwdriver

The *spiral pump-action* screwdriver, which can also be used as a ratchet, comes in three sizes of 343 mm, 358 mm and 711 mm lengths when released by the spiral lock. The spring release is fast and potentially dangerous unless controlled by holding the knurled sleeve at the front of the spiral shaft, next to the spiral lock. This sleeve should also be held between forefinger and thumb while pumping the screwdriver with the other hand, in a screwing operation. The 358 mm size pump is recommended, but the 711 mm size is very popular. A smaller version of the spiral pump screwdriver is available, in one size of 267 mm with a magazine handle holding two slotted bits, a Pozidriv bit and two drill bits. The use of drill bits in this compact-size pump is an attractive alternative for making speedy pilot holes. Interchangeable bits are supplied with the whole range of this type of screwdriver in different sizes of slotted and Pozidriv tips, and can be purchased separately.

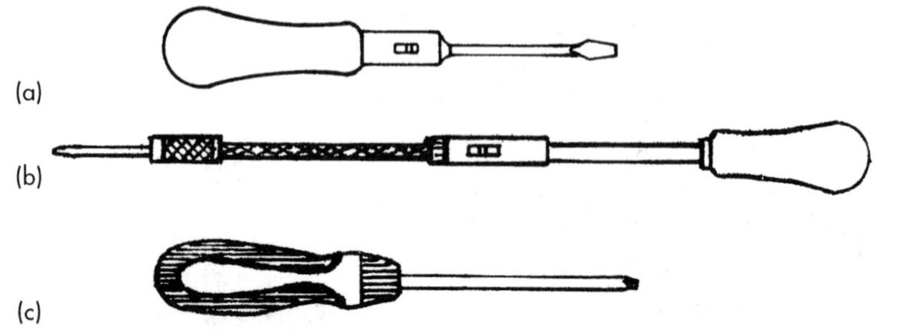

Figure 2.22(a) Ratchet screwdriver. (b) Spiral pump screwdriver. (c) Plastic-handled screwdriver

2.5.3 Plastic-Handled Screwdrivers

There is a large variety of these screwdrivers to choose from, each with its own feature and qualities, but some are not easy or comfortable to grip, often making it difficult to apply the required torque. The one illustrated in Figure 2.22(c) has a well-shaped polypropylene handle integrated with thermo-plastic elastomer inserts to provide improved grip and comfort in use. The size of the tip varies according to blade-length and these vary from 75 mm up to 300 mm, with flared slotted tips, and from 75 mm up to 200 mm with Supadriv/Pozidriv or Phillips' tips.

2.6 MARKING GAUGES

Figure 2.23: These tools may not have a use on first-fixing carpentry, but will be needed on second-fixing operations. Although still predominantly made of beech, the thumbscrews are made of clear yellow plastic and – although quite tough – if overtightened, may fracture. To protect the sharp marking-pin and for safety's sake, the pin should always be returned close to the stock after use. To use the gauge, it should be held as shown, with the thumb behind the pin, the forefinger resting on the rounded surface of the stock and the remaining fingers at the back of the stock, giving side pressure against the timber being marked. Always mark lightly at first to overcome grain deviations. The gauge is easier to hold if the face-edge arris – that rubs the inside of the outstretched thumb – is rounded off as shown. Also, to reduce wear and surface friction, plastic laminate can be shaped and bonded to the face of the stock.

2.7 CHISELS

2.7.1 Firmer and Bevelled-edge Chisels

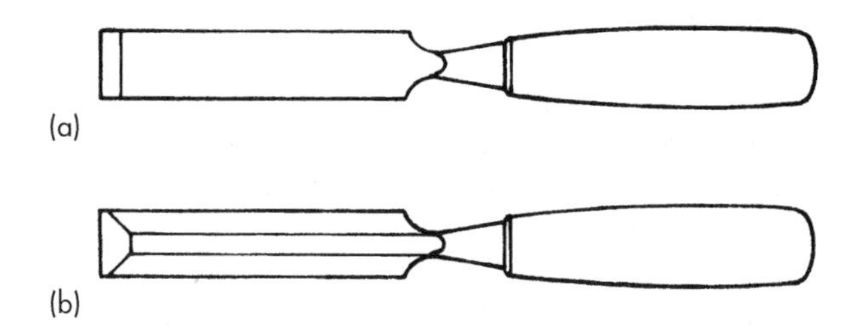

Figure 2.24 **(a)** Firmer chisel. **(b)** Bevelled-edge chisel

Figure 2.24: Firmer chisels are generally for heavy work, chopping and cutting timber in a variety of operations where a certain amount of mallet/hammer work and levering might be necessary to remove the chopped surface. Bevelled-edge chisels are generally for more accurate finishing tasks – such as paring to a gauged line – where mallet/hammer work, if any, is limited and levering should be avoided.

2.7.2 Types of Chisels

Although chisels with conventional wooden handles of boxwood or ash are still available, they are more suitable for bench joinery work, where they less likely to receive rough treatment. Because most site carpenters will hit a chisel with the side of a claw hammer, the modern range of chisels includes: splitproof handled

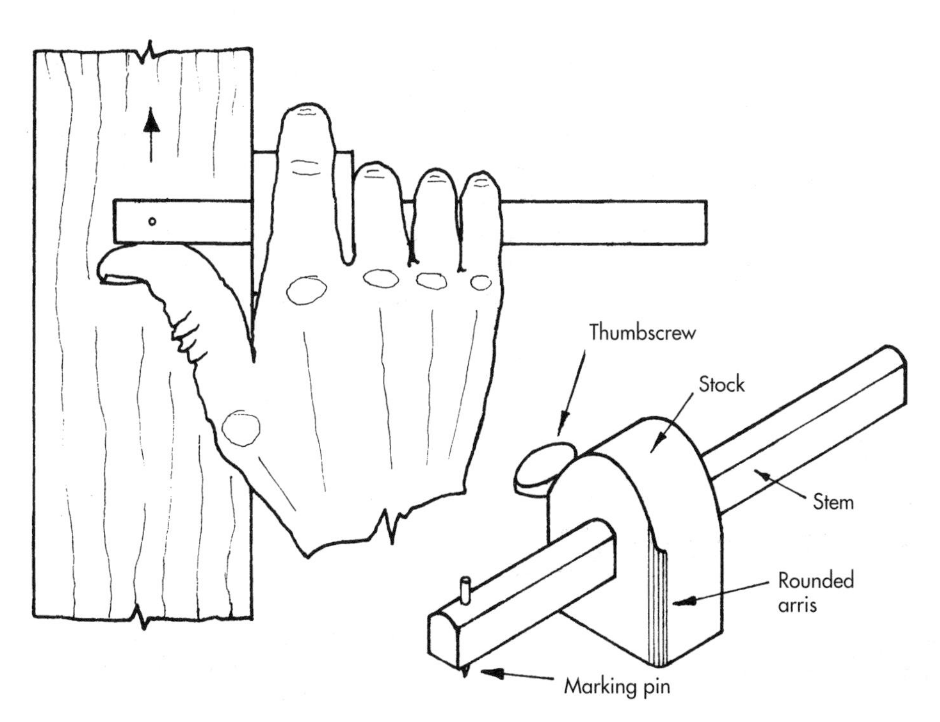

Figure 2.23 Marking gauge

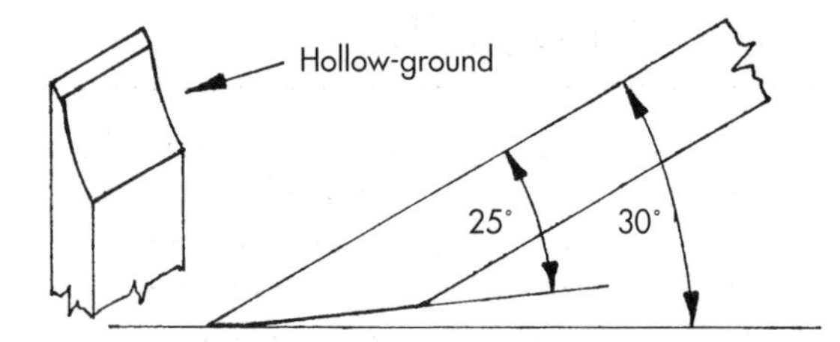

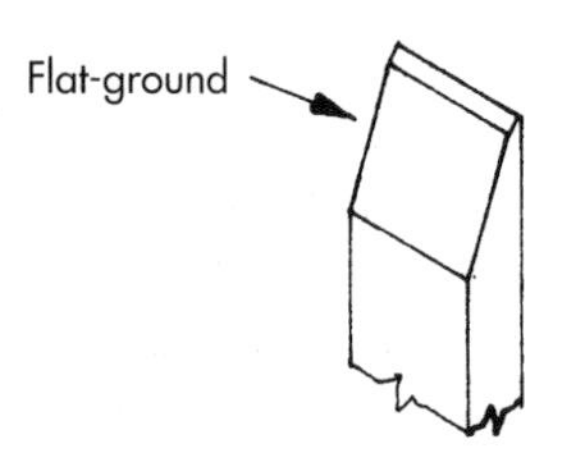

Figure 2.25 Grinding angle 25°; sharpening angle 30°

chisels with impact-resistant plastic handles, designed for use with a hammer and guaranteed for life; black non-slip polypropylene impact-resistant handled chisels; and heavy-duty shatterproof, cellulose acetate butyl handled chisels. Blade widths range from 3 mm to 50 mm. Recommended sizes for a basic kit are 6, 10, 12 and 25 mm in firmer chisels (Figure 2.24(a)) and 18 and 32 mm in bevelled-edge chisels (Figure 2.24(b)).

2.7.3 Grinding and Sharpening Angles

Figure 2.25: The cutting edge of chisels should contain a *grinding angle* of 25°, produced on a grindstone or grinding machine, and a *sharpening angle* of 30°, produced on an oilstone. The hollow-ground angle should not lessen the angle of 25° in the concave of the hollow. For extra strength, firmer and mortice chisels can be flat-ground.

2.8 OILSTONES AND THEIR USE

Artificially manufactured stones, made from furnace-produced materials, as opposed to natural stone, are widely used because of their constant quality and relative cheapness. Coarse, medium and fine grades are available. A *combination stone*, measuring 200 × 50 × 25 mm, is recommended for site work. This stone is coarse for half its thickness, and fine on the alternate side for the remaining half thickness. As these stones are very brittle, they should be housed in purpose-made (or shop-purchased) wooden boxes for protection.

2.8.1 Oiling the Stone

When sharpening, use a thin grade of oil, animal or mineral, but not vegetable oil, which tends to solidify on drying, so clogging the cut of the stone. Lubricating oil is very good. Should the stone ever become clogged, giving a glazed appearance and a slippery surface, soak it in petrol or paraffin for several hours, then clean it with a stiff brush or sacking material and allow it to dry before re-using.

2.8.2 Sharpening

Figure 2.26: When sharpening chisels or plane irons, first apply enough oil to the stone to cover its surface and help float off the tiny discarded particles of metal, then hold the tool comfortably with both hands, assume the correct angle to the stone (30°), then move back and forth in an even, unaltering movement until a small sharpened (or honed) edge is obtained. This action produces a metal burr which is turned back by reversing the cutter to lay flat on the stone, under finger-pressure, and by rubbing up and down a few times. Any remaining burr can be removed by drawing the cutter across the arris edge of a piece of wood.

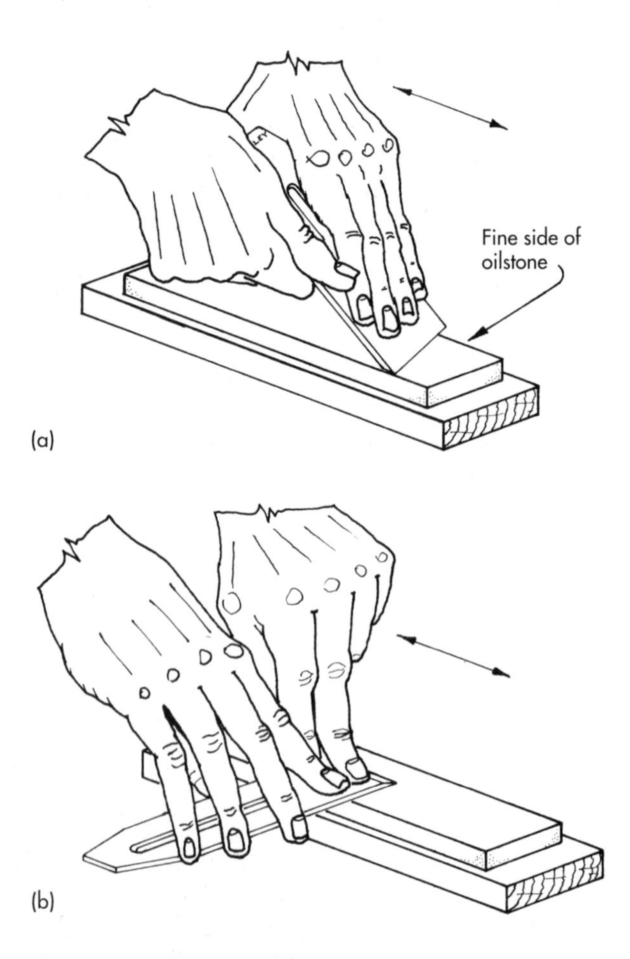

Figure 2.26(a) Recommended hand-hold for sharpening a plane iron. **(b)** Removing metal burr and polishing underside of cutting edge

2.8.3 Use of Oilstone

The stone should always be used to its maximum length, the cutter lifted occasionally to bring the oil back into circulation. Narrow cutters, such as small chisels, whilst traversing the length of the stone, should also be worked across the stone laterally to reduce the risk of dishing (hollowing) the stone in its width.

2.8.4 Oilstone Box

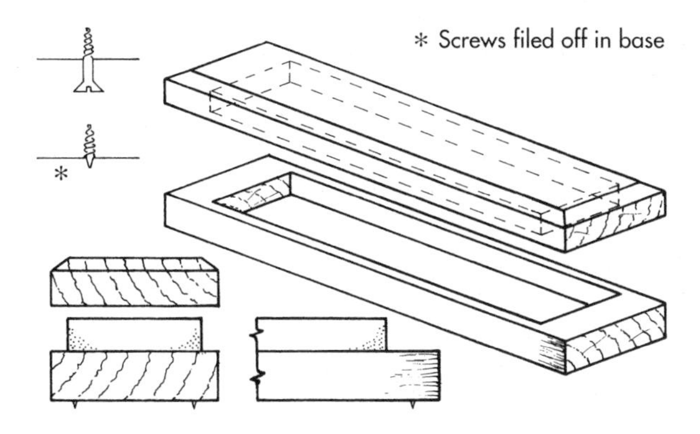

Figure 2.27 Oilstone box

Figure 2.27: Although now available in tool shops, this was traditionally a hand-made item, usually of hardwood, required to protect the stone from damage and the user from contact with the soiled oil. It is easily made from two pieces of wood, each measuring a minimum of 240 × 62 × 18 mm, to form the two halves of the box. With the aid of a brace and bit and chisel (or a router, if available), recesses are cut to accommodate the stone snugly in the base and loosely in the part which is to be the lid. To stop the box from sliding while sharpening, two 12 mm × 4 gauge screws can be partly screwed into the underside of the base and filed off to leave dulled points of about 2 mm projection.

2.9 HAND PLANES

The two planes to be recommended as most useful for site work are the No. 4½ smoothing plane with a cutter width of 60 mm and a base length of 260 mm and the No. 5½ jack plane, also with a cutter width of 60 mm, but a base length of 381 mm. Narrower smoothing and jack planes with 50 mm cutter widths and less length and weight, notably the No. 4 and the No. 5, are thought to be more suitable for bench joinery work.

2.9.1 Knowledge of Parts

Figure 2.28 shows a vertical section through a smoothing plane to identify the various parts which need to be named and known for reference to the plane's usage. These named-parts also apply to the jack plane and other planes of this type.

2.9.2 Planing and Setting-up Details

These planes are also available with corrugated (fluted) bases to reduce surface friction, especially when planing resinous or sticky timbers; if not fluted, the sole of the plane can be lightly rubbed with a piece of beeswax or candlewax. When planing long lengths of timber, like the edge of a door, lift the heel slightly at the end of the planing stroke to break the shaving. On new planes, the cutter has been correctly ground to 25°, but not sharpened. To sharpen, remove the cutter from the back iron and carry out the sharpening procedure outlined in the text to Figure 2.26. When re-assembling, set the back iron within 1–2 mm of the cutter's edge and, if necessary, adjust the lever-cap screw so that the replaced lever cap is neither too tight nor too loose.

2.9.3 How to Check Cutter-projection

Always check the cutter projection before use, by turning the plane over at eye-level and sighting along the sole from toe to heel. The projecting cutter will appear

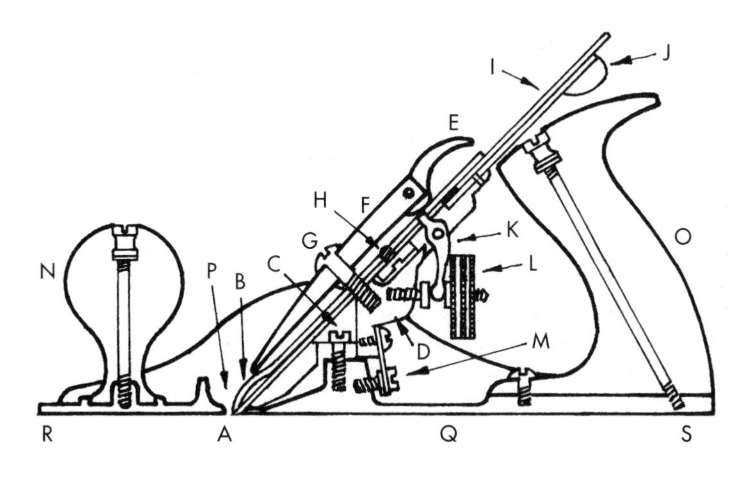

Figure 2.28 Vertical section through smoothing plane. The parts are: **A** mouth, **B** back iron, **C** frog-fixing screws, **D** frog, **E** lever, **F** lever cap, **G** lever-cap screw, **H** back-iron screw, **I** cutting iron (cutter), **J** lateral-adjustment lever, **K** cutter-projection adjustment lever, **L** knurled adjusting-nut, **M** mouth-adjustment screw, **N** knob, **O** handle, **P** escapement, **Q** sole (base), **R** toe and **S** heel.

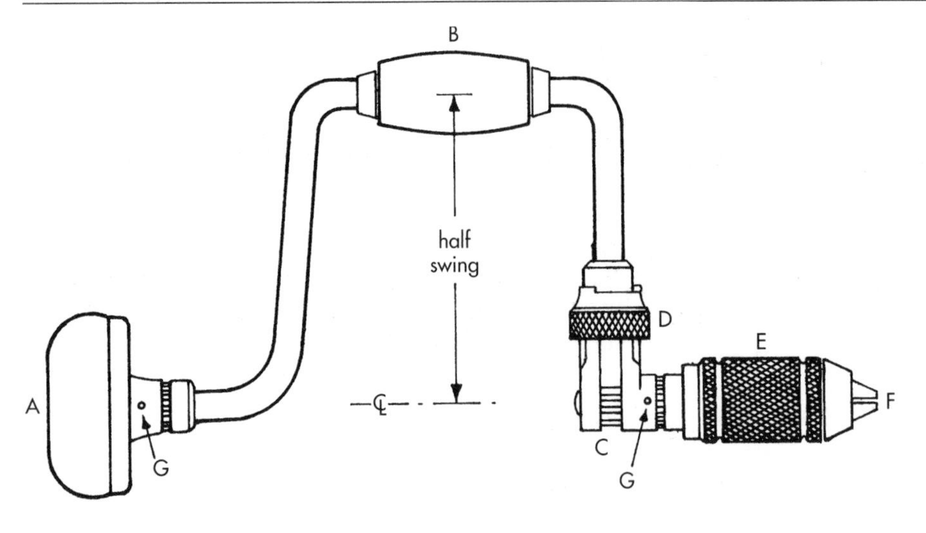

Figure 2.29 Ratchet brace. Parts of the brace are named as follows: **A** head, **B** handle, **C** ratchet, **D** ratchet-conversion ring, **E** jaws-adjustment shell, **F** universal jaws, and **G** oil hole

as an even or uneven black line. While sighting, make any necessary adjustments by moving the knurled adjusting-nut and/or the lateral-adjustment lever. For safety and edge-protection, always wind the cutter back after final use of the plane. Bear in mind that the body is made of cast iron, and if dropped, is likely to fracture – usually across the mouth. Keep planes dry and rub occasionally with an oily rag.

2.10 RATCHET BRACE

Figure 2.29: This item should be carefully chosen for its basic qualities and any saving in cost could prove to be foolish economy. Essentially, the revolving parts – the head and the handle – should be free-running on ball bearings, the ratchet must be reliably operational for both directions and the jaws must hold the tapered-tang twist bits, and the dual-purpose combination auger bits with parallel shanks, firmly and concentrically. The recommended *sweep* (diameter of the handle's orbit) is 250 mm. Braces with a smaller or larger sweep are available. The advantages of the ratchet are gained when drilling in situations where a full sweep cannot be achieved, such as against a wall or in a corner – or when using the screwdriver bit under intense pressure and sustaining the intensity by using short, restricted ratchet-sweeps.

2.11 BITS AND DRILLS

2.11.1 Twist Bits

Figure 2.30: These are also referred to as *auger bits* and traditional types are spiral-fluted, round shanked with tapered tangs. Their disadvantage nowadays is that they will only fit the hand brace and not the electric or cordless drill. However, a set of modern bits, without this

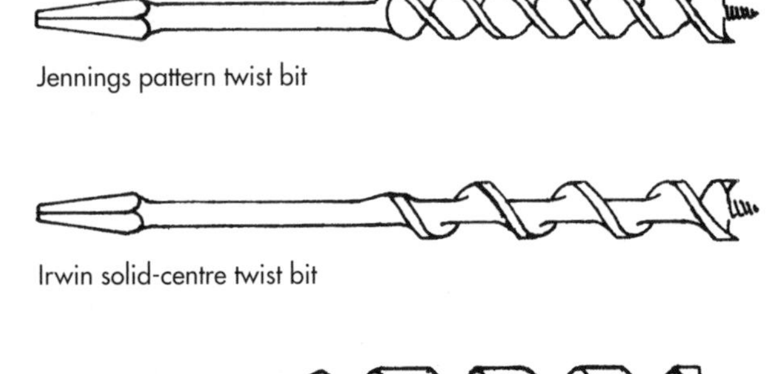

Sandvik combination auger bit

Figure 2.30 Twist bits/auger bits

disadvantage, is now an option. These bits are also spiral-fluted and round shanked, but are minus the tapered tang and will fit either the electric/cordless drill or the ratchet brace. They are known as combination auger bits. All of these bits are for drilling shallow or deep (maximum 150 mm) holes of 6–32 mm diameter. *Jennings pattern* twist bits have a double spiral and are used for fine work; *Irwin solid-centre* twist bits have a single spiral and are more suitable for general work; *Sandvik combination* auger bits have a wide single spiral with sharp edges and give a clean-cut hole suitable for fine or general work. Seven combination bits are recommended for the basic kit, these being sizes 6, 10, 13, 16, 19, 25 and 32 mm.

2.11.2 Drilling Procedure

If the appearance of a hole has to be considered, then care must be taken not to break through on the other side of the timber being drilled. This is usually achieved by changing to the opposite side immediately the point of the bit appears. Alternatively, drill through into a piece of waste timber clamped onto or seated under the blind side.

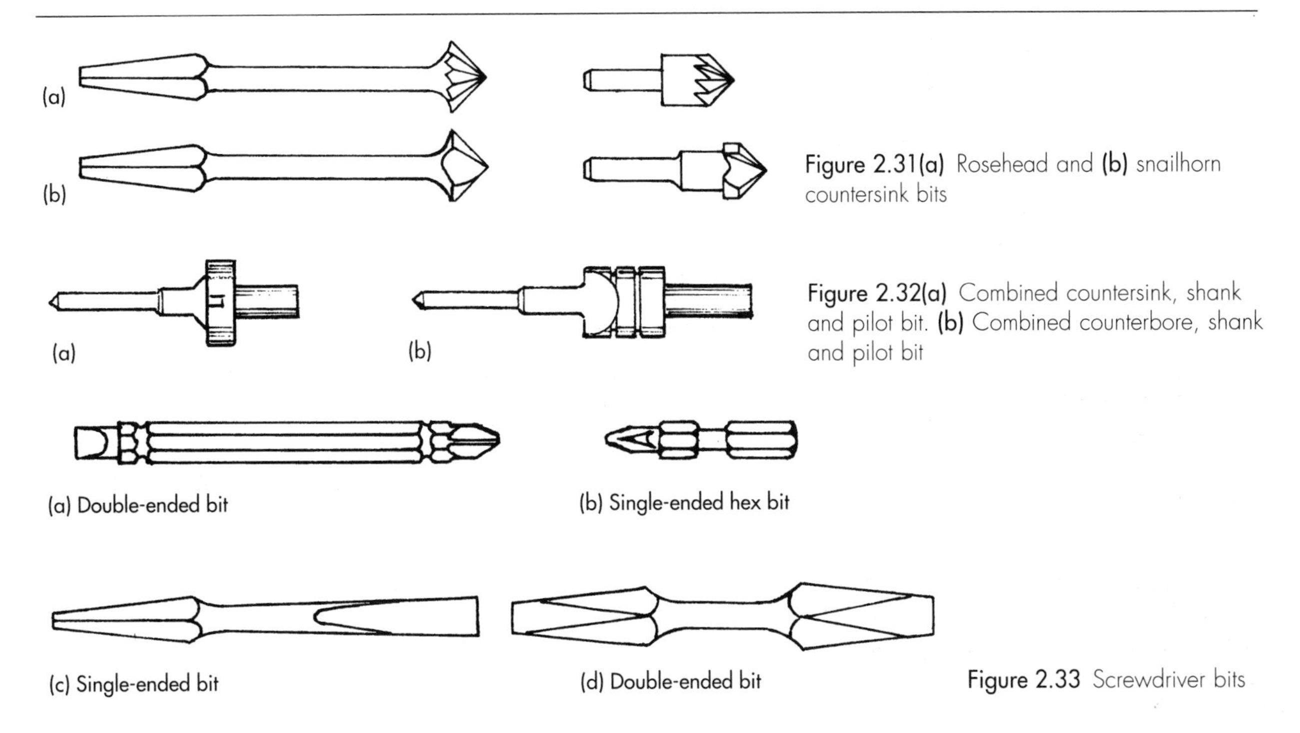

Figure 2.31(a) Rosehead and **(b)** snailhorn countersink bits

Figure 2.32(a) Combined countersink, shank and pilot bit. **(b)** Combined counterbore, shank and pilot bit

Figure 2.33 Screwdriver bits

2.11.3 Sharpening Twist Bits

Always avoid sharpening twist bits for as long as possible, but when you do, sharpen the inside edges only, with a small flat file; never file the outer surface of the spur cutters. Always take care, when drilling reclaimed or fixed timbers, not to clash with concealed nails or screws, as this kind of damage usually ruins the twist bit.

2.11.4 Countersink Bits

Figure 2.31: These are for screw-head recessing in soft metal and timber. The *rosehead pattern* type is for metals, such as brass or aluminium, although it is also used for softwood (and can be used for hardwood). The *snailhorn pattern* type is used just for hardwood. As illustrated, these are available with a round shank and traditional tapered tang for use with the ratchet brace, or with short or long, round shank only, for use with the hand drill or the electric or cordless drill.

2.11.5 Combined Countersink and Counterbore Bits

Figure 2.32: These two modern drill-bits are useful on certain jobs and although they can be used in traditional hand drills, they will of course be more efficient in electric or cordless drills. The combined countersink bit is available in seven different sizes and is for drilling a pilot hole, shank hole and countersink for woodscrews in one operation. The combined counterbore bit is available in 12 different sizes and is for drilling a pilot hole, shank hole and a counterbored hole (the latter receives a glued wooden pellet after screwing) also in one operation.

2.11.6 Screwdriver Bits

Figure 2.33: There is a wide range of modern bits suitable for electric or cordless drills, in different lengths (Figures 2.33(a) and (b)) to fit different gauges of screws and different types, such as screws with Supadriv/Pozidriv inserts, Phillips' inserts and slotted inserts. The two traditional screwdriver bits shown in Figures 2.33(c) and (d) have tapered tangs for use with the ratchet brace. These bits are still very useful, mainly for the extra pressure and leverage obtained by the brace and occasionally required in withdrawing or inserting obstinate or long screws. The double-ended bit (d) has a different size slotted tip at each end, in the shape of – and to act as – a tang.

2.11.7 Twist Drills

Figure 2.34: The twist drill is another of those tools adopted from the engineering trades and put to good

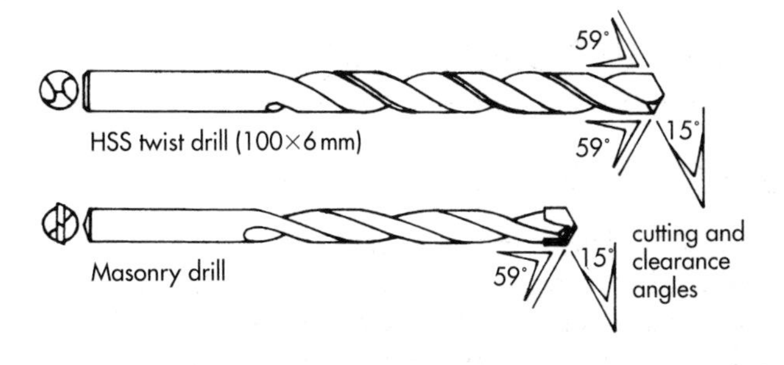

Figure 2.34 Twist drills and masonry drills

use in drilling holes in timber. Their round shanks will fit the chuck of the electric, cordless, or hand drill. A set of these twist drills, of high-speed steel (HSS), varying in diameter by 0.5 mm and ranging from 1 mm to 6 mm, is essential for drilling pilot holes and/or shank holes for screws in timber. They may also be used of course for drilling holes in metals such as brass, aluminium, mild steel, etc. (after marking the metal with a *centre punch*). When dull, these drills can be sharpened on a grinding wheel, but care must be taken in retaining the cutting and clearance angles at approximately 60° and 15°, respectively.

2.11.8 SDS Drills and Masonry Drills

Drills of various diameter for drilling plug holes in brick, block and concrete, or similar materials, have improved in recent years and masonry drills with heavy-duty carbide tips are now an option. Also, for use with the powerful SDS-Plus range of hammer drills, SDS-Plus-shank hammer-drill bits are available.

2.12 INDIVIDUAL HANDTOOLS

2.12.1 Bradawls

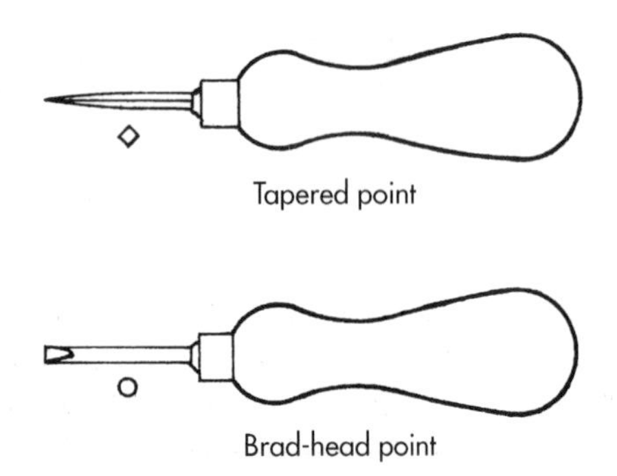

Figure 2.35 Bradawls

Figure 2.35: These are used mainly for making small pilot or shank holes when starting screw fixings. Two different sizes are available and different types. One type of awl has a flat brad-head point which should always be pushed into the timber at right angles to the grain before turning; the other type of awl has a square-sectioned, tapered point which acts as a reamer when turned – ratchet fashion – into the timber. Although it seems to be less common, the square-tapered awl is very much recommended.

2.12.2 Pincers

Figure 2.36: These are used for withdrawing small nails and pins, not fully driven in. Although these are usually

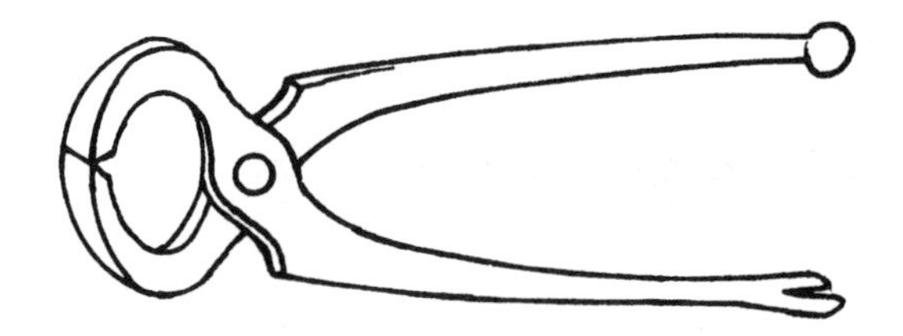

Figure 2.36 Pincers

extracted by the claw hammer, occasionally a pair of pincers will do the job more successfully. When levering on finished surfaces, a small piece of wood or thin, flat metal placed under the fulcrum point will reduce the risk of bruising the surface. Different sizes are available, but the 175 mm length is recommended.

2.12.3 Wrecking bar

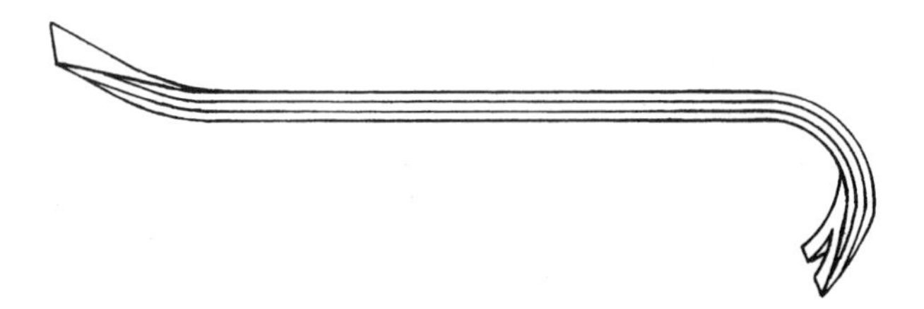

Figure 2.37 Wrecking bar

Figure 2.37: This tool is also referred to as a *crowbar*, *nail bar* or *pinch bar*. It is not essential, but is useful in construction work for extracting large nails and for general leverage work. There is a choice of five sizes: 300, 450, 600, 750 and 900 mm. If necessary, leverage can be improved by placing various-size blocks under the fulcrum point. The 600 mm length is recommended.

2.12.4 Hacksaw

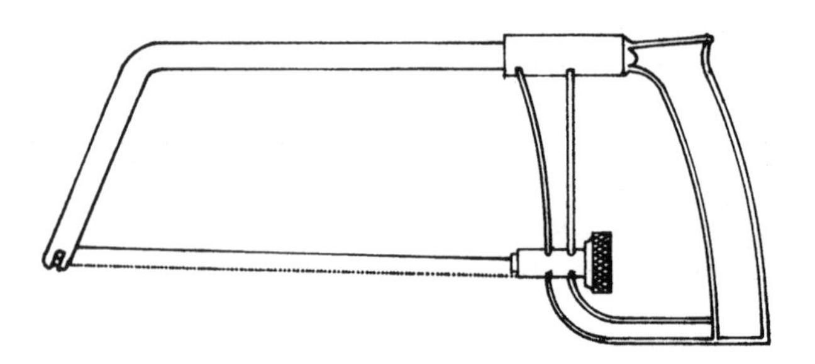

Figure 2.38 Junior hacksaw

Figure 2.38: This is another useful addition occasionally required. Carpenters do not normally need a full-size hacksaw for cutting large amounts of metal objects, so the *junior hacksaw*, with a blade length of 150 mm, is recommended for the limited amount of use involved.

2.12.5 Nail Punches

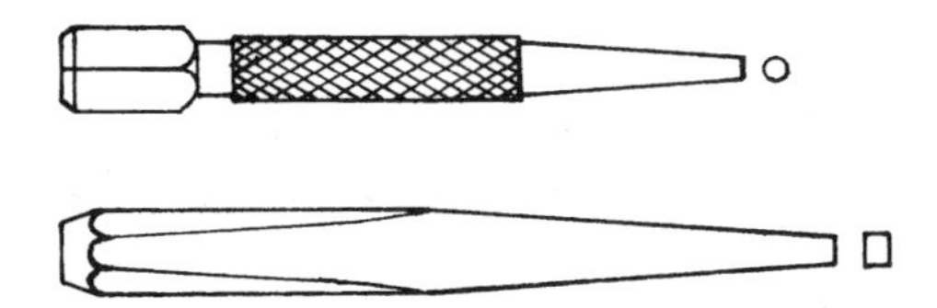

Figure 2.39 Nail punches

Figure 2.39: These are essential tools, especially in second-fixing carpentry, when used to sink the heads of nails or pins below the surface of the timber, by about 2 mm, to improve the finish when the hole is *stopped* (filled) prior to painting. They are also of use occasionally, assisting in driving a nail into an awkward position, or in skew-nailing into finished timbers, switching to a punch to avoid bruising the surface with the hammer. At least three nail punches are recommended for the basic kit, say 1.5, 2.5 and 5 mm across the points. Some points have a concave shape to reduce the tendency to slip.

2.12.6 Cold Chisel and Bolster Chisel

Figure 2.40: A 19 × 300 mm cold chisel and a bolster chisel with a 75 mm wide blade are recommended additions to the tool kit, for use on odd occasions when brick or plasterwork requires cutting. Carbon steel chisels are commonly used, although nickel alloy chisels are more suitable. Keep the cutting edges sharp, by file for nickel alloy and by grinding wheel for carbon steel.

Safety Note

Figures 2.39 and 2.40: For safety's sake, grind the sides of the heads before they become too mushroom-shaped from prolonged usage. Also, develop the technique of holding the tool as you would the barrel of a rifle, so that the palm of the hand, not the knuckles, faces the hammer blows. Alternatively, plastic grip hand guards are available separately to fit onto chisels.

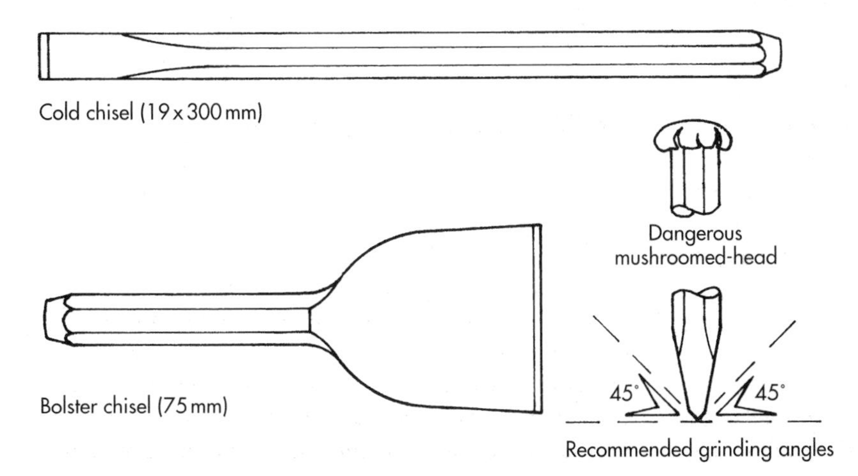

Figure 2.40 Cold chisel and bolster chisel

2.13 PORTABLE POWERED CIRCULAR SAWS

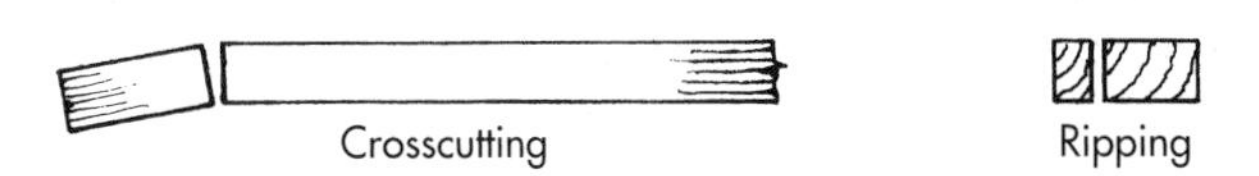

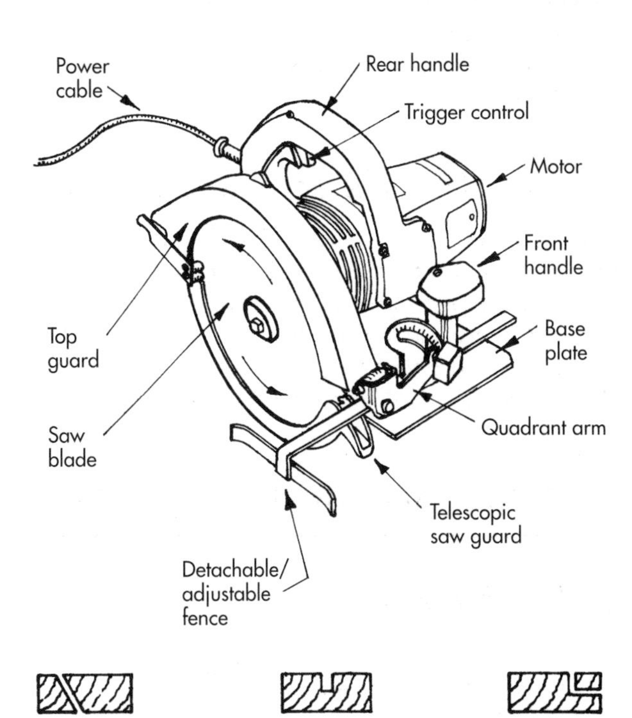

Figure 2.41 Portable powered circular saw

Figure 2.41: These saws are widely used nowadays to save time and energy spent on handsawing operations. Although basically for ripping and crosscutting, they can also be used for bevel cuts, sawn grooves and rebates. Models with saw blades of about 240 mm or more diameter are recommended for site work.

2.13.1 Adjusting the Base Plate

Before use, the saw should be adjusted so that when cutting normally, the blade will only just break through the underside of the timber. This is easily achieved by releasing a locking device which controls the movement of the base plate in relation to the amount of blade exposed. For bevel cutting, the base tilts laterally through 45° on a lockable quadrant arm. The telescopic saw guard, covering the exposed blade, is under tension so that after being pushed back by the end of the timber being cut, it will automatically spring back to give cover again when the cut is complete. For ripping, a detachable fence is supplied.

2.13.2 Using the Saw

Before starting to cut, the saw should be allowed to reach maximum speed and should not be stopped or restarted in the cut. Timber being cut should be securely held, clamped or fixed – making certain that any fixings will not coincide with the sawcut and that there are no metal obstacles beneath the cut. Always use both hands on the handles provided on the saw, so reducing the risk of the free hand making contact with the cutting edge of the blade. At the finish of the cut, keep the saw suspended away from the body until the blade stops revolving.

2.13.3 Safety Note

Additional safety factors include: keeping the power cable clear of the cutting action; not overloading the saw by forcing into the material; drawing the saw back if the saw-cut wanders from the line and carefully re-advancing to regain the line; wearing safety glasses or protective goggles; disconnecting the machine from the supply while making adjustments, or when it is not in use; keeping saw blades sharp and machines checked on a regular basis by a qualified electrical engineer; checking voltage and visual condition of saw, power cable and plug before use; and working in safe, dry conditions.

2.14 ELECTRIC AND CORDLESS DRILLS AND SCREWDRIVERS

Figure 2.42: There is nowadays a wide range of dual- and triple-purpose drills to choose from, starting with the basic *rotary-only drill* and ending with the advanced *electro pneumatic hammer drill*. In the electric-powered range, of either 110 or 240 volts, the following combinations are available: the *drill/screwdriver*, *drill/impact (percussion) drill/screwdriver*, *drill/rotary hammer drill/screwdriver*; and combinations of battery-powered models such as the *cordless screwdriver*, *drill/screwdriver*, *drill/impact drill/screwdriver*, and the *drill/rotary hammer drill/screwdriver*. Careful consideration needs to be given in choosing a particular model in relation to the type of work to be done and its location regarding whether power is readily available and if so, whether it is 110 or 240 volts.

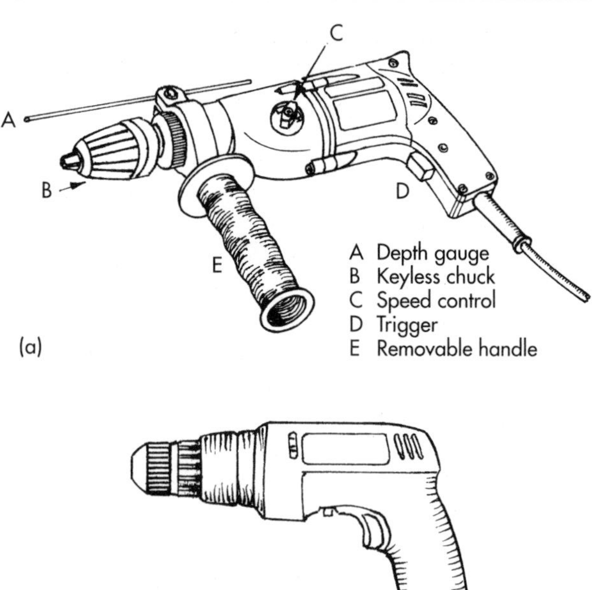

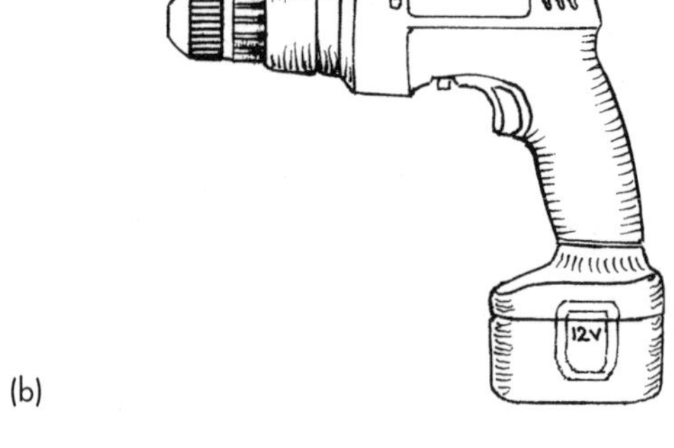

Figure 2.42(a) Electric drill. **(b)** Cordless drill/screwdriver

2.14.1 Technical Features of Drills and Drivers

- *Drills with rotary impact or percussion action* create a form of hammer action generated by a ridged washer-type friction bearing and should only be used for occasional drilling into masonry or concrete.
- *Drills with rotary hammer action* are designed with an impressive impact mechanism involving a reciprocating piston and connecting rod, which strikes the revolving drill bit at between 0 and 4000 blows per minute on certain models. This makes drilling of masonry or concrete much easier and faster and is recommended when drilling into these materials frequently.
- *Drills with electro pneumatic rotary hammer action* are designed with an impact mechanism which converts the power into a reciprocating pneumatic force to drive a piston and striker at between 0 and 4900 blows per minute on certain models. It develops eight times the power of a normal hammer drill and is recommended for ease, speed and efficiency when drilling frequently or constantly into masonry or concrete.
- *Safety clutches* are torque-limiting mechanisms fitted

to impact and hammer drills to protect the user if the drill bit becomes jammed in the material being drilled. They also protect the gears of the machine and the motor from short-term overloading.

- *Chucks* either have three jaws and require a separate chuck key, or are hand-operated and known as *keyless chucks*. Chuck sizes are usually 10, 13 and 16 mm. Some combination drills have an SDS chuck system, using special quick-release keyless chucks that accept only SDS drill bits.
- *The Fixtec system* on some SDS drills enables an SDS chuck to be replaced quickly by a standard three-jaw chuck without the use of tools.
- *Hex in spindle* means that a drill with this facility has a 6.35 mm ($\frac{1}{4}$in) hexagon recess in the drive spindle (accessible after removing the chuck) to take hexagon shanked screwdriver bits.
- *Drills with variable electronic speed* have an accelerator function for gentle start-up during drilling and screwdriving and, on certain jobs, can be used for driving in screws without first drilling a pilot hole.
- *Drills with adjustable torque control* can be preset and gradually adjusted to achieve precise screwdriving tightness.
- *Reversing rotation* is available with some drills and drivers, which is necessary for removing screws – and desirable when bits get jammed in a drilling operation.

2.14.2 Safety Note

Small pieces of loose timber being drilled should be firmly held or clamped; reliable step ladders, trestles, orthodox platforms or scaffolds should be used when drilling at any height above ground level; safety glasses or protective goggles should be worn; before making adjustments, or when not in use, a machine should be disconnected from the power supply; bits should be resharpened or renewed periodically; machines should be maintained on a regular basis by a qualified electrical engineer or by the manufacturer; always check visual condition of machine, power cable and plug or socket before use; always keep a proper hold on the drill until it stops revolving; always work in safe, dry conditions. On certain jobs, wear hard hats/helmets.

2.15 POWER PLANERS

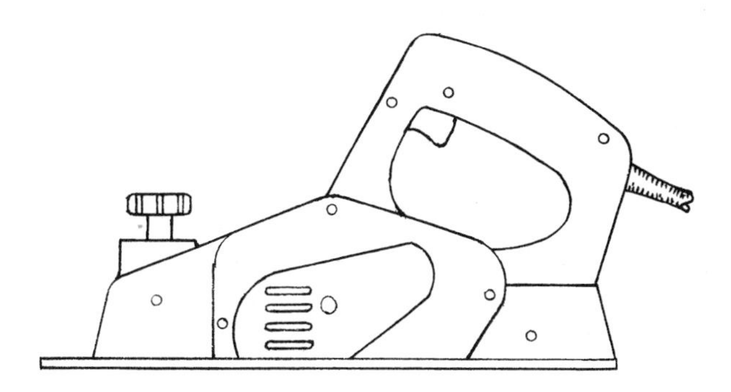

Figure 2.43 Power planer

Figure 2.43: Power planers are often used nowadays in conjunction with traditional planes such as the jack and the smoothing plane. They are sometimes preferred on such jobs as door-hanging, to lessen the strenuous task of 'shooting-in' the door by planing its edges. However, on this particular job, it is good practice to finish off with a hand plane, to remove the unsightly rotary-cutter marks – which can be very pronounced if the planer is pushed along at too great a feed-speed.

2.15.1 Making the Choice

When choosing a planer for frequent or constant use, avoid models classified in a DIY range; choose one with a professional/industrial classification. The model illustrated in Figure 2.43 is one in a range of three by this particular manufacturer. It has a TCT (tungsten carbide tip) cutter with a width of 82 mm, planing depth of 0–3 mm, rebating depth of 0–22 mm, a vee groove in the base to facilitate chamfering, a safety switch catch to prevent unintended starting and an automatic pivoting guard enclosing the cutter block for increased safety. This latter feature allows the planer to be put down safely before the revolving cutter block has come to a stop. One of the other models of this industrial threesome has an increased cutter width of 102 mm, but a reduced planing depth of 0–2.5 mm. Both models mentioned run at a no-load-speed of 13 000 rpm.

3

Making a Carpenter's Tool Box

3.1 INTRODUCTION

The large number of tools that carpenters need to perform their trade, demands some kind of box or bag in which to store them between jobs, or in which to transport them when moving around. Nowadays, bags (basses) and holdalls are more popular for fitting into vans and the boots of cars, and tool boxes, as such, are rarely seen on site. However, the carpenter's tool box is still a better container for tools, if only to be left at home for storage, while a holdall conveys required-tools back and forth to the workplace.

3.2 CONSTRUCTION

Traditionally, the design and construction of this box was made up as follows.

3.2.1 Carcase

Figure 3.1: This comprises the top, bottom and side material, which should be of selected softwood, straight-grained and free from large knots and other defects. To keep the weight of the box to a minimum, the finished thickness should be 13 mm, the width – ex 175 mm – finished to 170 mm. The corners of the carcase should be formed with through-dovetail joints, glued together.

3.2.2 Cladding

The front and back of the box should have 4 mm thick plywood glued and pinned to its edges.

3.2.3 Tray

A shallow tray or drawer, for holding small tools – especially edge-tools such as chisels – is made to fit the inside top of the box. This is supported by 20 × 9 mm finished hardwood side runners, glued and pinned or screwed to the sides of the box with $\frac{3}{4}$ × 6 countersunk screws, two each side. The tray, with dovetailed corners and 6 mm thick cross-divisions, is made up from 70 × 9 mm finished softwood and a 4 mm plywood base.

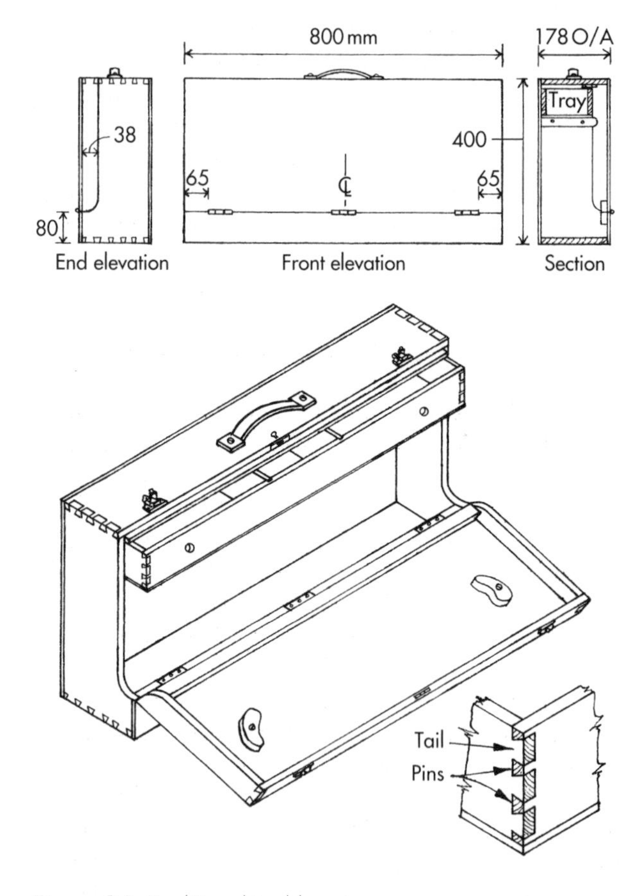

Figure 3.1 Traditional tool box

3.2.4 Hinge Fillets

These are glued and pinned to the inside faces of the lid and box to complete the structure and accommodate the hinges. They should be made from at least 28 × 16 mm finished softwood.

3.2.5 Fittings

These comprise $1\frac{1}{2}$ pairs of 50 or 63 mm butt hinges or a continuous strip hinge, sometimes referred to as a piano hinge, a case handle with small bolts, case clips (one pair) and box lock or, alternatively, a padlock and hasp and staple of 75 mm safety pattern type.

3.2.6 Construction Details

There is a traditional method for setting out dovetails, but the one described here, developed through an attraction to geometry some years ago, is preferred and recommended for its simplicity and speed when making dovetails by hand.

After cutting up the carcase material squarely with a panel saw to the length and height of the box, with an allowance of 2 mm (1 mm each end) on each of the four pieces, mark the thickness of the material, plus the allowance, 13 + 1 mm, in from each end on two pieces only, one long and one short. Square these around the material (face, edge, face, edge) with a sharp pencil. Mark the other two pieces of carcase from the two already marked. These marks are called shoulder lines.

3.2.7 Forming Through-dovetails

Figure 3.2: The sides of the box are now ready for dovetailing. First, select the best face-side and mark a centre line across the width, between the shoulder and the end of the timber. This line is used to plot the dividing points for the tails. Now decide how many dovetails are required and obtain a pair of sharp dividers or a pair of compasses. If you decide – on this width – to have 5, as illustrated, the dividers must be stepped out by trial-and-error stepping, $5\frac{1}{2}$ steps from one edge and $5\frac{1}{2}$ from the other, i.e. left to right along the centre line, then right to left. The sixth stepping, when one point of the dividers is off the timber, over the edge, allows visual judgement to be made as to whether half a step, more or less, has been achieved. If under half-a-step is over the edge, try again; if slightly over, you could try again or let it go. (Letting it go will result in slightly wider dovetails.)

3.2.8 Dovetail Angle and Setting-out Method

The dovetail angle for softwood is usually set to a ratio of 1:6. This can be set up on a sliding bevel by alignment to a right-angled line across a board which has been marked 6 cm across and 1 cm along the edge. The bevel is now used to pick up the divider-points along the centre line and the dovetail angles are marked with a sharp pencil.

The easiest way to think of the setting-out method, is to remember that $\frac{1}{2}$ a divider step more than the

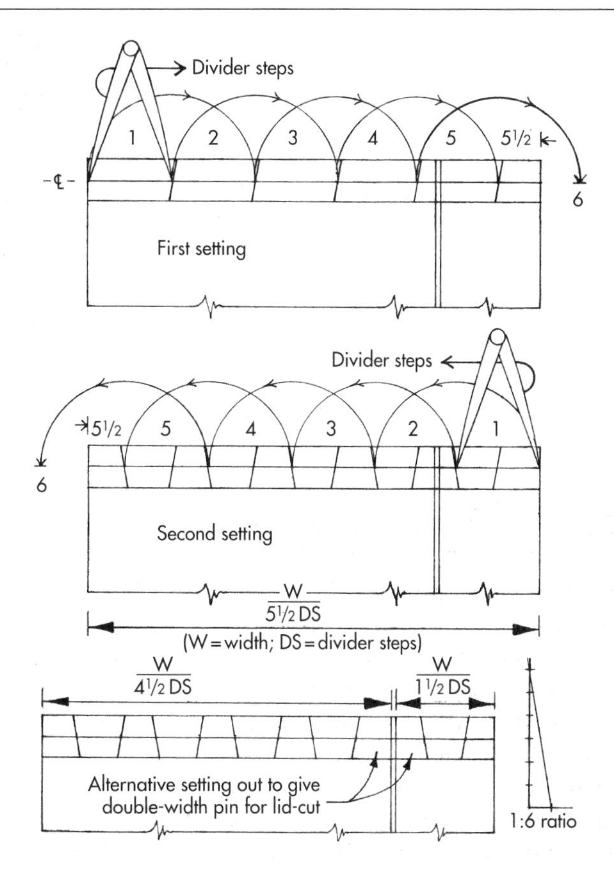

Figure 3.2 Setting-out details

number of dovetails is required, i.e. 1 dovetail, $1\frac{1}{2}$ divider steps (or spacings) across the timber; 2 dovetails, $2\frac{1}{2}$ spacings; 3 dovetails, $3\frac{1}{2}$ spacings, etc. If larger dovetails than pins are required by this method, as shown in Figure 3.6, simply step out the dividers until much less than half a step remains at the edge.

3.2.9 Cutting the Tails and Pins

Next, square the tails across the top edge of the end grain, cut carefully with a fine saw (preferably a dovetail saw, but a tenon saw will suffice), remove the outer shoulders with the same saw, the inner shoulders with a coping saw and bevelled-edge chisels. Hold each joint together and mark the tails onto the end-grain. Square these lines onto the faces of the bottom and top of the box. Remember that it is the dovetail shapes that are now removed, and repeat the cutting operation for the pins as for the tails.

3.2.10 The Need for Speed of Assembly

The thin carcase material is very prone to distortion across its width, known as cupping, especially if the growth rings are tangential to the face, as shown in Figure 3.3. For this reason, the joints should be formed as quickly as possible and the carcase glued together and checked diagonally for squareness, as in Figure 3.4. If the dovetails are a snug fit, there should be no need for the box to be held together with cramps while the glue is setting.

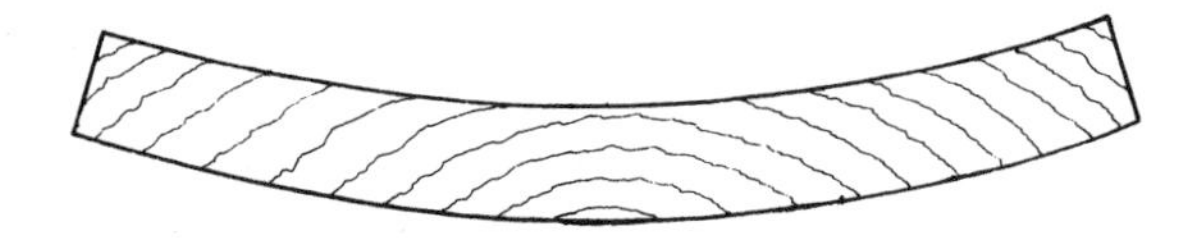

Figure 3.3 Cupping of carcase material

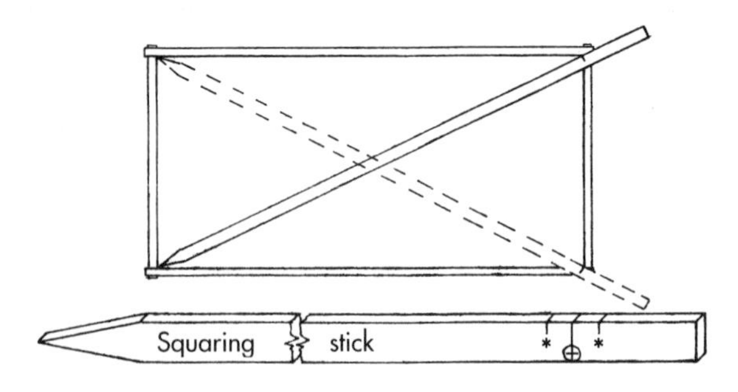

Figure 3.4 Checking the box for squareness: check both diagonals with stick; if two marks * are recorded, distort the structure until the centre ⊕ of these marks registers on both checks

3.2.11 Preparation for Cladding

Figure 3.5: After the glue has set, a gauge line representing the eventual edge of the lid, can now be marked – or may have been marked earlier to assist in setting out the dovetails – and should terminate as a quadrant shape 80 mm up from the base. This quadrant shape *only*, should be cut on each side with a coping saw before cladding.

3.2.12 Applying the Cladding

The 4 mm thick plywood for the front and back, having been cut slightly oversize by a few millimetres, is then glued and pinned with 18 mm panel pins at

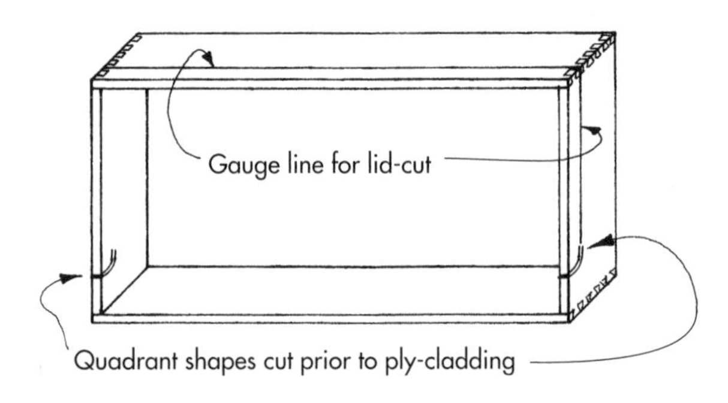

Figure 3.5 Quadrant shapes cut prior to ply-cladding

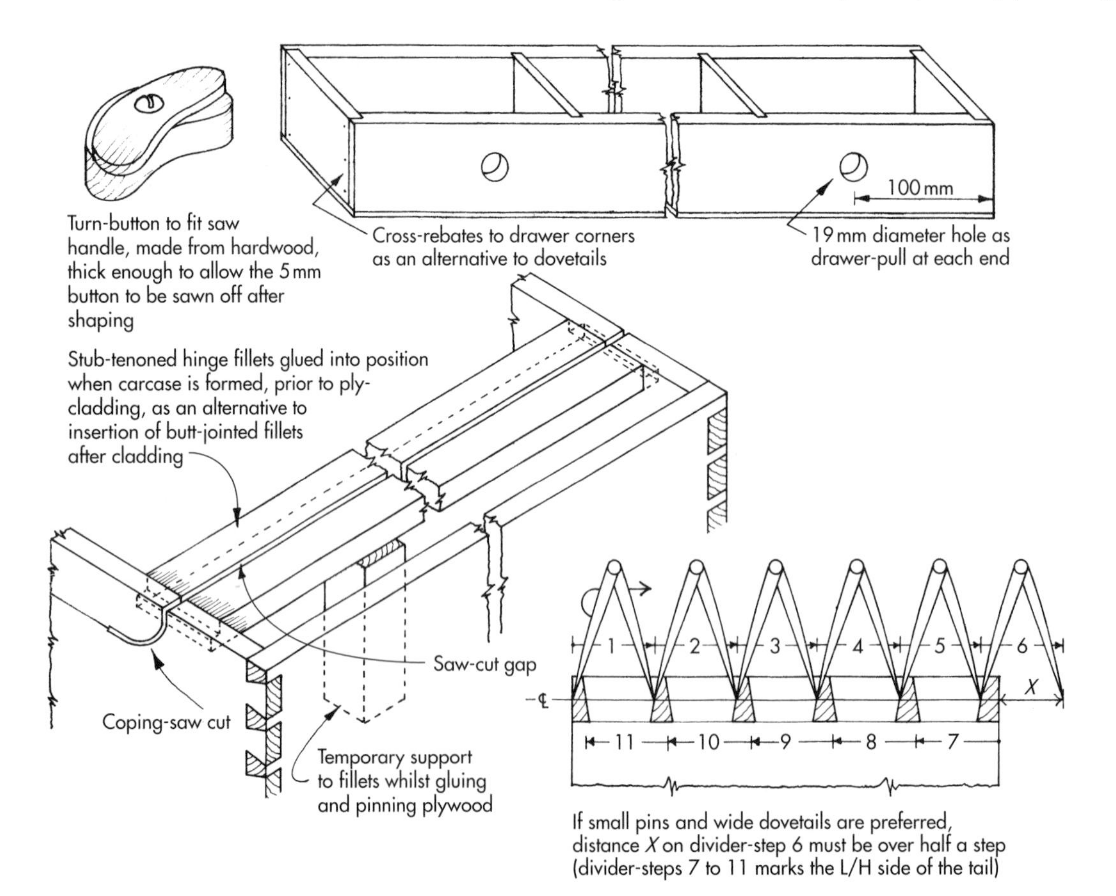

Figure 3.6 Alternative details

approximately 75 mm centres (c/c) into position, transforming the assembly into an inaccessible box. The box is then cleaned up with a smoothing plane to a flat finish all round and sanded.

3.2.13 Releasing the Lid

By careful use of a tenon saw (or a finely set panel saw) working from each top corner, across the top and down the sides to meet the pre-cut quadrants, then across the plywood front, the lid is cut and released. The hinge fillets, including the end-grain abutments, are then glued into position and pinned through the face of the ply (Figure 3.6). When set, the lid can be hinged to the box, after cleaning up all the sawn edges with glass paper only – no planing!

3.2.14 Adding Fittings and the Tray

The handle, bolted not screwed, case clips and hasp and staple can now be fitted. First, to help transfer the weight of a full box to the lid, when lifted, a minimum 35 × 7 mm hardwood fillet should be glued and screwed up to the underside lock-edge of the box, projecting about 6–9 mm into the lid area, as seen in Figure 3.1.

Finally, again using through-dovetails, or simple cross-rebates (Figure 3.6), the tray is made with one or two cross-divisions, to the inside length minus 2 mm for sliding tolerance. Turn-buttons for saw handles (Figures 3.1 and 3.6) can be made and glued to the lid to hold at least two saws.

4

Making Builder's Plant for Site Use

4.1 SAW STOOL

4.1.1 Introduction

A carpenter's saw stool, sometimes called a horse or trestle, has a few variations in design, but the one shown in Figure 4.1 is most commonly used. The length and height can also vary, although the height should not be less than that shown, otherwise on hand rip-sawing operations (still done occasionally on short lengths of timber) – when the saw should be at a steep angle of about 60–70° – the end (toe) of the saw may hit the floor.

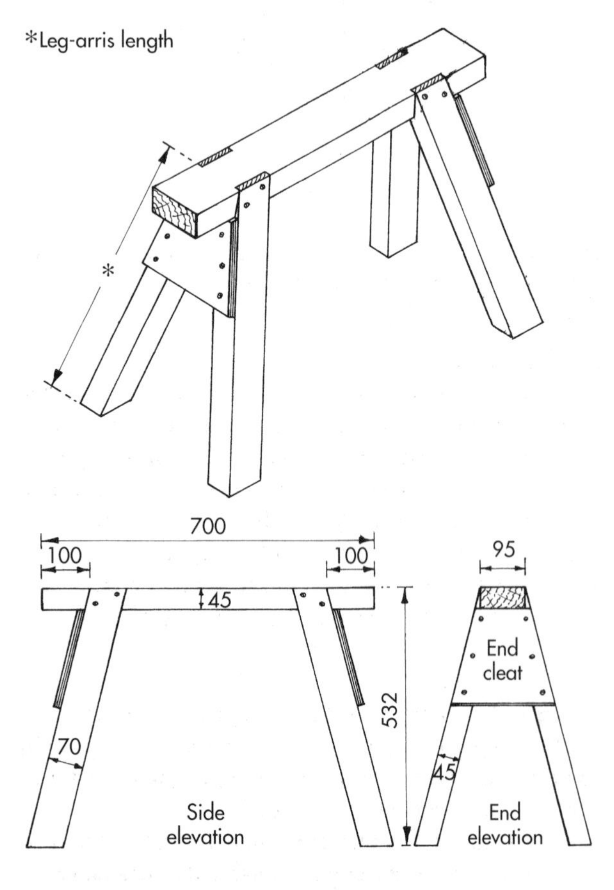

Figure 4.1 Isometric view of a typical carpenter's saw stool

4.1.2 Material Required

The material used is usually softwood and can be a sawn finish or planed all round (par). The latter reduces the risk of picking up splinters when handling the stool. Material sizes also vary, often according to what may or may not be available on site, or, for design reasons, in consideration of the weight factor of the stool.

Typical sizes for a sturdy stool are given in the illustrations, showing the top as ex. 100 × 50 mm, legs ex. 75 × 50 mm, end-cleats 6, 9 or 12 mm plywood or MDF board.

4.1.3 Angles of Legs

The angles of the legs are not critical to a degree and are usually based on the safe angle-of-lean used on ladders, which is to a ratio of 4 in 1 (4:1). This refers to a slope or gradient measured by a vertical rise of four units over a horizontal distance of one unit. This ratio works out to give approximately 76°.

4.1.4 Optional Nail-box Facility

Some tradespeople used to incorporate a tray within the leg structure, to act as a nail box. The tray, which was sub-divided to contain a variety of nails, screws, etc., had shallow sides of about 50–75 mm depth, plywood base and was usually fixed halfway up the legs on cross or longitudinal bearers. The advantages of this combined stool/nail box were outweighed by the increased weight factor, restricted access to nails, trapping of sawdust and cleaning-out difficulties.

4.1.5 Need For Geometry

By following very basic criteria and guesswork, saw stools are often roughly made without any regard for the geometry involved. This is partly to save time, of course, but also reflects a lack of knowledge of the subject. Making a saw stool should be a quick and simple operation, joinery and geometry-wise (it can be done in about an hour). However, it might appear complex and lengthy here, because additional methods have been shown regarding the geometry involved. The reason for this is because the inclination of saw-stool legs is like the inclination of hip rafters – and therefore emphasis is on the geometry to help introduce the reader to understanding the roofing chapter.

4.1.6 Length of the Leg

Before metrication of measurement in industry, the height of saw stools ranged between 21 and 24 in. The approximate metric equivalent is 532 and 610 mm. The lesser height is used here for preference. Once the vertical height has been decided, the length of the legs has to be worked out. Basically, without the necessary additional allowances, this is a measurement along the outside corner of the leg (indicated in Figure 4.1 with an asterisk) known as the *leg-arris length* (*arris* is a French/Latin-derived word used widely in the trade to define sharp, external angles).

4.1.7 Determining Leg-arris Length

The leg-arris length can be worked out in three different ways and it will help build up an understanding of the geometry to explore each of these before proceeding. These different methods are (A) by practical geometry, (B) by drawing-board geometry and (C) by a method of calculation.

Method A, By Practical Geometry

Finding the leg-arris length by this method is based on the fact that a piece of timber leaning at an angle, like the hypotenuse of a right-angled triangle, has a vertical height and a base length. Once the measurement of the base and height are known, the length of the timber (on the hypotenuse) can be worked out. These theoretical triangles must first be visualized on the side and end elevations, as illustrated in Figure 4.2. Knowing that the vertical height (vh) is to be 532 mm and the leg-angle is 4 in 1, the base measurement is worked out by dividing vh by 4, i.e.

$$\frac{vh}{4} = \frac{532}{4} = 133\,\text{mm}$$

Using Base Measurements

Figure 4.3: Now that the base measurement of the side and end elevation triangles is known to be 133 mm, this information can be used to find the unknown base measurement of the leg-arris triangle. Because the leg is leaning at the same angle in both elevations, the leg-arris base must be a 45° diagonal within a square of 133 × 133 mm.

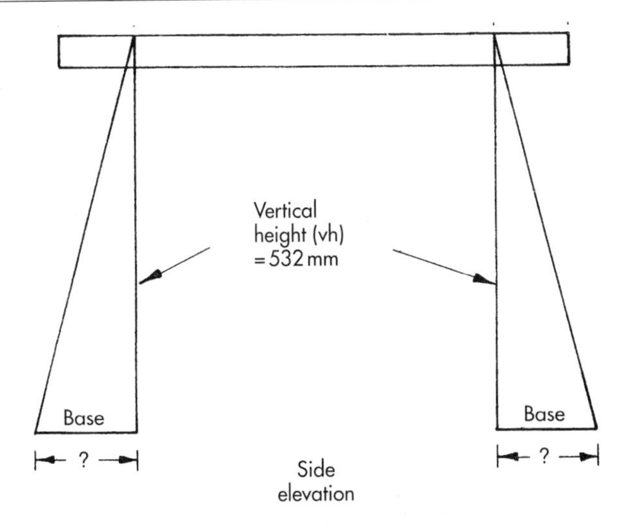

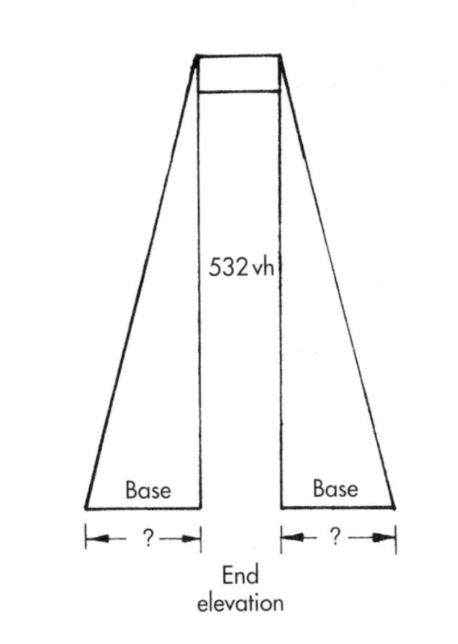

Figure 4.2 With 4 in 1 leg-angle, the base equals vh ÷ 4 = 532 ÷ 4 = 133 (base measurement = 133 mm)

Finding Leg-arris Length

All that needs to be done now, therefore, is to form a true square with these measurements, draw a diagonal line from corner to corner and measure its length. The product is the base measurement of the leg-arris triangle. This measurement is then applied to the base of another right-angled setting out, the stool's vertical height is added, and the resultant diagonal on the hypotenuse measured to produce the true leg-arris length.

Practical Method

Figure 4.4: Two practical ways of doing this are to use a steel roofing square (now called a metric rafter square by some manufacturers), or to use the right-angled corner of a piece of hardboard or plywood.

As illustrated, the base measurement *b* of 133 mm is set on each side of the angle to enable the diagonal *c*, the base of the leg-arris triangle, to be measured. This

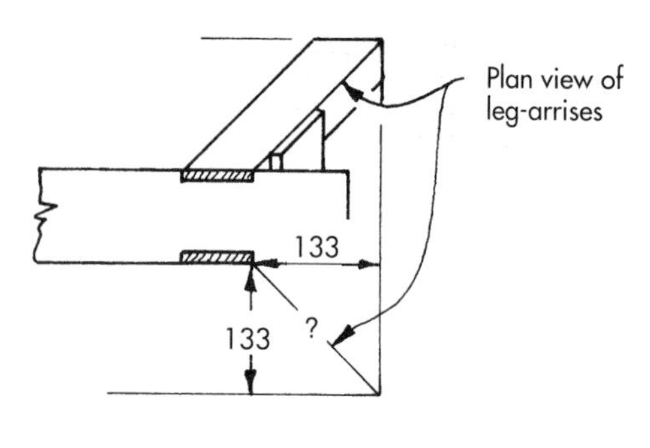

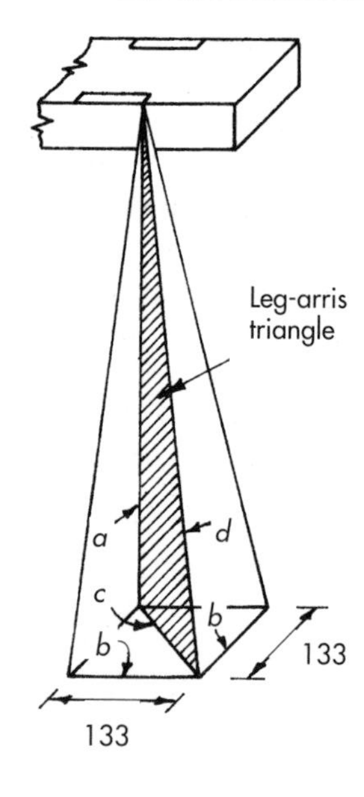

Figure 4.3 Base measurements, 133 × 133 mm, forming a square, give required diagonal base-measurement of leg-arris triangle c

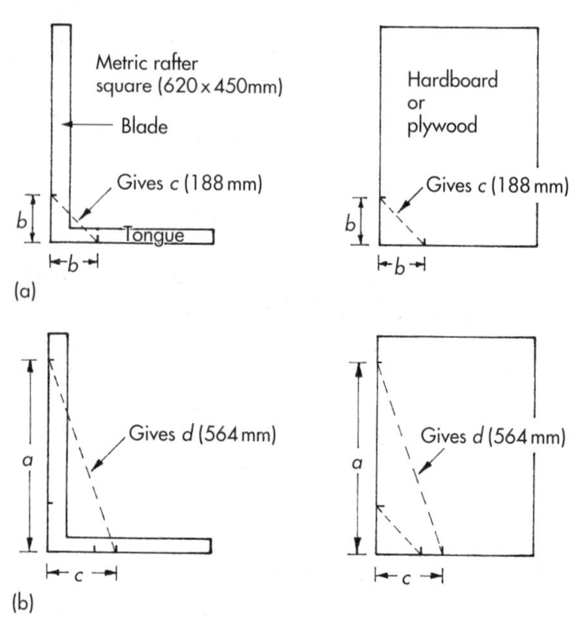

Figure 4.4(a) Set *b* on each side and measure the diagonal *c*. (b) Set *c* and *a* on each side to find length of *d*

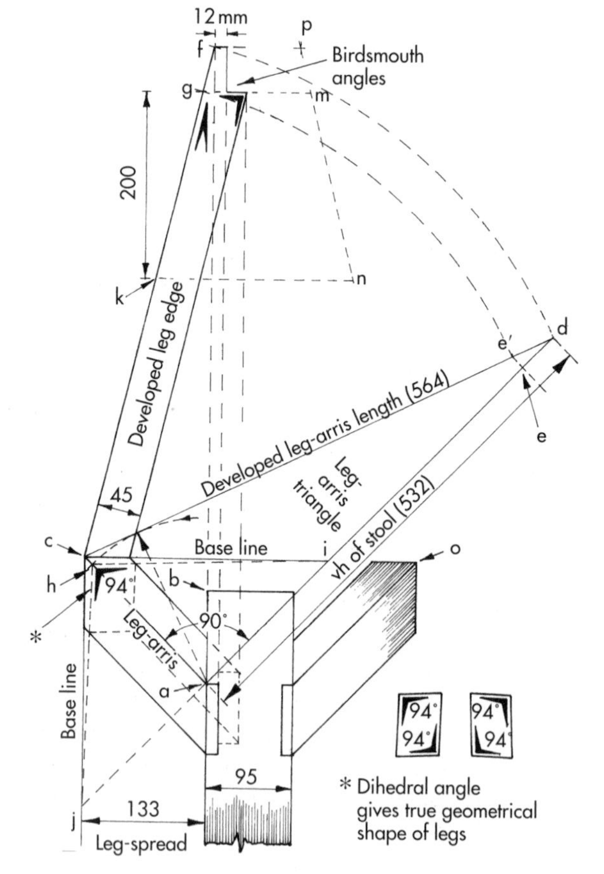

Figure 4.5 Setting out

measurement of 188 mm, forms the base of another setting out on the roofing square or hardboard, in relation to the vertical height of the stool *a*, of 532 mm, being placed on the opposite side of the angle. The diagonal *d* is then measured to produce the true leg-arris length of (to the nearest millimetre) 564 mm.

Method B, By Drawing-board Geometry

Finding the leg-arris length by this method is a useful exercise in geometry, but does not lend itself to site (or workshop) application.

Setting Out

Figure 4.5: The setting out can be full size or scaled down to a half or quarter full size. As illustrated, the first step is to draw a part-plan view of the stool's top, 95 mm (ex. 100 mm) wide, with the top of the leg-arris line marked at a, 100 mm in from the end, b. Point a is then squared across the top to locate the start of the leg-housings.

Establishing Base Lines

Next, the legs' theoretical base lines, shown here on two sides only, are established. This is determined by the legs leaning in two directions at an angle of 4 in 1, so that the vertical height divided by four, equals base measurement of leg-spread, i.e. vh ÷ 4 = 532 ÷ 4 = 133 mm. This measurement fixes the base line to the side of the stool as shown and at the return end, measured from point a, to form a right-angle with the side. The leg-arris line, a to c, is then drawn. At 90° to this, rising from point a, another line is drawn to point d, set at the vertical height of the stool. Now join c to d to establish the developed leg-arris length. Mark point e from d, equal to stool-top thickness, 45 mm (ex. 50 mm), and extend out squarely to cut developed leg-arris at e′.

Forming Birdsmouth Angles

Using point c as centre for the compass, transfer d to f and e′ to g, to intersect with line b–f. Join f to c and establish leg-edge thickness, 45 mm, parallel to f–c. Extend horizontal lines from f to g to form birdsmouth angles and developed leg edge.

Developing the End-cleat Shape

To develop the shape of the end cleat (one required at each end), mark its depth (say, 200 mm) vertically down from g and form line k through n, parallel to line c–o. Establish point p, equal to the plotting of point f, and strike a line from p, in line with o at the base of the legs, to form m–n, in relation to a final line from g through m, parallel to k–n. Angles g–m–n–k show the true cleat shape.

Dihedral Angle

This is to do with the true geometrical, sectional shape of the legs and is explained in Figure 4.6. To find this angle geometrically, as illustrated in Figure 4.5, with a as centre for the compass, describe an arc tangential (at 90°) to line c–d to cut line c–a at h. Extend line i–a to form point j. Join j–h–i to produce a dihedral angle of 94°.

True Sectional Leg-shape

Figure 4.6: The true geometrical, sectional shape of the legs is rhomboidal (rectangular, with opposite sides equal in length and parallel to each other, but with all angles out of square). This shape is to meet the dihedral angle created by the legs' angles-of-lean. Perhaps an appreciation of the dihedral angle can best be understood by holding a try square (better still, a roofing square) in a truly horizontal position, against a square section of timber positioned to lean at angles similar to the saw-stool leg. Then, as illustrated at Figure 4.6, tapered gaps showing on the face-side and edge of the leg will prove that the dihedral angle has changed from 90° for a leg in the upright position, to 94° for a leg leaning at angles of 76° (4 in 1) on two elevations.

Accepted Leg-shape

In practice, to simplify the work, the legs are left in a true rectangular shape – but it must be realized that this causes a slight misfit of the birdsmouthed legs into the

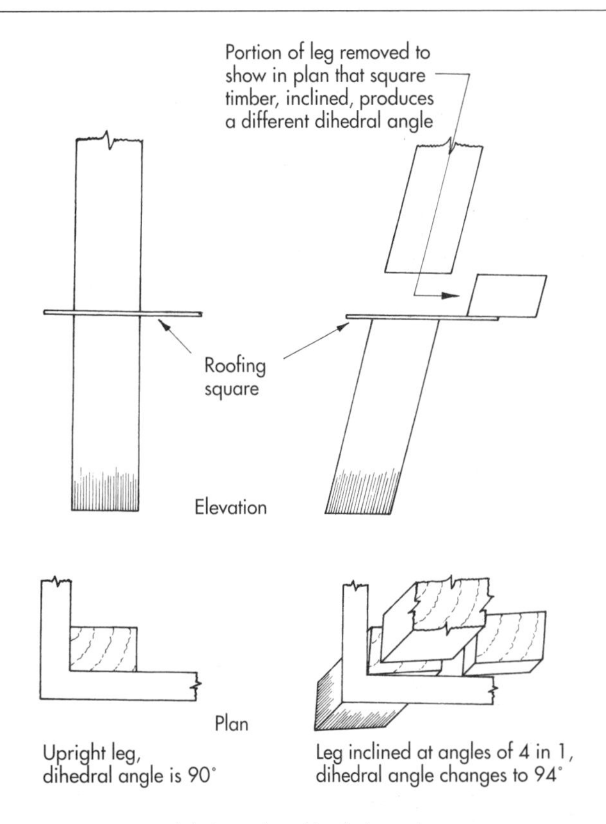

Figure 4.6 Establishing the dihedral angle

housings and an unequal seating of the end cleats onto the leg edges. The latter is the most noticeable, if viewed from the underside.

These slight irregularities are not enough in themselves to warrant the extra time and effort in making the legs to a rhomboidal shape. The strength of the stool is not impaired to any appreciable degree, nor the visual standard of finish.

Method C, By Calculation

Figure 4.7: Finding the leg-arris length by a method of calculation, entails using Pythagoras' theorem, the

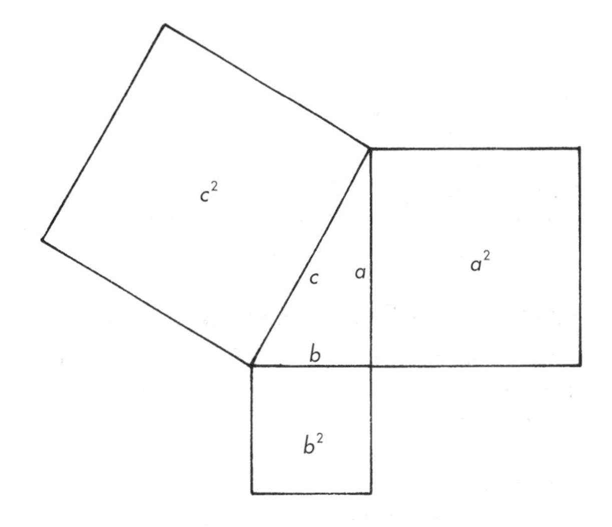

Figure 4.7 Pythagoras' theorem: $a^2 + b^2 = c^2$

square on the hypotenuse c of a right-angled triangle is equal to the sum of the squares on the other two sides (a and b), i.e.

$$a^2 + b^2 = c^2$$

Thereby it is possible to find the length of the hypotenuse if the length of sides a and b are known. Once the sum of the *square* (represented by the superscript '2' after the number, meaning that the number is to be multiplied by itself) on the hypotenuse has been worked out, the square root of that sum will give the length of the hypotenuse c, i.e. $\sqrt{c^2} = c$.

Base Measurement and Abbreviations

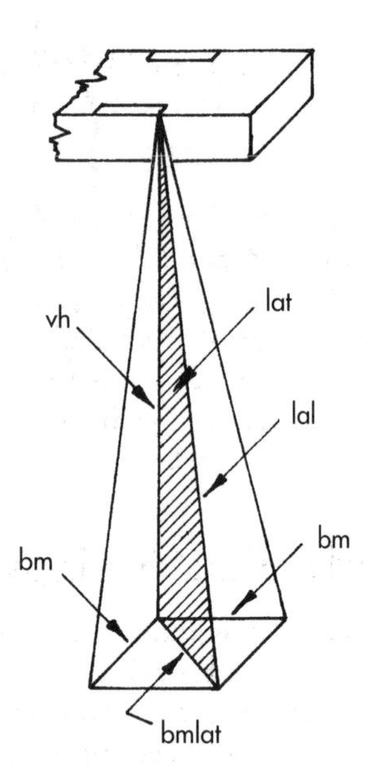

Figure 4.8 Triangular formation of leg-spread

Figure 4.8: First, as in the other methods, the base measurement of the leg spread must be determined by dividing the vertical height by four, i.e. vh ÷ 4 = 532 ÷ 4 = 133 mm base measurement. The following abbreviations have been used in the illustration showing the triangular formation of the leg spread:

vh = vertical height
lat = leg-arris triangle
lal = leg-arris length
bm = base measurement (4 in 1)
bmlat = base measurement of leg-arris triangle

Finding the Base Measurement of Leg-arris Triangle

Figure 4.9: By using Pythagoras' theorem, the base measurements are now applied to the adjacent and opposite sides of the leg-spread triangle to find the length of the hypotenuse, namely the base measurement of the leg-arris triangle (bmlat), as follows:

$$c^2 = 133^2 + 133^2 \quad (c^2 = 133 \times 133 + 133 \times 133)$$
$$c^2 = 17\,689 + 17\,689$$

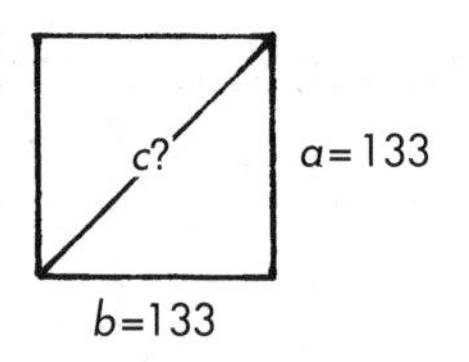

Figure 4.9 Leg-spread triangle

$$c^2 = 35\,378$$
$$c = \sqrt{35\,378} \quad (c = \text{the square root of } 35\,378)$$
$$c = 188.09 \text{ (say 188)}$$

Therefore, bmlat = 188 mm.

Finding the Leg-arris Length

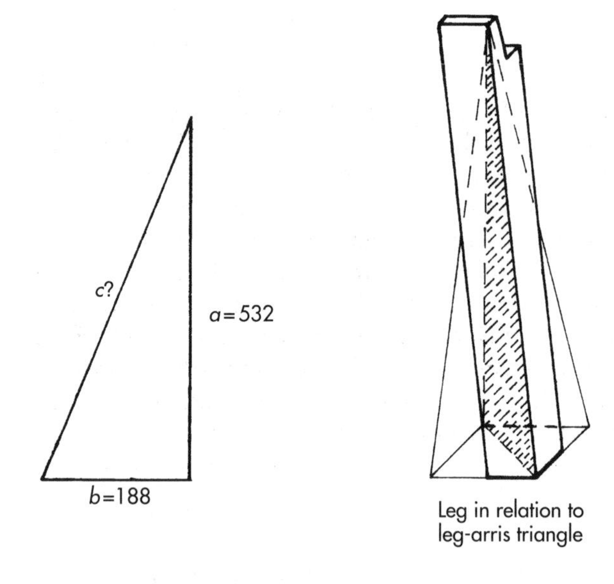

Figure 4.10 Leg-arris triangle

Figure 4.10: Finally, the leg-arris length (lal) is found by applying vh and bmlat to the same formula:

$$c^2 = 532^2 + 188^2 \quad (c^2 = 532 \times 532 + 188 \times 188)$$
$$c^2 = 283\,024 + 35\,378$$
$$c^2 = 318\,402$$
$$c = \sqrt{318\,402} \quad (c = \text{the square root of } 318\,402)$$
$$c = 564.271 \text{ (say 564)}$$

Therefore, leg-arris length = 564 mm.

4.1.8 Adding Cutting Tolerances

Now that the true leg-arris length has been determined by one of the foregoing methods, at least another, say 35 mm at the top, plus say 11 mm at the bottom, cutting tolerance has to be added to the leg as an allowance for the angled setting out involved. The length of legs, therefore, equals the true leg-arris length, 564 mm, plus 46 mm tolerance, equals 610 mm.

4.1.9 Setting Up the Bevel(s)

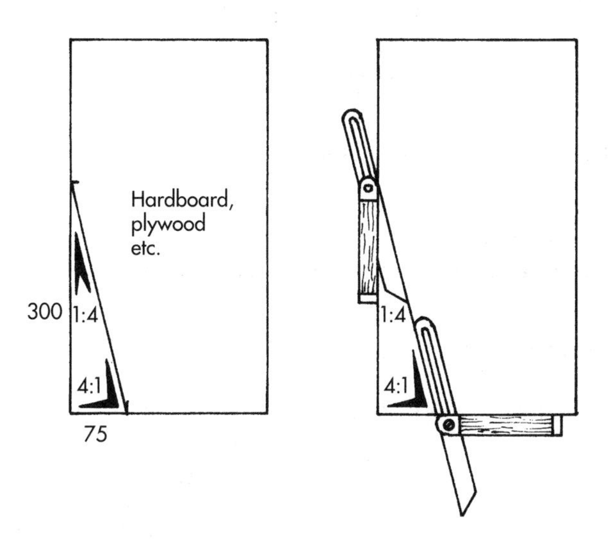

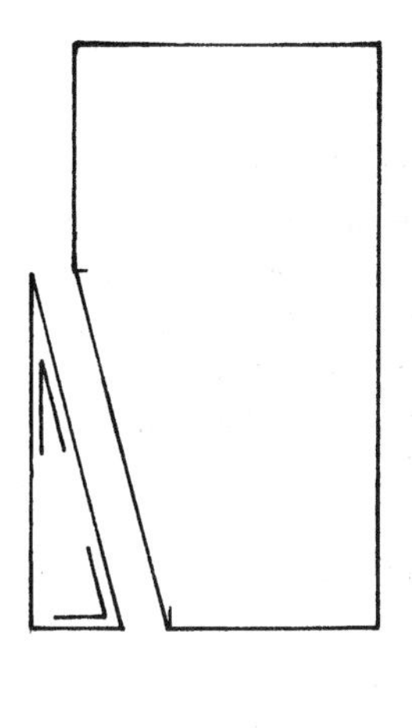

Figure 4.11 Mark 4 in 1/1 in 4 angle, set up bevels, or cut out as template

Figure 4.11: The legs and stool top can now be cut to length, ready for marking out. The marking out can either be done with a carpenter's bevel (or bevels), or with a purpose-made template. As illustrated, the 4 in 1/1 in 4 angles are set out against the square corner of a piece of hardboard, plywood or MDF board, to a selected size. Sizes of 300 mm and 75 mm are advisable as a minimum ratio if the setting out is to be cut off and used as a template in itself. If carpenters' bevels are to be used, set up one at 4 in 1 and the other at 1 in 4, as illustrated. If only one bevel is available, set and use at the 4 in 1 (76°) angle before resetting for use at the 1 in 4 (14°) angle.

4.1.10 Alternative Setting Out

Figure 4.12: This shows an alternative way of setting out the two required angles, to enable a carpenter's bevel to be set up. First, mark a square line from point a on the face of a straightedge or on the underside-face of the stool-top material. Then, from point a, mark point b at 4 cm (40 mm) and point c at 1cm (10 mm). Draw line c through b to establish the first required angle of 4 in 1 (76°). From point a, mark point d at 1 cm and point e from either side of a, at 4 cm. Draw line e through d to establish the second angle of 1 in 4 (14°). The advantage of setting out on the underside of the stool's top, is that the setting out will remain visible for years, ready at any time for making the next stool.

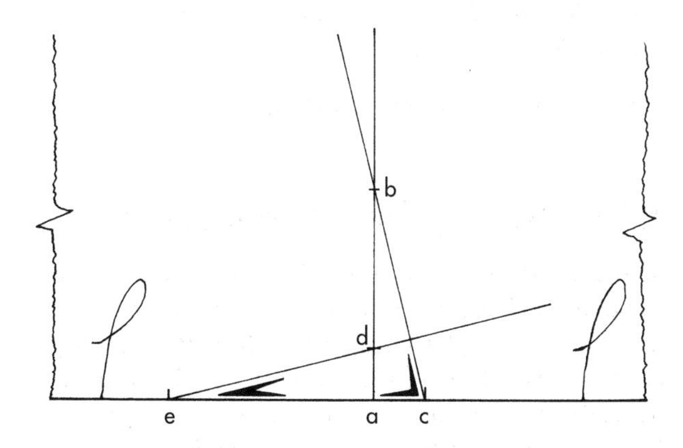

Figure 4.12 Alternative setting out

4.1.11 Marking Out the Legs

Figure 4.13(a): Before attempting to mark out the birdsmouth cuts, it is essential to identify and mark the starting point on each leg to avoid confusion. As illustrated, this point is on each inside-leg edge, about 6 mm down, and relates to the uppermost point of the compound splay cut in relation to the top surface of the stool. This mark is referred to as zero point.

4.1.12 Two Slopes Down, Two Slopes Up

Figure 4.13(b): From zero point, the 4 in 1 angle always slopes down on edge A, turns the corner and, likewise, slopes down on face-side B, turns again and this time slopes up on edge C, and up on back-face D, to meet the first point (zero) on face-edge A. So, on the four sides of each leg, the marking sequence leading from one line to the other, is two slopes down, two slopes up, back to zero point on leg-edge A.

4.1.13 Waste Material

Note that in Figure 4.13(b) the shaded areas represent the timber on the waste side of the cut. Graphic demarcation such as this – but roughly marked – is often used by carpenters and joiners to reduce the risk of removing the wrong area of material.

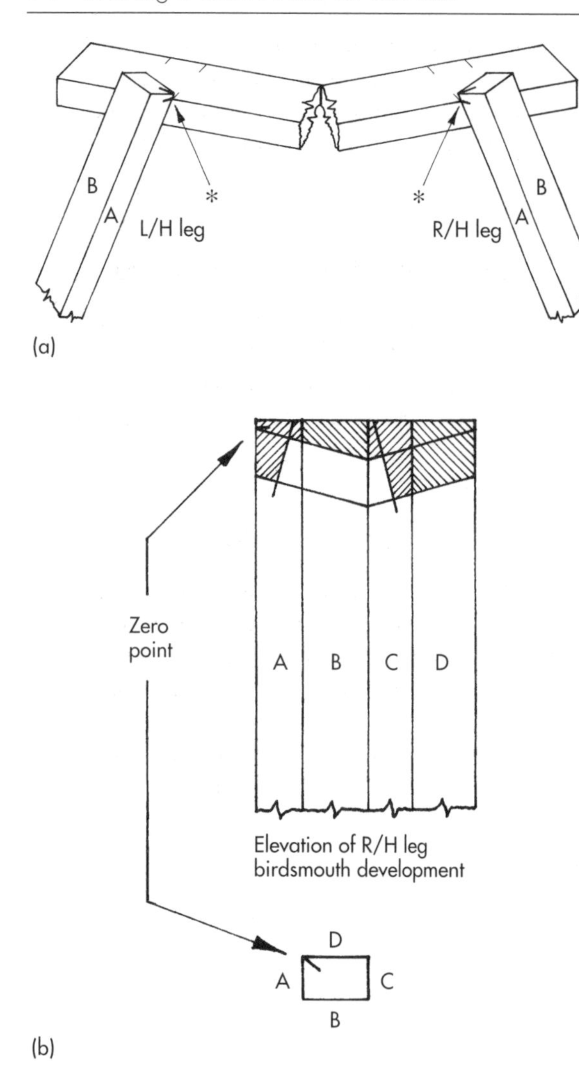

Figure 4.13(a) Marking out of legs starts from the points marked *. (b) Plan view of the marking sequence

4.1.14 Marking Zero-point Arrises

The zero-point arris can be quickly determined on each leg by placing the four legs together, with the inside-leg edges and back-faces (A and D) touching, and by marking the top corner of each leg on the middle intersection, as illustrated in Figure 4.14.

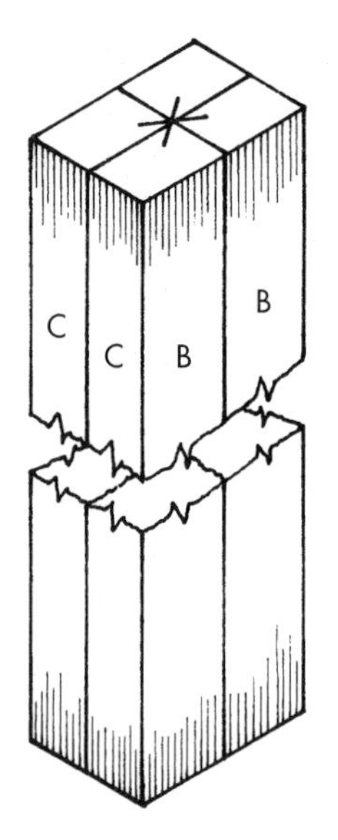

Figure 4.14 Marking zero-point arrises

4.1.15 Use of Template – First Stage

Figure 4.15: By using the template positioned on its 4 in 1 (76°) angle, surfaces A, B, C and D of the leg are first marked as illustrated here and as described in Figure 4.13(b).

4.1.16 Use of Template – Second Stage

Figure 4.16: Returning to surface A, the template is again positioned and marked on the 4 in 1 angle, set at 48 mm down, measured at right-angles to the first line. This represents the stool-top thickness, 45 mm, plus 3 mm to offset the fact that the stool-top thickness registers geometrically at a greater depth on the 4 in 1 angled legs. Next, measure 12 mm in on the top line to represent the beak of the birdsmouth-cut and use the 1 in 4 (14°) angle of the template to mark the line as shown.

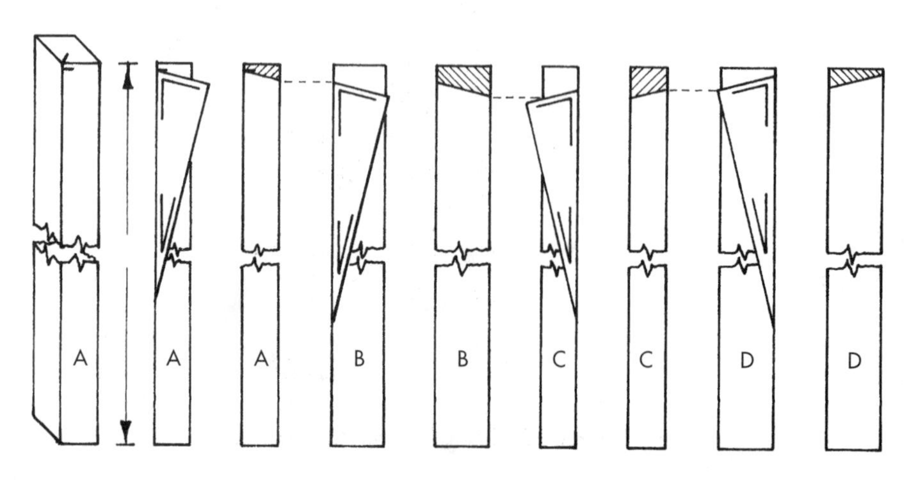

Figure 4.15 Marking out of the legs – first stage

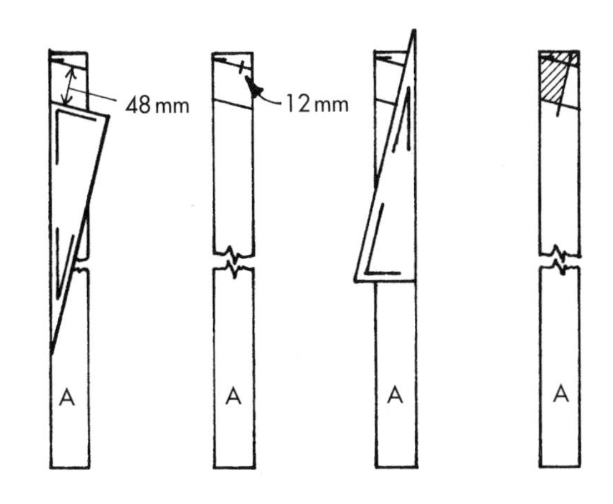

Figure 4.16 Marking out of the legs – second stage

4.1.17 Use of Template – Third Stage

Figure 4.17: To complete the marking out of the first leg, the lower line on surface A, representing the stool-top thickness, is picked up on surface B by the template and marked parallel to the line above. It is picked up on surface C and again marked parallel to the line above, then the 1 in 4 (14°) angle is marked in at 12 mm to complete the beak of the birdsmouth (this birdsmouth-marking is on the opposite side to the one marked on surface A). Finally, the lip of the birdsmouth is picked up from C and marked on surface D, yet again parallel to the line above.

4.1.18 Complete the Marking Out

Figure 4.18: As illustrated, complete the marking out of the four legs, being careful to follow the marking sequence already explained, i.e. slope down from zero point on surface A of each leg to produce two opposite pairs of legs.

4.1.19 Cutting Birdsmouths

Figure 4.19: By using a suitable handsaw (a sharp panel saw is recommended, as hardpoint saws are ineffective for ripping), the four birdsmouths should be carefully cut to shape by first ripping down from the beak and then by cross-cutting the shoulder line on the lower lip. This can be done on a stool, if one is available, or on any convenient cutting platform – failing access to such

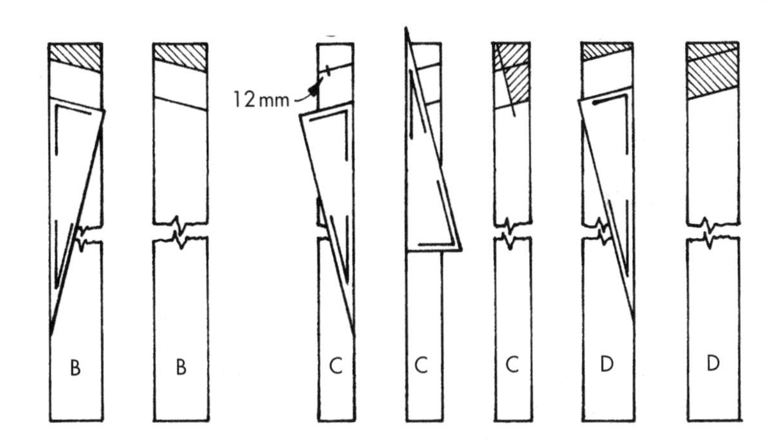

Figure 4.17 Marking out of the legs – third stage

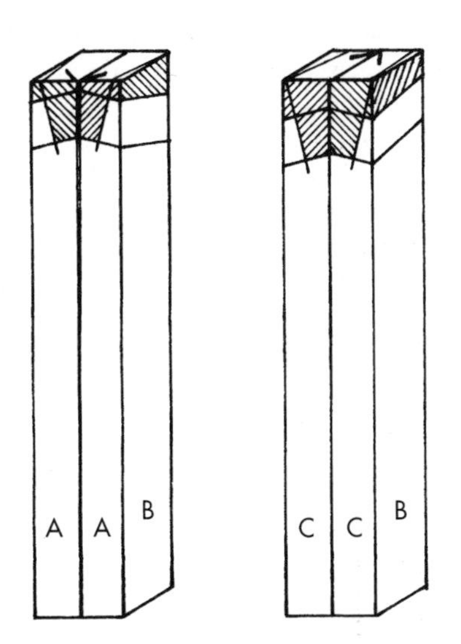

Figure 4.18 The four legs marked out

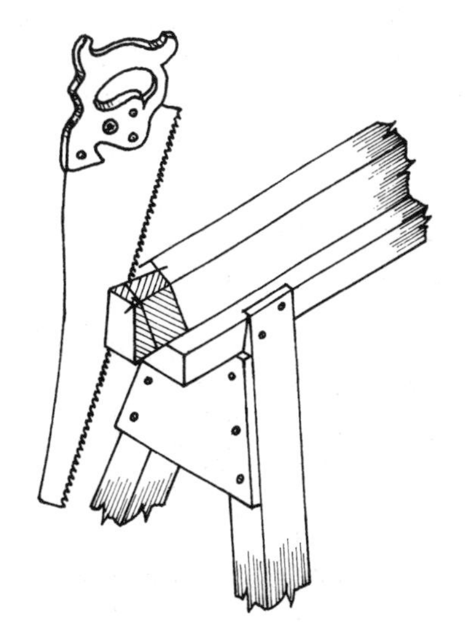

Figure 4.19 Cutting birdsmouths to shape

equipment as a joinery-shop vice, etc. The cross-cutting of the shoulders can also – more accurately – be cut with a tenon saw.

4.1.20 Checking Matched Pairs

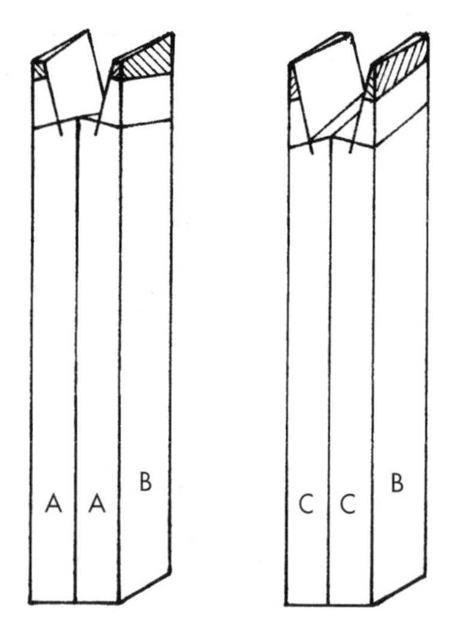

Figure 4.20 The completed, matched pairs

Figure 4.20: After cutting, check that the legs match in pairs and hold each birdsmouth against the side of the stool-top material at any point to check the cut for squareness. If necessary, adjust by chisel-paring; inaccuracies, if any, are usually found on the inner surface of the end grain of the lower lip.

4.1.21 Marking Leg-housings

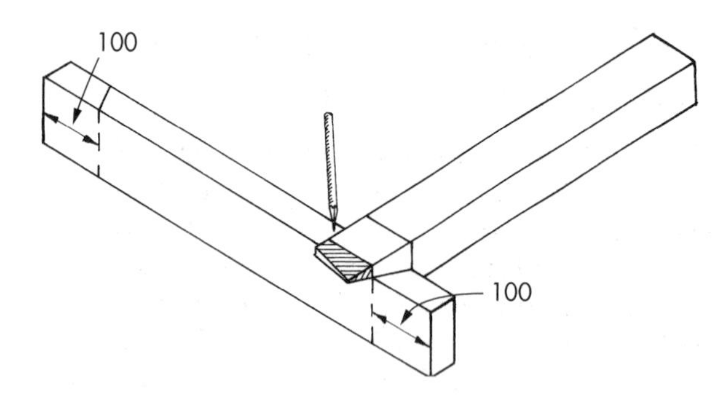

Figure 4.21 Mark leg-housings

Figure 4.21: Next, as illustrated, the birdsmouthed legs are held in position on the side of the stool top, 100 mm in from the ends, and the sides marked to indicate the housings.

4.1.22 Cutting Leg-Housings

Figure 4.22: When marked and squared onto the face-sides and underside, the housings are then gauged to

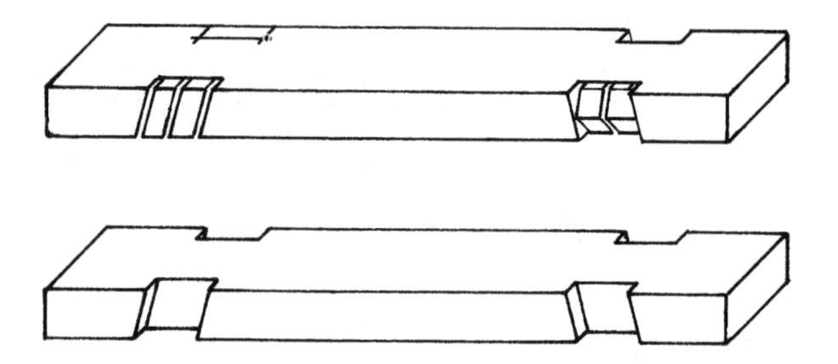

Figure 4.22 Gauge housings to depth, cut and pare out with wide bevel-edged chisel

12 mm depth on both sides of the stool top, ready for housing. The angled sides of the housings can be cut with a tenon saw or panel saw and pared out with a wide bevel-edged chisel. The recommended chiselling technique, for safety and efficiency, is to pare down (known as vertical paring) – either by hand pressure or with the aid of a mallet – onto a solid bench or boarded surface from alternate faces of the stool top.

4.1.23 Alternative Leg-fixings

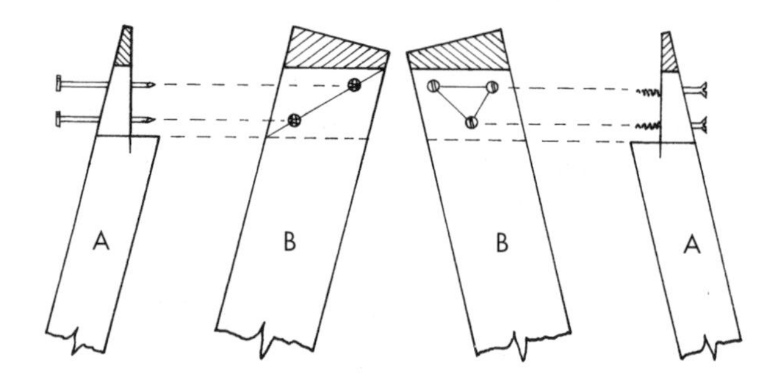

Figure 4.23 Alternative fixings

Figure 4.23: The stool top can now be set aside and the legs prepared for fixings – if screws are to be used. As illustrated, two or three fixings may be used and these are usually judged rather than marked for position. However, for strength, the fixings should neither be too near the edges nor too close together. Shank holes should be drilled and countersunk for, say, 2 in × 10 gauge screws. If nails are used instead of screws, 63–75 mm round-head wire nails are preferred.

Take care when screwing or driving-in fixings, to ensure that they are parallel to the stool-top's surface, i.e. at a 4 in 1 angle to the surface of the legs.

4.1.24 End Cleats

Figure 4.24: To enable the assembly of the stool to flow without interruption, it will now be necessary to prepare the end cleats. By using the template as illustrated, these can be set out economically on board material or plywood and cut to shape with the panel or hardpoint saw. The size and shape of the cleats is critical, as the

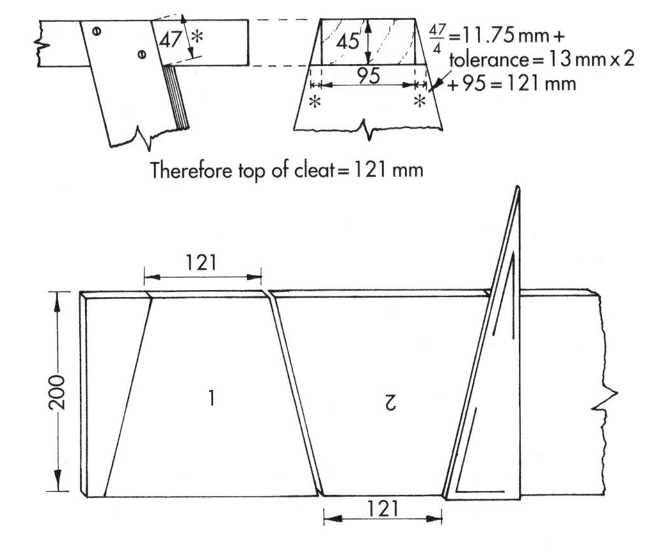

Figure 4.24 End cleats

assembled legs are unlikely to assume the correct leg-spread on their own and will rely on the end cleats to correct and stabilize their posture.

Common Error

It is a common error in stool-making to fix the legs and mark the shape of the cleats unscientifically from the shape of the distorted leg assembly. This is done by marking the outer shape of the stool's end onto the cleat material laid against it.

4.1.25 Working Out Cleat Size

The setting out of the cleats, therefore, should be marked from the template (or bevel) in relation to a measurement at the top of the cleat. This measurement is made up by the width of the stool top and the visible base of the birdsmouth beak on each side. This can either be measured in position or determined by dividing the 4 in 1 diagonal thickness of the stool top by four. That is, as illustrated in Figure 4.24, 47 ÷ 4 = 11.75 mm × 2 = 23.5 mm, plus width of stool top (95 mm) giving 118.5 mm. Add 2.5 mm as a working tolerance and for final cleaning up, therefore, top of cleat equals 121 mm.

4.1.26 Fixing the Legs

Figure 4.25: Now ready for assembly, the legs are fitted and screwed or nailed into the housings, making sure that the lower lip of the birdsmouth cut fits tightly to the underside of the stool top.

4.1.27 Fixing the Cleats

Figure 4.26: Next, the top edges of the cleats are planed off to a 4 in 1 angle, to fit the underside of the

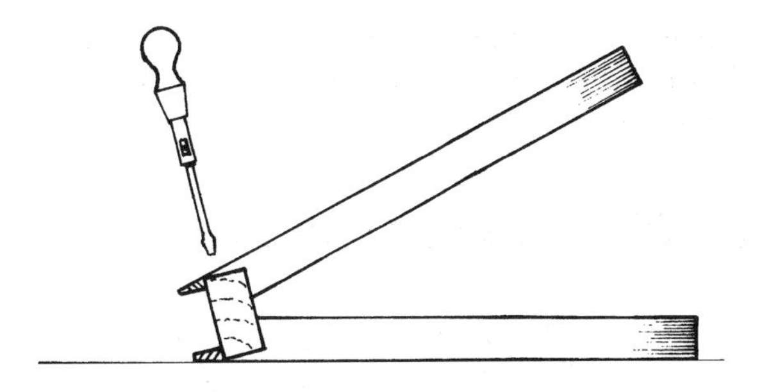

Figure 4.25 Legs screwed into housings

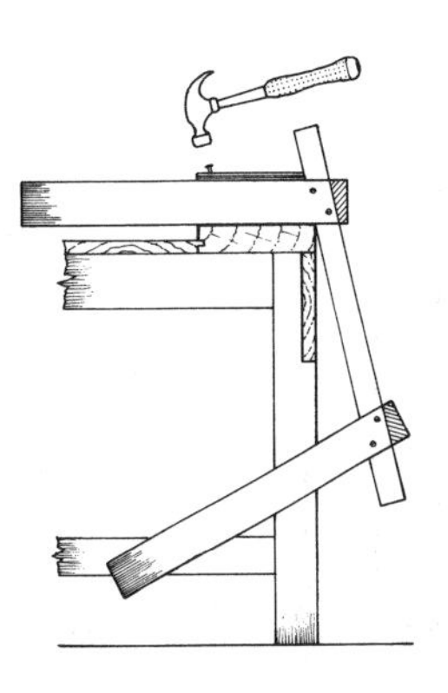

Figure 4.26 Fix plywood cleats with 50 mm round wire nails or $1\frac{1}{2}$ in × 8 gauge screws

stool top, then the cleats are nailed or screwed into position, using 50 mm round-head wire nails or $1\frac{1}{2}$ in × 8 gauge screws. During the fixing operation, to avoid weakening the leg connections, the stool should be supported on a bench, as illustrated, or any other suspended platform.

4.1.28 Removing the Ears and Marking the Leg-waste

Figure 4.27: Next, the projecting ears are cut off from the legs, with a slight allowance, say 1 mm, left for planing to an even finish with the stool top. After cleaning-up the top to a flat finish, measure down each leg and mark the leg-arris length (previously determined) to establish the stool's height. Join these marks with a straightedge and mark the line of feet on the ends and sides of the stool – or, if preferred, use the template or bevel instead of the straightedge.

4.1.29 Removing Leg-waste

Figure 4.28: Finally, lay the stool on its alternate sides, and alternately end for end, up against a wall, bench, etc., and carefully cut off the waste from the legs with a panel or hardpoint saw. Clean off waste material from the sides of cleats and clean off any splintering arrises to finish.

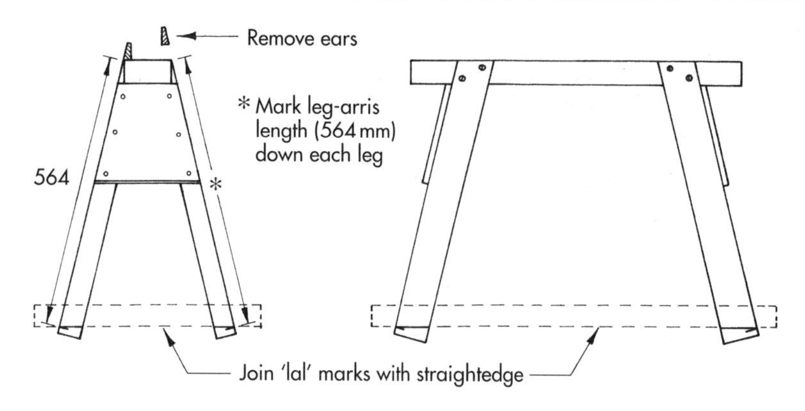

Figure 4.27 Join 'lal' marks with straightedge (or use template or bevel) to establish line of feet

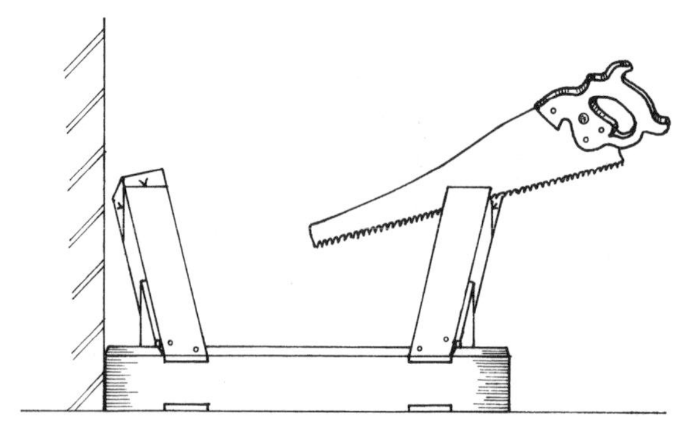

Figure 4.28 Lay the stool on its side, end against the wall, and remove leg-waste

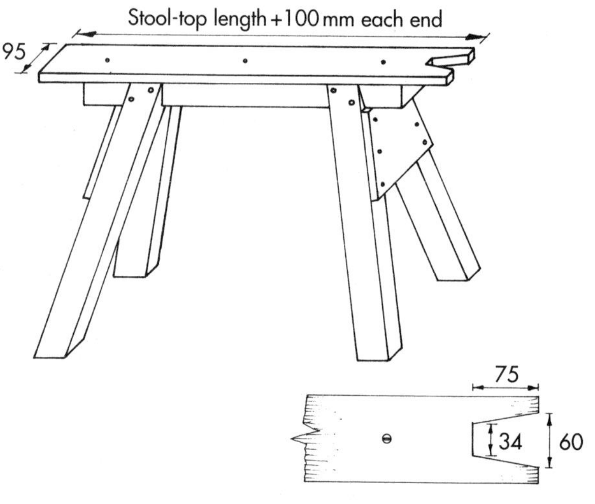

Figure 4.30 Optional vee-ended top

Uneven Legs

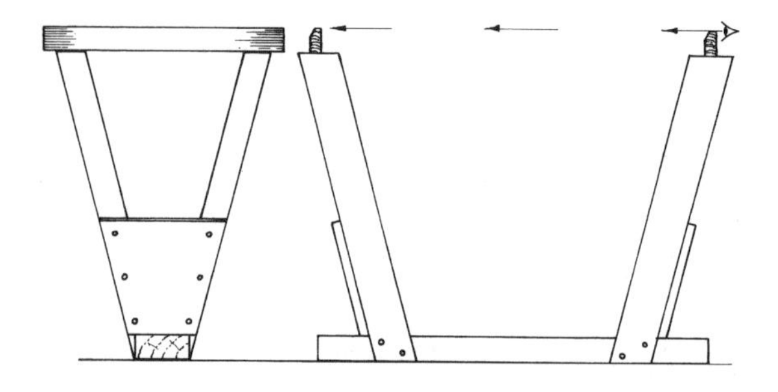

Figure 4.29 Using winding sticks on uneven legs

Figure 4.29: If it is discovered that the stool is uneven and wobbly, turn it upside down and, as illustrated, place *winding sticks* (true and parallel miniature straightedges for checking twisted material) on the feet at each end and sight across for alignment. If in line, then the floor is uneven; if out of true, plane fractions off the two high legs and re-sight and re-plane, if necessary, until the legs are even.

4.1.30 Vee-ended Top

Figure 4.30: Sometimes, a vee-shape is cut in the end of a stool to facilitate the holding of doors-on-edge when they are being 'shot-in' (planed to fit a door-opening). To accommodate this, the legs ought to be set in 150 mm from each end of the stool top, instead of 100 mm.

This allows the top of the stool to touch the wall at one end, while the other holds the door in the vee cut without touching the bottom of the cleat. However, this does create a potentially dangerous stool if used as a 'hop-up', and you step along the stool towards an over-extended end.

Alternatively, as illustrated, a separate vee-ended board, of ex. 25 mm material, can easily be made and fitted to the top of the stool – and removed again, when required.

4.2 SAW HORSE

4.2.1 Introduction

Figure 4.31: Special vices were used for saw sharpening, but more commonly wooden frames known as *saw horses* or *saw stocks* are used. The material is usually softwood and can be of a prepared or sawn finish. The wooden jaws, made from ex. 75 × 25 mm material, which hold the saw in the frame are known as *saw chops*. The frame is made up of two ex. 75 × 50 mm (or ex. 100 × 50 mm) stiles or legs, ex. 75 × 25 mm cross rails on each side, acting as foot and knee rails, optional top rails above the knee rails and an optional diagonal brace.

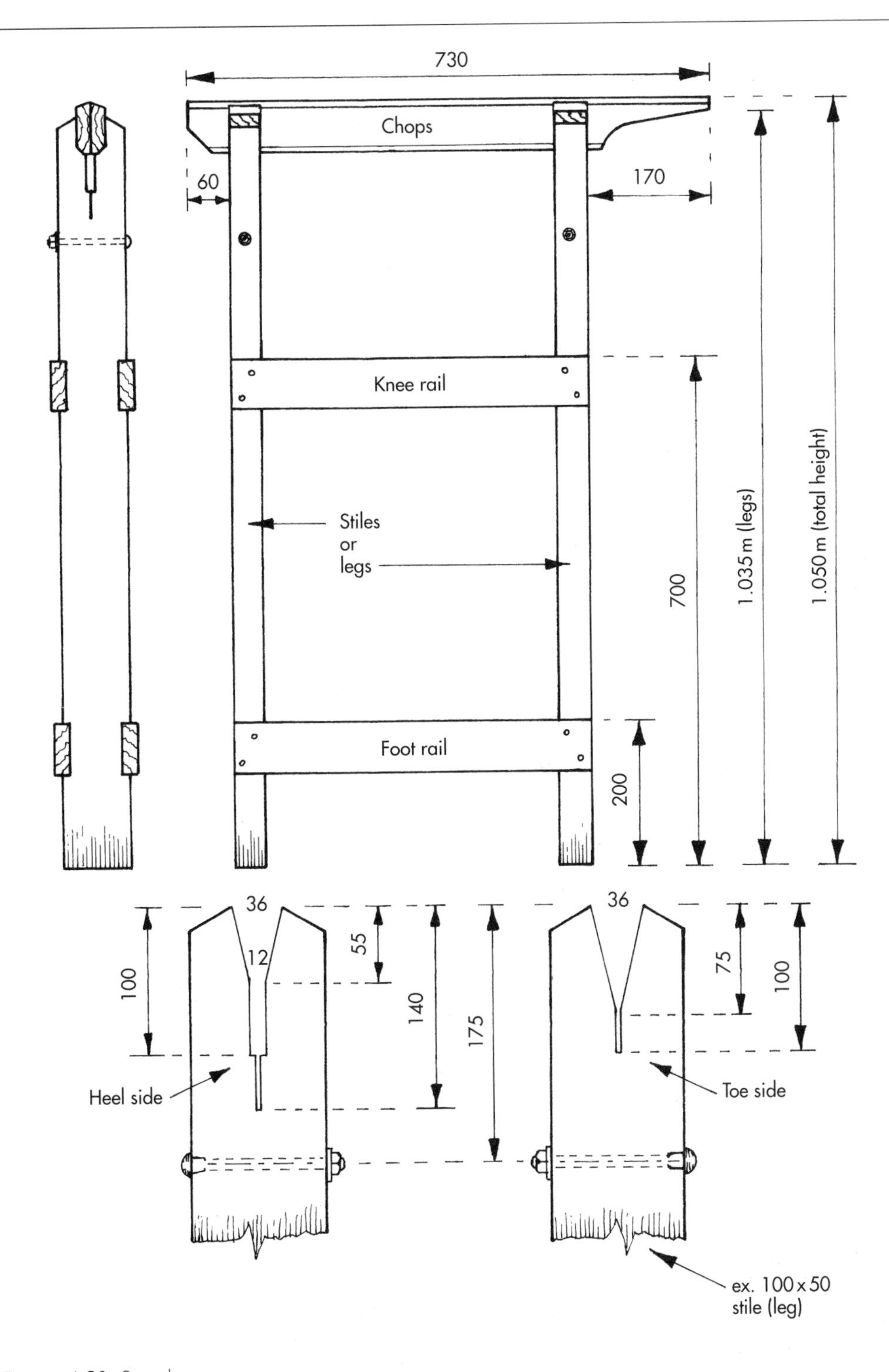

Figure 4.31 Saw horse

Shop- or Site-made Horses

The diagonal brace was usually seen on site-made horses, to strengthen the simple construction of the surface-nailed rails. On workshop-made horses, the rails are usually housed and screwed into the stiles and so there is no need for a brace. Also, with shop-made horses, the saw chops are often made of hardwood and a coach bolt is inserted through the edge of each stile, near the top, to offset the tendency for the legs to split when the saw chops are driven in to the vee-cuts to hold the saw.

4.2.2 Preparing the Stiles

The first step in making a saw horse is to prepare the stiles. Because it is important that a person takes up the correct posture at the horse, the height is critical and ideally should be to suit the individual. The total height of 1.050 m given here would be suitable for a tall person of about 1.830 m (6 ft).

4.2.3 Provision for Saw Chops

The vee-cut to receive the saw chops should be marked out in relation to the vee-shaped housings to be cut in the chops themselves. As illustrated in Figure 4.31, the vee-shape should promote a slow (gradual) wedge action, as opposed to a less acute angle that would not take such a good grip on wedge shapes driven into it. At the base of the vee-cuts, a saw cut is made to the depths shown, to house the upturned blade of the saws being sharpened. One stile, on the side chosen to take the heel of the saws, must have a further slot of about 12 mm width to accommodate back-saws such as tenon saws.

4.2.4 Alternative Rail Fixings

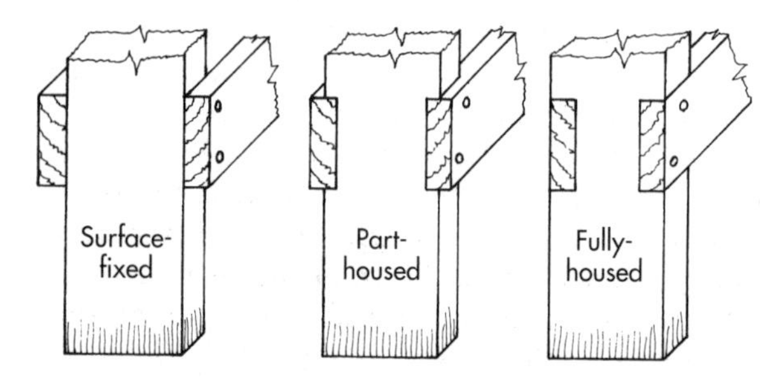

Figure 4.32 Alternative rail fixings

Figure 4.32: Next, the rails are cut to length and either surface-fixed, part-housed or fully housed onto the stiles, using 50 mm round-head wire nails or $1\frac{3}{4}$ in × 10 gauge countersunk screws. Their position, to act as foot and knee rails, is an important feature of the saw horse as a means of holding the frame steady and leaving the carpenter's hands free for the sharpening operation.

4.2.5 Tapered Housings in Saw Chops

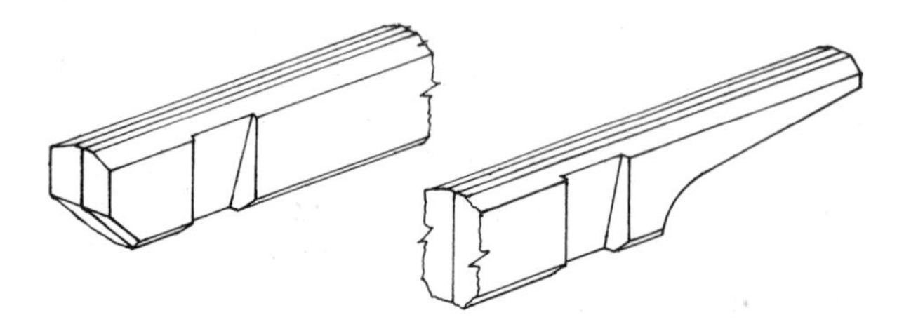

Figure 4.33 Tapered housings cut in saw chops

Figure 4.33: Following the rail-fixing, the chops are prepared to fit the stiles and must be of sufficient length to accommodate the longest saw. Tapered, vee-shaped housings are marked and cut to fit the vee-cuts already established in the stiles. These housings must be unequal distances from the ends to allow for a greater projection of the saw chops from the stile that was slotted to take back-saws.

4.2.6 Allowance for Handles

As illustrated in Figures 4.31 and 4.33, the projecting chops are cut out to a shape that will house the handles of the various types of saws that might require treatment. This shape can vary, or be varied, according to the range of saws that need attention.

4.2.7 Optional Saw Chops

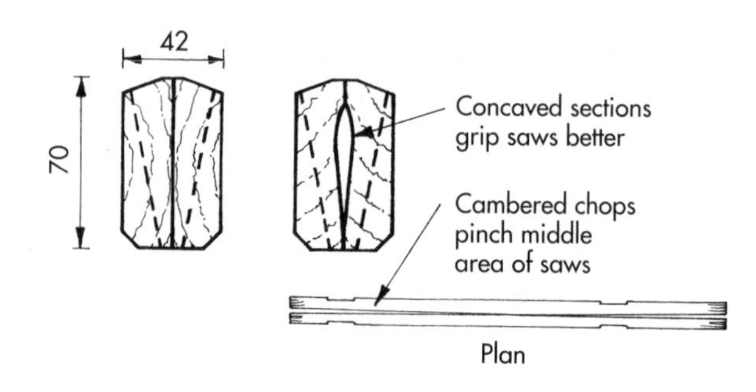

Figure 4.34 Details of saw chops

Figure 4.34: As illustrated, the sectional shape of the chops can vary between a site-horse and a shop-horse. A concave shape, as seen in the second drawing, helps to pinch the saw just below the gullets of the teeth, thereby eliminating distracting movement of the blade during sharpening. Also, a further refinement of the saw chops is achieved if each inner surface is planed very slightly out of true in length, to a convex or camber shape. When the chops are tightened into the horse, this will pinch the middle area of the blade lacking the support of the outer stiles.

4.2.8 Optional Leg Bolts

Finally, if required, the holes are drilled for 9 mm diameter coach bolts – after insertion and tightening of the nuts onto the washers, any surplus bolt should be cut off with a hacksaw and filed or hammered to remove any dangerous burrs or sharp edges.

4.3 NAIL BOXES

4.3.1 Introduction

Nail boxes appear in all shapes and sizes and vary between very simple – and often very rough – constructions where all joints are butted and nailed, to more elaborate forms with dovetailed corners, housed cross-divisions and shaped handles.

4.3.2 Preferred Nail Box

The dovetailed type are not now seen in industry, but serve as a very useful jointing exercise for apprentices, trainees or students and can also provide a presentable nail box for one's own workshop. Apart from this, the simple, easier-constructed box serves its purpose well enough – whether in the workshop or out on site. This purpose is to provide a manageable means of transporting a sufficient supply of nails of different size and type from one work-location to another.

4.3.3 Compartment Variations

Generally speaking, on first-fixing operations, the compartments in the box can be fewer and therefore larger, to house such nail sizes as 75 and 100 mm round-head wire nails, whereas, on second-fixing operations, more compartments are usually needed to house a greater variety of smaller nails such as 38 and 50 mm oval nails, etc.

4.3.4 Dovetailed Nail Box

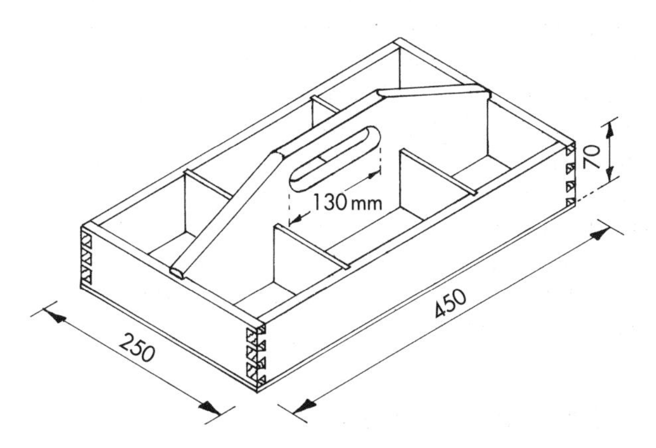

Figure 4.35 Dovetailed nail box

Figure 4.35: The one illustrated is built up of 70 × 12 mm finish sides and ends, jointed together with through-dovetails on each corner. It has a 145 × 21 mm finish handle-division housed into the ends, to one-third the end-material thickness, and solid timber or 6 mm plywood cross-divisions housed 4 mm into the sides and handle division. The 130 × 32 mm slot for the handle is drilled out with a 32 mm diameter centre bit, chisel-pared and chamfered to a clean finish. The assembly is then glued together and checked for squareness before the 4 mm plywood base is glued and pinned (with 15–18 mm panel pins) into position.

4.3.5 Modern Nail Box

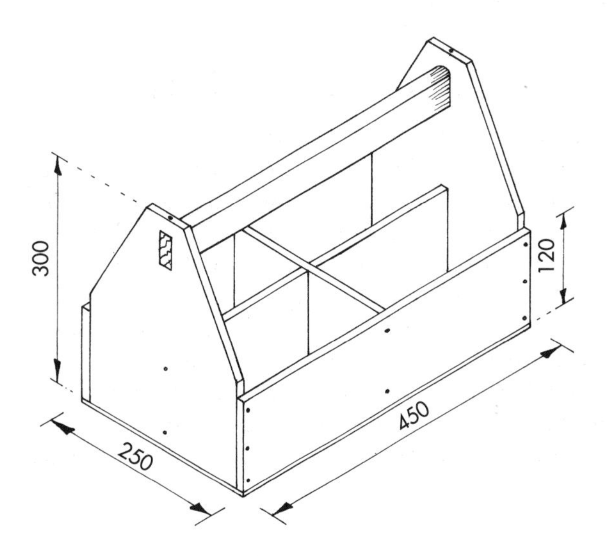

Figure 4.36 Modern nail box

Figure 4.36: As illustrated, this is built up of 10–12 mm plywood sides, ends and cross-divisions, an ex. 50 × 25 mm handle morticed into each end, and a 4 or 6 mm plywood base. The assembly is nailed together unglued with 38 or 50 mm round-head wire nails. The cross-halved plywood divider is dropped loosely into the box, without side housings, and fixed through the sides and ends with 38 mm oval or wire nails. Finally, the handle is inserted and nailed or screwed through the top of each end – after making two pilot holes.

4.4 HOP-UPS

4.4.1 Introduction

Even though 'wet plastering' has diminished since dry lining became popular, hop-ups are still used by plasterers when rendering/floating and setting walls. In these situations, the plasterer uses a floating or skimming trowel, a handboard (hawk) with which to repeatedly carry the plaster to the wall, and a 'board and stand' from which to feed the material onto the hawk. Because the loaded hawk is in one hand and the trowel in the other, he cannot easily – or safely – climb step-ladders if plastering to a height beyond his reach. Furthermore, step-ladders inhibit the plastering action. So, there is a need for an easily movable (repositioned by foot), easily ascendable (stair-like rise), and non-restrictive piece of equipment such as a hop-up.

4.4.2 Traditional Hop-up

Figure 4.37: As illustrated, this is simple in design structure, built up of square-edged or tongued-and grooved boarding, cleated and clench-nailed (protr

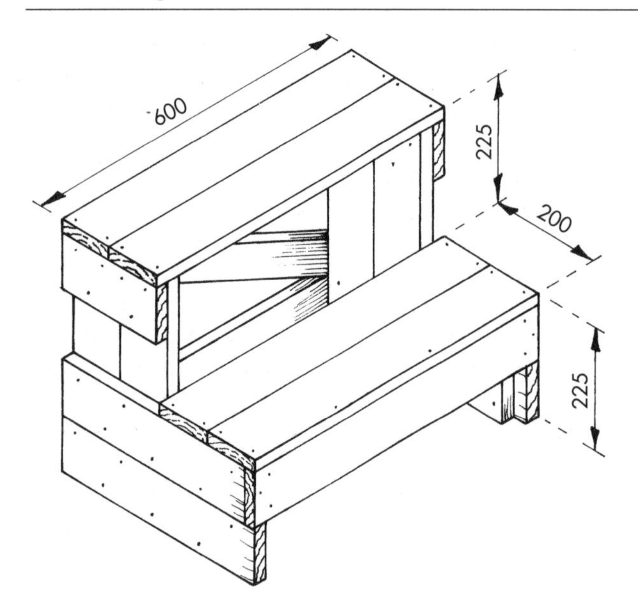

Figure 4.37 Traditional hop-up

nail-points bent over for a stronger fixing). The 100 × 25 mm sawn boarding, shown here, is arranged to form a two-step hop-up with a step rise of 225 mm.

First, two side-frames are constructed of vertical board and horizontal cleats clench-nailed together with 56 mm round-head wire nails or cut, clasp nails. These frames are then joined together by the tread boards being nailed into position and two cross-rails at low level, one at the front, the other at the back. These should be fixed with 63 mm round-head wire nails or oval nails. Finally, a diagonal brace of 50 × 25 or 75 × 25 mm section is also fixed at the back.

4.4.3 Modern Hop-up

Figure 4.38: The modern version of a hop-up shown here is simplified by using 18 mm plywood or Sterling board as side frames, traditional boarded steps, cross-rails and diagonal bracing. On a hop-up of this width (600 mm) 18 mm plywood or MDF board could also be used as tread boards.

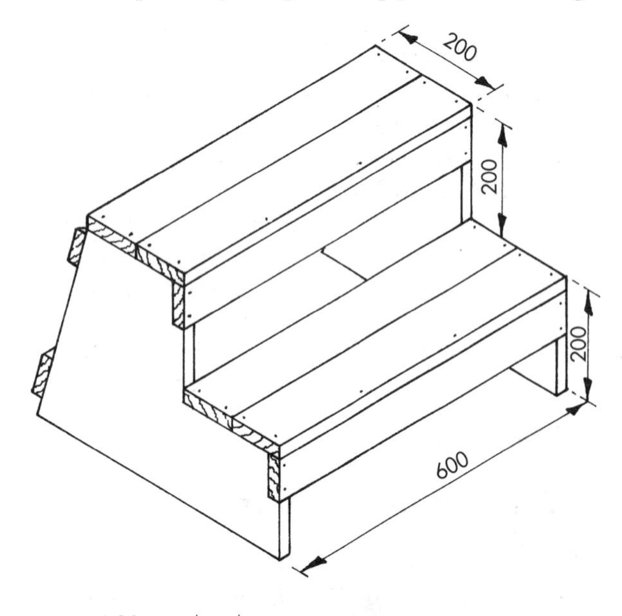

Figure 4.38 Modern hop-up

4.5 BOARD AND STAND

4.5.1 Introduction

A board and stand is also a piece of site equipment still required in a wet-plastering operation. It acts as a platform upon which the mixed plastering materials are deposited and are more easily trowel-fed onto the hawk placed under the board's edge. There are two types of stand that can be made, one being rigid in construction, the other, folding. Both types have a similar, separate mortar board which lays in position on top without any attachment to the stand.

4.5.2 Rigid Stand

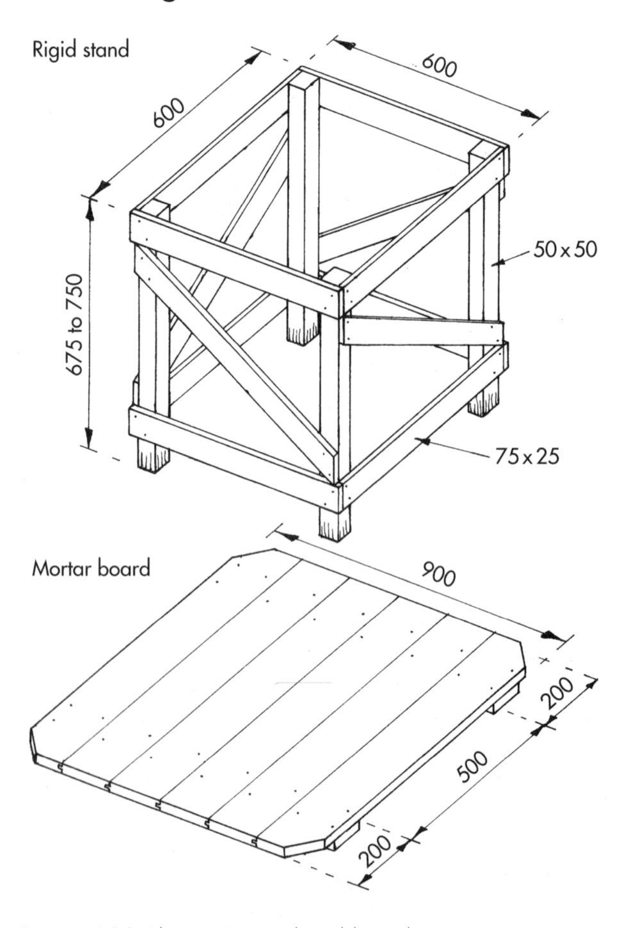

Figure 4.39 Plasterer's stand and board

Figure 4.39: The height of this can vary from 675 to 750 mm and the width and depth is usually about 600 × 600 mm. This allows the board to overhang the stand to facilitate the loading of the hawk. The material used can be sawn or prepared softwood. First, 50 × 50 mm legs and 75 × 25 mm rails are cut to length to form two frames. On each frame, one rail is nailed to the top of the legs, the other is nailed about 100 mm clear of the bottom. Each frame is then braced diagonally with 50 × 25 mm or 75 × 25 mm bracing material and the two frames joined with the remaining cross-rails at top and bottom. Finally, the two remaining braces are fixed; 50–63 mm round-head nails are used throughout.

4.5.3 Folding Stand

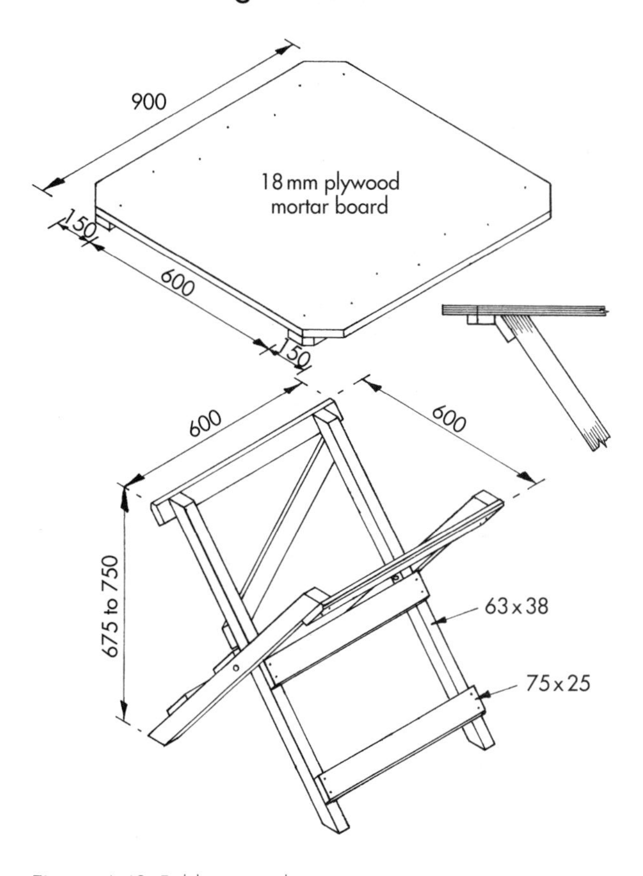

Figure 4.40 Folding stand

Figure 4.40: When in the open position, the height, width and depth can be similar to those given for the rigid stand. The leg-length can be worked out as already explained for the saw stool, either by practical geometry or by calculation (by using, say, 750 mm as vertical height and 600 mm as base measurement and adding at least 70 mm allowance for splay cuts). Alternatively, the side elevation showing the crossed legs could easily and quickly be set out at half-scale or full-size, so that the exact details of legs were available.

Construction Details

The legs, of 63 × 38 mm or similar section, with central holes drilled for 9–12 mm diameter coach bolts, are fixed together with 75 × 25 mm rails to form two interlocking frames. The inner frame must be minus a tolerance allowance in width to enable it to fit easily into the outer frame. Ideally, the allowance, say 2–3 mm, should, on assembly, be taken up with a washer each side, between the frames.

Once the frames are bolted together, they are partly retained in the open position by the middle rails on each side, but mostly rely on the top rails fitting between the cleats of the mortar board. Accordingly, the cleats on the board must be so positioned as to leave a clear middle area of 600 mm. Before bolting the frames together, diagonal braces should be added for extra strength as indicated. Nails or screws can be used for fixing the rails and braces, although with this type of construction screws are advisable. Finally, the ends of the bolts, if protruding more than a reasonable amount, should be close-cut and burred over.

4.5.4 Mortar Boards

These are made from 900 mm to 1 m square, either from tongued-and-grooved boarding, cleated and clench-nailed together (cut, clasp nails are ideal for this), or made from sheet materials such as resin-bonded plywood, MDF board, Sterling board or similar. As indicated, the sharp corners are usually removed. Whether made from sheet material or T&G boards, cleats, as indicated, will still be required to retain the board's position on the stand. Cleat material is usually 75 or 100 × 25 mm. Mortar boards used by bricklayers – often referred to as *spot boards* – only vary by their size, which is usually between 600 and 760 mm square.

4.6 BUILDER'S SQUARE

4.6.1 Introduction

Figure 4.41: Builders' squares are large wooden try-squares, made by the carpenter or joiner, for use on site during the early stages of setting out walls and foundations, etc. Their size varies from about 1 to 2 m and the two blades, forming the square, may be of equal length or have one blade longer than the other. They are used mostly by bricklayers as an aligning tool rather than an instrument, to help establish internal right-angles of walls or partitions. The initial setting-out of right-angled walls nowadays is usually done with an instrument such as an optical site square.

4.6.2 Material and Construction Details

The material, from ex. 75 × 25 mm for the smaller size, up to ex. 125 × 32 mm for the larger squares, should be

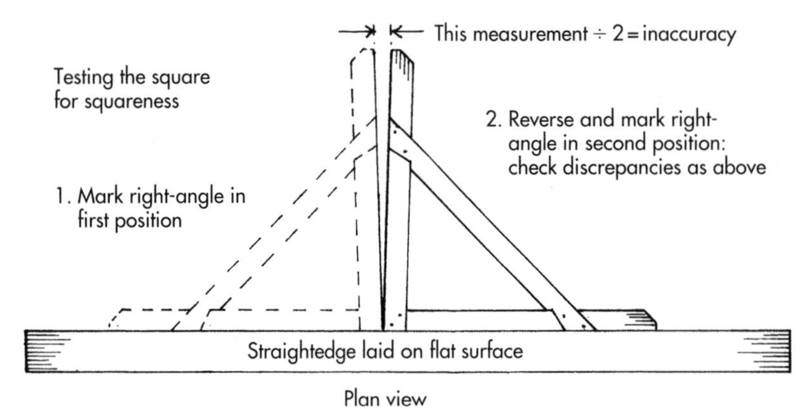

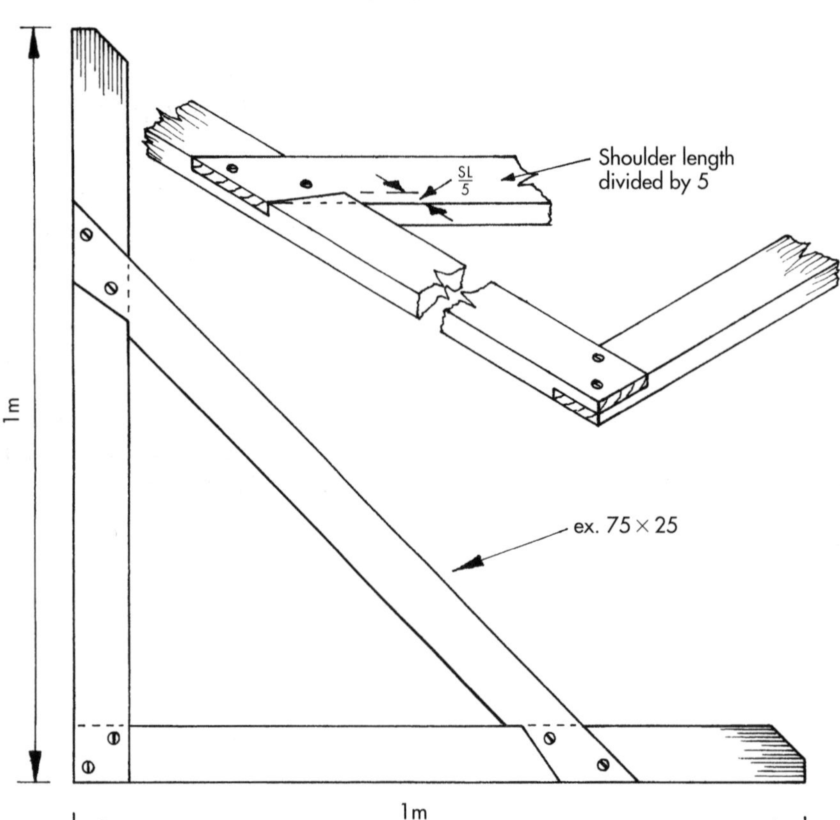

Figure 4.41 Builder's square

prepared softwood, carefully selected for density, straight grain and freedom from large knots. A corner half-lap joint is used on the connection of the two blades and should be screwed together. Alternatively, this joint can be formed with a haunched mortice and tenon. Single-splay dovetail halving joints, as illustrated, are used to connect the diagonal brace to the blades – and these should also be screwed; joints are not usually glued.

4.6.3 Assembling the Square

First, make the right-angled corner joint and fix the two blades together temporarily with one screw. Now test for squareness with either a roofing square or by using what is known as the 3–4–5 method. This refers to a ratio of units conforming to Pythagoras' theorem of the square on the hypotenuse being equal to the sum of the squares on the other two sides. For example, using 300 mm as a unit, mark three units (3 × 300 = 900 mm) accurately along one face-side edge from the corner, four units (4 × 300 = 1.2 m) along the adjacent edge from the corner and then check that the diagonal measures five units (5 × 300 = 1.5 m). Adjust the blades, if necessary, until the diagonal measures exactly five units. Lay the brace carefully into position and mark for halving joints. Now dismantle the corner joint, form the other two joints, then reassemble, screw up and test for squareness.

4.6.4 Testing for Squareness

Figure 4.41: As illustrated, this can be done by laying a straightedge on a flat surface such as a sheet of plywood or hardboard, squaring a mark from this with the builder's square in a left-hand position, reversing the square to a right-hand position and marking another line close to the first – then checking for any discrepancies in the lines. If necessary, true up any small inaccuracies by planing.

4.7 STRAIGHTEDGES AND CONCRETE-LEVELLING BOARDS

4.7.1 Straightedges

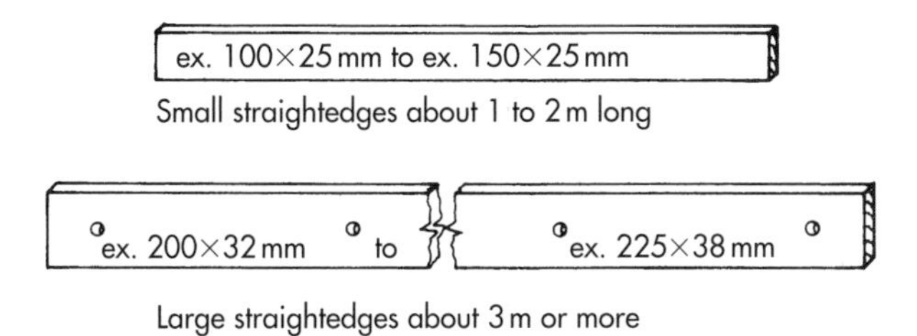

Figure 4.42 Straightedges

Figure 4.42: These are boards used by various tradespeople for setting out straight lines, checking surfaces for straightness, and levelling and plumbing with the addition of a spirit level. As the name implies, the essential feature of these boards is that the two edges are straight and parallel to each other. This can, of course, be done by hand-planing, but is best achieved on surface planer and thicknessing machines.

Varying Sizes

Lengths of straightedges vary from 1 to 2 m, with sectional sizes of ex. 100 × 25 mm to ex. 150 × 25 mm. Large straightedges of 3 m or more in length, are usually made from ex. 200 × 32 mm, or 225 × 38 mm boards. Holes of about 38 mm diameter were traditionally drilled through the straightedge, at about 900 mm centres along its axis. This was done to establish the board visually as a proper straightedge and to discourage anybody on site from claiming it for other uses. Prepared softwood, of similar quality to that selected for the 'building square', should be used. Periodic checks for straightness are advisable.

4.7.2 Concrete-levelling Boards

Figure 4.43: These are usually about 5–6 m long, made from 225 × 38 mm sawn softwood. As illustrated, a handle arrangement is formed at each end to assist in easier control and movement of the board during the levelling operation. If a bay of concrete was to be laid within a side-shuttered area, or within the confines of a brick upstand, the levelling board would have to be long enough to rest on the shutters or brickwork each side. As the concrete was being placed, a person at each handle would tamp the concrete and pull the board, zig-zagging back and forth across the surface.

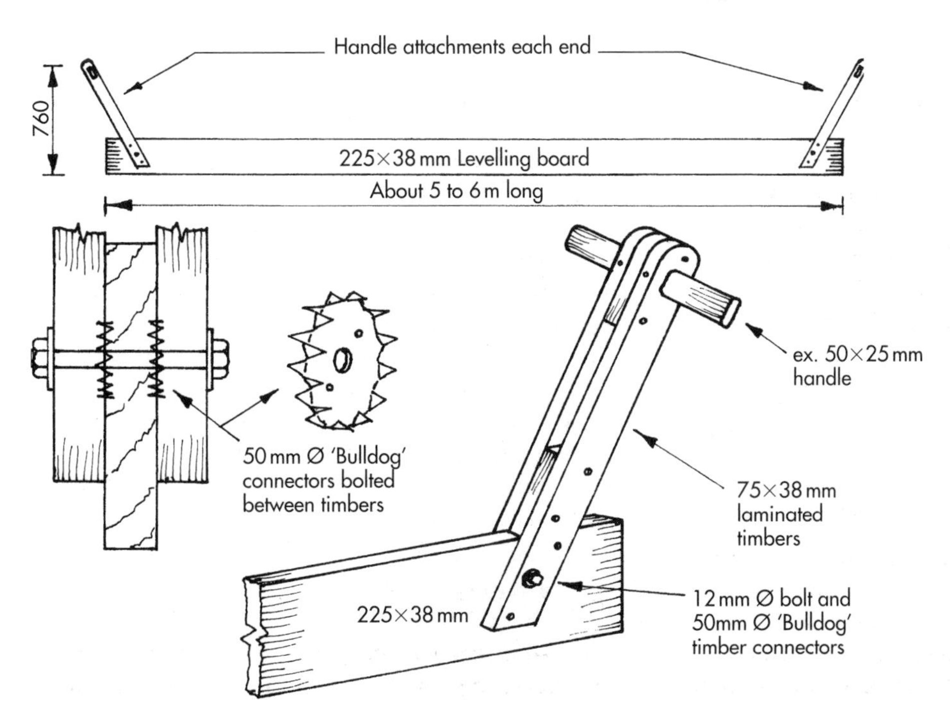

Figure 4.43 Concrete-levelling boards

4.8 PLUMB RULES

4.8.1 Traditional Plumb Rule

Figure 4.44 Traditional plumb rule and plumb bobs

Figure 4.44: This traditional piece of equipment, in its original form as illustrated, is here for reference, as it is obsolete nowadays, even though plumb bobs themselves are still very useful in some carpentry operations. When used with a plumb rule, waiting for the plumb bob to settle against the gauge line, to indicate plumbness, was a tedious and slow operation – although very accurate, if carefully done.

4.8.2 Modern Plumb Rule

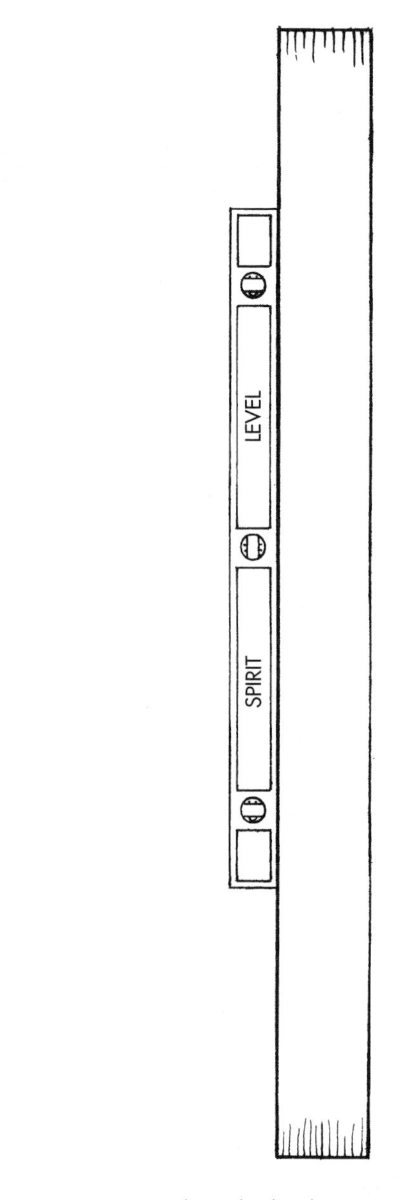

Figure 4.45 Modern plumb rule

Figure 4.45: Although extra-long spirit levels are now obtainable to replace plumb rules, the modern plumb rule consists of a straightedge of about 1.675 m length of ex. 125 × 25 mm selected softwood, with a shorter (say 750 mm) spirit level placed on its edge when in use. When fixing door linings, or striving for accuracy in plumbing on any similar operation, the straightedge and level combined are preferable to the relatively short spirit level on its own. This is because any inaccuracies in the shape of hollows or rounds in the surface of the item being plumbed/fixed, show up easily because of the plumb rule or straightedge's greater length.

5

Sharpening Traditional Saws

5.1 INTRODUCTION

Most carpenters seem to use hardpoint, throwaway saws for the following reasons:

- they are relatively cheap to buy;
- they retain their original sharpness for a long time if used without hitting metal objects such as hidden nails;
- when the saw is blunt or damaged, there is no need to lose time in sharpening or paying for the saw to be sharpened.

However, traditional saws, that require to be sharpened when blunt, are still in use, still being sold to a lesser extent, and recent research showed that some carpenters still regard them as their 'good saws', even though they used mostly hardpoint saws.

Therefore, this chapter is for those people who want to master the high degree of skill and judgement required in successful sharpening of traditional saws.

5.2 SEQUENCE OF OPERATIONS

There are four separate operations involved for saws in a bad condition; these are known as *topping*, *shaping*, *setting* and *sharpening* – they must be performed in that sequence. If a saw is in good condition, has not been neglected or abused, but has lost its edge through normal use, then the action needed is less drastic and it will only require sharpening.

5.2.1 Topping

Figure 5.1: Normally, the points of the saw teeth conform to a straight line, or – with some saws – a slightly segmental curve in the length of the saw. If, through lack of the carpenter's time or skill, the saw is sharpened roughly on a number of occasions, the line or camber will lose its original shape and the teeth will become

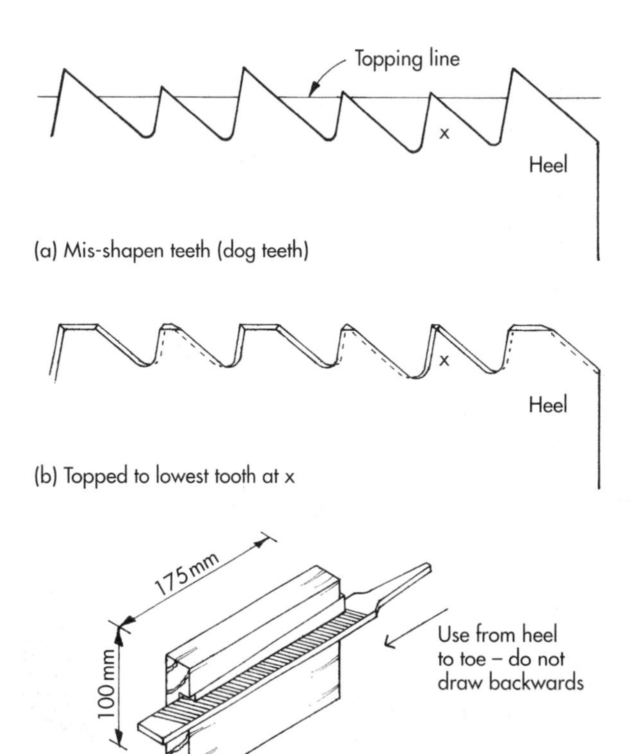

Figure 5.1 Topping

mis-shapen and unequal in size and height. Such teeth are known as *dog teeth*. To remedy these faults, the teeth must be reshaped and the first step is known as topping. This means running a flat mill file over the points of the teeth until the lowest tooth has been 'topped' by the file (Figure 5.1(a) and (b)) and the overall shape regained in the length of the saw. To achieve this, periodic sightings – looking along the saw at eye-level – must be made while filing.

Topping Tool

To assist the filing operation, a wooden block, grooved to take the file and a wooden wedge, should be made.

The complete assembly, illustrated in Figure 5.1(c), is known as a *topping tool*. Its main advantage lies in keeping the file, by virtue of the block being pressed against the saw blade, at right angles to the saw. Finally, it must be borne in mind that excessive topping creates extra work in the next operation, shaping.

5.2.2 Shaping

Files Needed

Figure 5.2(a): This operation is carried out with a saw file. Such files are equilaterally triangular (60°), slim-tapered in various lengths, single or double-ended and fitted into plastic or wooden saw-file handles. A 150 mm double-ended file is recommended for tenon and panel saws and a 200 or 225 mm double-ended file for crosscut and rip saws.

Technique of Shaping

Figures 5.2(b) and (c): When shaping, the filing action is always square across the saw blade and follows in every consecutive gullet from the heel on the left to the toe on the right. The idea is to eliminate the dog teeth and create evenly shaped teeth leaning towards the toe of the saw at the correct pitch and points per 25 mm. *Pitch* refers to the angle-of-lean given to the front cutting edges of the teeth, as shown in the chapter on tools. The recommended pitch angle for rip saws is 87°, for crosscut saws is 80°, and for panel and tenon saws is 75°. Until experience is gained in angle-judgement, the required angle can be set on a metal template or sliding bevel to test the degree of accuracy in initial shaping.

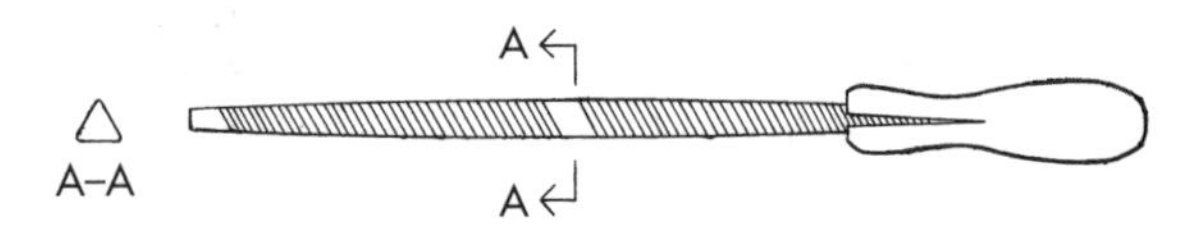

(a) Double-ended saw file

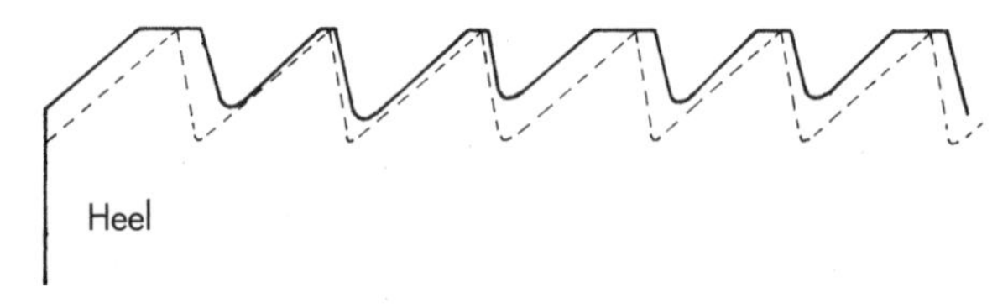

(b) Shaping outline to be judged

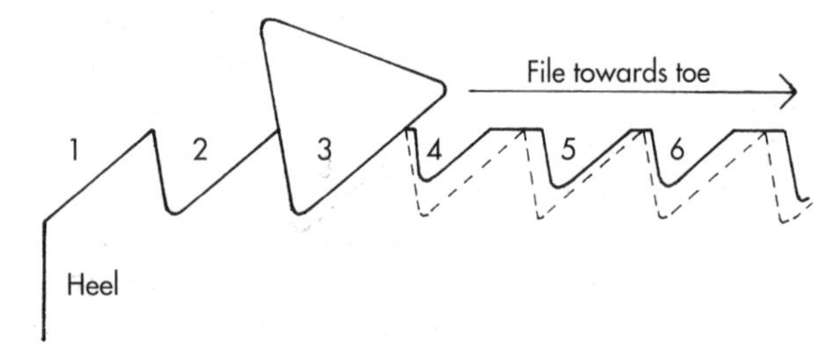

(c) Recommended saw position for shaping sequence

(d) Finished appearance

Figure 5.2 Shaping

Filing Action

Figures 5.2(c) and (d): When filing, the handle should be held firmly in one hand and the end of the file steadied with the thumb and first two fingers of the other hand. All file strokes should be forward-acting and *not* drawn backwards. Take care to establish the correct angles on the first few teeth and then familiarize yourself with the feel of the file resting in a corrected gullet. Maintain this feel as you continue the shaping operation and combine it with visual appraisal of the shape and pitch of the teeth in relation to the shiny, flat areas on the tips. These are produced in the topping operation and will gradually diminish as the gullets deepen, and will indicate that the shaping must stop immediately the shiny flat spots are removed. Finally, it helps to rub chalk on the file periodically to reduce the tendency of the file to become clogged with metal particles.

5.2.3 Setting

Figure 5.3: This next operation, known as *setting*, refers to the bending of the tips of the teeth, every other one, out from the face of the saw on one side and then setting the alternate row of teeth on the other side. This is done with a pliers-type tool known as a *saw-set*. The idea is that the cut – or kerf – made by the saw is slightly wider than the thickness of the saw blade, to facilitate an easy sawing action. However, too much *set* can be a disadvantage, as the saw tends to run adrift in an oversize saw kerf. For this reason, it is advisable to set the saw slightly less than the normal setting indicated on the saw set and only set the saw when it is really necessary – not every time the saw is sharpened.

Saw-Set Tool

The saw-set (Figure 5.3(a)) has a knurled hand-screw controlling a wheel-shaped, bevelled anvil – the edge of which is numbered with different settings, relative to points per 25 mm (pp25) and not *teeth* per 25 mm (which is always one less than pp25 and if used, therefore, creates more set). The anvil has to be adjusted so that the required pp25 numeral (6 pp25 setting shown in Figure 5.3(b) is set exactly opposite the small plunger which ejects when the levered handles are squeezed.

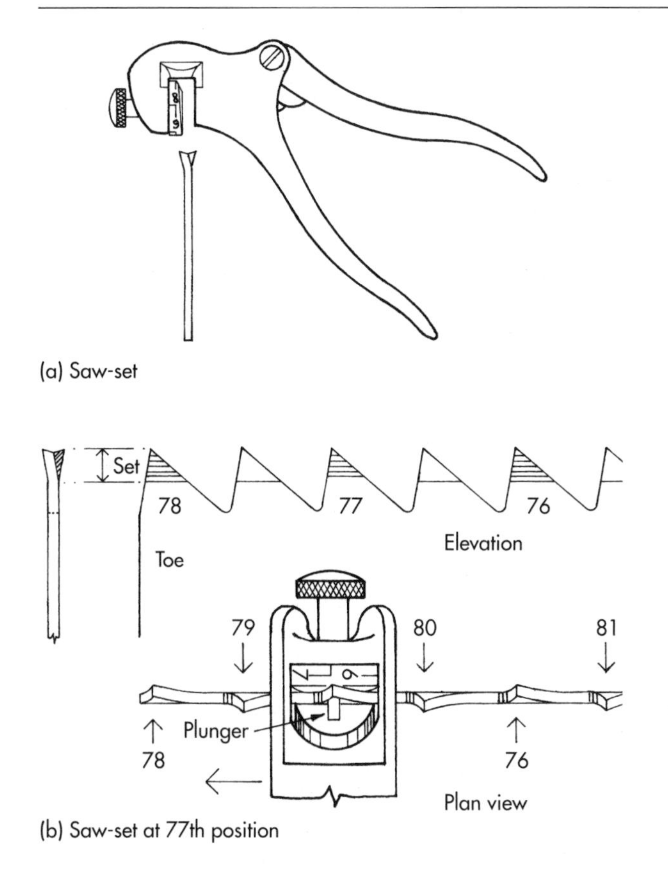

(a) Saw-set

(b) Saw-set at 77th position

Figure 5.3 Setting

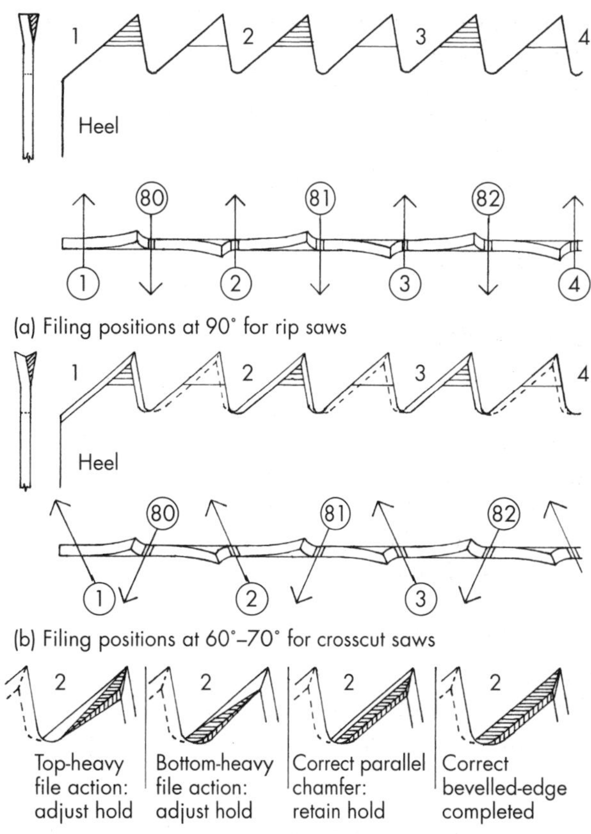

(a) Filing positions at 90° for rip saws

(b) Filing positions at 60°–70° for crosscut saws

(c) Sharpening technique

Figure 5.4 Sharpening

Technique of Setting

To set the saw, hold it under your arm, with the handle in front, the teeth uppermost, and place the saw-set on the first tooth facing away from the plunger. Looking down directly from above, squeeze the saw-set firmly and then carefully repeat this operation on every other tooth thereafter from the heel towards yourself and the toe at the rear. Turn the saw around so that the handle is now at the rear, under your arm, and set the alternate row of teeth from the toe towards yourself and the heel – which is gradually worked out from the under-arm position. The saw is now ready for sharpening.

Note that when setting old saws, squeeze the saw set very gently, as the metal becomes brittle with age and teeth can be easily broken off in the setting operation.

5.2.4 Sharpening

Figure 5.4: This final operation is concerned with creating sharp edges and points to the outer tips of the teeth, by filing every other gullet at an angle to the face of the saw on one side and then the alternate row of gullets at an opposing angle to the face of the saw on the other side, as shown in Figure 5.4(b).

Sharpening Theory

The method of sharpening shown in Figure 5.4(b) is for saws designed to cut across the grain, such as crosscut, panel and tenon saws, the theory being that the sharp pointed outer tips of the vee-shaped teeth act as knives cutting two close lines across the timber. Short-grained pieces of fibre between the lines break up as the saw moves forward; the broken fibres are collected in the gullets as *sawdust* and released when the saw passes through the timber. Rip saws are similarly sharpened on alternate sides, but the angle of sharpening is square or almost square across the saw, as indicated in Figure 5.4(a). This eliminates the pointed outer tips required for crosscutting and produces square-tipped teeth which provide a scraping/shearing action necessary for effective down-grain ripping.

Technique of Sharpening

When sharpening, take care not to lose the basic shape of the teeth; this is best achieved by gaining the feel of the file and by keeping your eye on the back edges of the teeth. The first stroke of the saw file, at an angle of 60–70° to the saw face, should show a parallel chamfer on the back edge of the tooth – if not parallel, then adjust the file accordingly on subsequent strokes and stop filing immediately the chamfer-edged tooth becomes completely bevel-edged (Figure 5.4(c)).

This normally takes from two to four strokes and, once established, each gullet should receive the same number of strokes thereafter. This promotes a rhythmic filing action necessary for speed and accuracy. If a saw is only being resharpened, whereby the edges of the teeth are already bevelled, it helps to top the saw lightly – repeat *lightly* – with the topping tool. The idea is to split the shiny flat spots in half when sharpening from one side and then remove the remaining halves when sharpening from the other side.

5.3 SAW HORSE

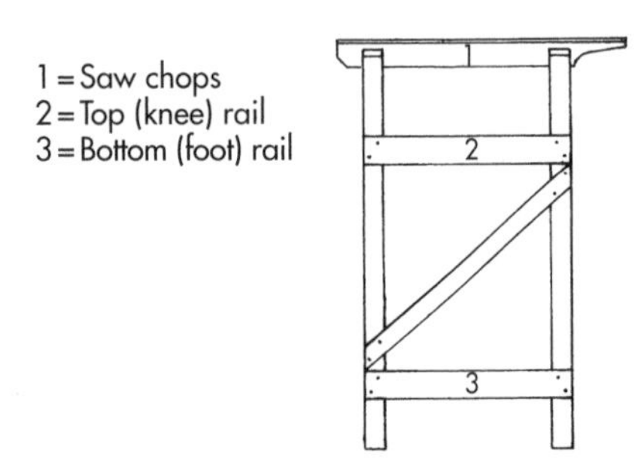

Figure 5.5 Saw horse

Figure 5.5: Special vices can be used for saw sharpening, but more commonly wooden frames known as *saw horses* or *saw stocks* are used. The wooden jaws that hold the saw in the frame are known as *saw chops*. Saw horses are purpose-made and the design and construction is shown in the chapter dealing with making small plant on site.

Sharpening Procedure (for Right-handed Persons)

Figure 5.6(a): Cramp the saw high in the saw chops and top lightly with the topping tool. Reposition the saw so that only about 4 mm remains between the top of the saw chops and the base of the gullets (Figure 5.6(b)). With the saw handle to your left, rest the saw horse against a bench or window sill, etc., with good light in front and above the saw.

Starting Position

Figure 5.6(c): Take up your position against the horse by resting your right foot on the bottom rail, with your knee pressing against the top rail to keep the saw horse steady. Start to file at the heel (near the saw handle; Figure 5.6(d)) in the gullet affecting the back of the first tooth leaning away – noting that the file, at an angle of 60–70° to the saw face, *always points towards the saw handle*. After two or three forward strokes, aimed at 'splitting the shiner', repeat the action in every other gullet thereafter, moving rhythmically towards the toe of the saw on your right.

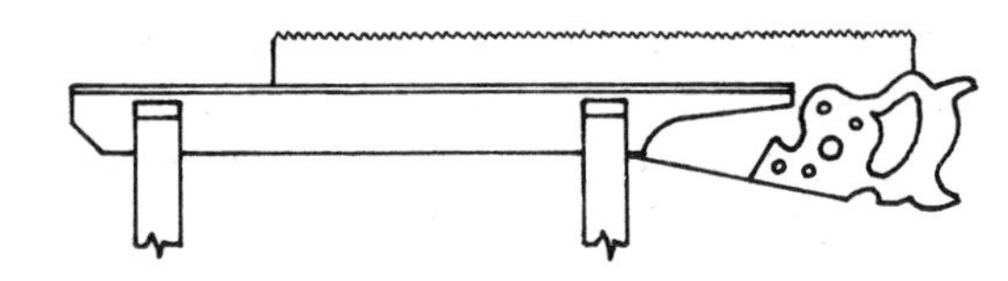

(a) Topping position

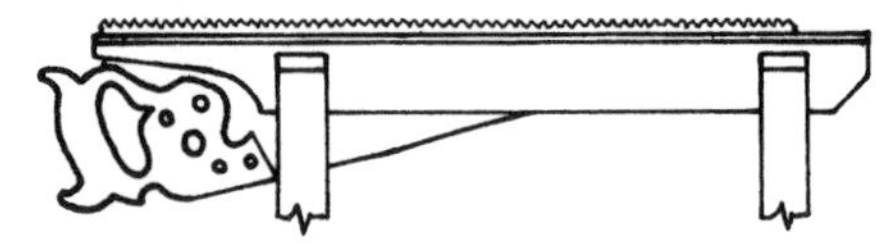

(b) Sharpening position

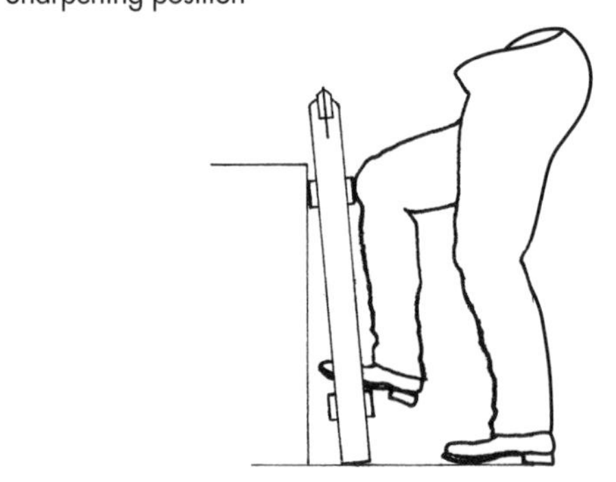

(c) Starting position

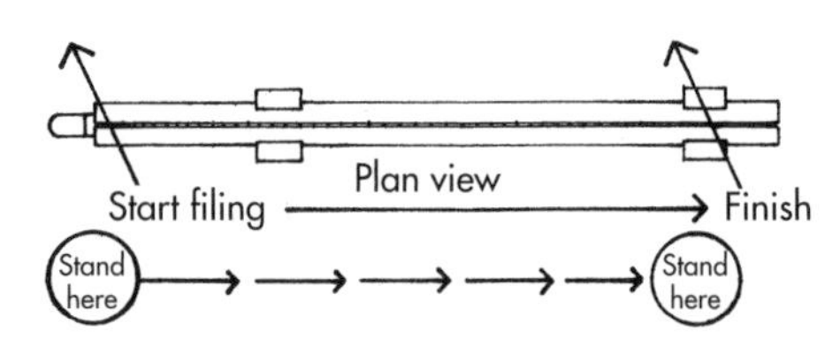

(d) Filing

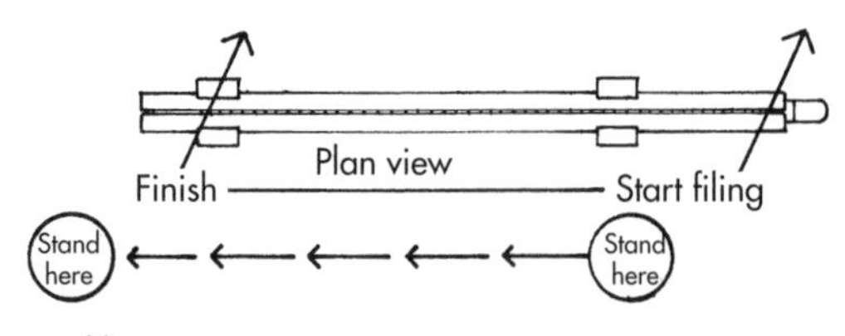

(e) Change position

Figure 5.6 Sharpening procedure

Changing Position

When the last quarter of the saw's length is reached, it will be found easier to switch the leg position and support the horse with the left foot and knee. When completed, turn the horse around so that the saw handle is now to your right (Figure 5.6(e)). Support the horse with your left foot and knee and start to file at the heel again (near the handle), in the gullet affecting the back of the first tooth leaning away, again remembering that the file *always points towards the saw handle*. With the same amount of file strokes used on the first operation, aim to remove the remaining half of the 'shiner' and produce sharp points while moving rhythmically towards the toe of the saw on your left, indicated in Figure 5.6(e). In the last quarter, switch the leg position to give support with the right foot and knee.

6

Fixing Door Frames, Linings and Doorsets

6.1 INTRODUCTION

Wooden frames and linings are fixed within openings to accommodate doors which are to be hung at a later stage in the second-fixing operation. This is necessary where wet trades, such as bricklaying and plastering and early-stage rough building operations, are involved. However, where dry methods of construction are used (or achieved, as described later), *doorsets* are sometimes used. These are made up of linings or frames with pre-hung doors attached, locks or latches and architraves in position. This reduces site work by eliminating the conventional second-fixing door-hanging operation.

6.1.1 Lining Definition

Figure 6.1: A door lining, by definition, should completely cover the reveals (sides) and soffit (underside of the lintel) of an opening, as well as support and house the door. Nowadays, this coverage only usually occurs where the opening is within a block or stud-partition wall.

6.1.2 Frame Definition

Figure 6.2: A door frame should be of sturdier construction, strong enough to support and house the door without relying completely on the fixings to the structural opening. Unlike linings, which have loose doorstops, frames are rebated to receive the door and usually have hardwood sills.

6.1.3 Internal or External

Generally speaking, frames are used for external entrance doors and linings for internal doors. Apart from this, door frames are usually set up and built in at the time the opening is being formed, whereas a lining, because of its thinness (usually 21–28 mm) and flexibility, should only be fixed after the opening is formed. Linings also require a greater number of fixings and a more involved fixing technique than door frames, as described later.

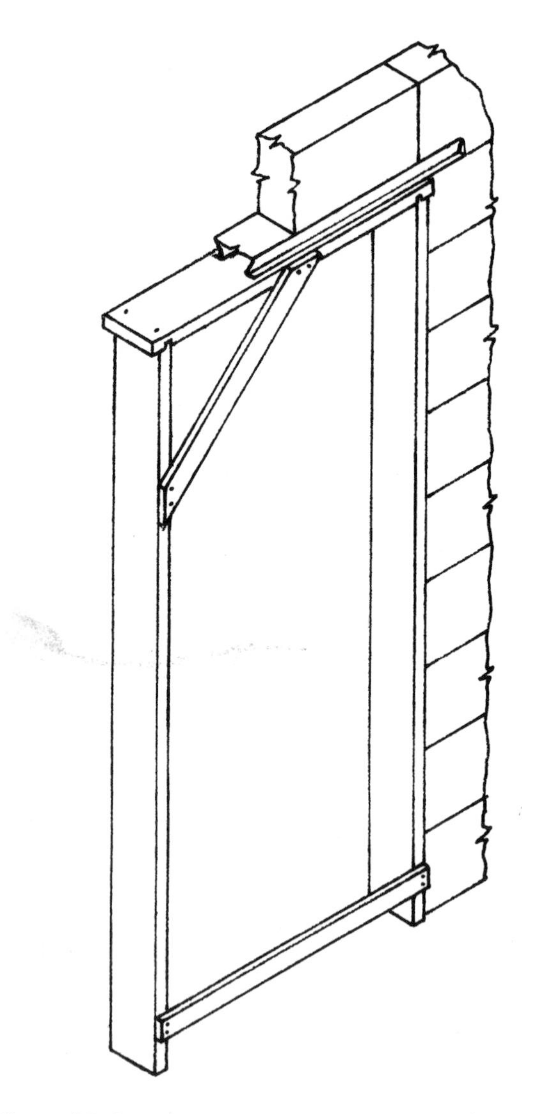

Figure 6.1 Door lining

6.1.4 Protecting External Frames

Figure 6.3: To inhibit initial moisture penetration into the timber from wet trades and to offset any future risk

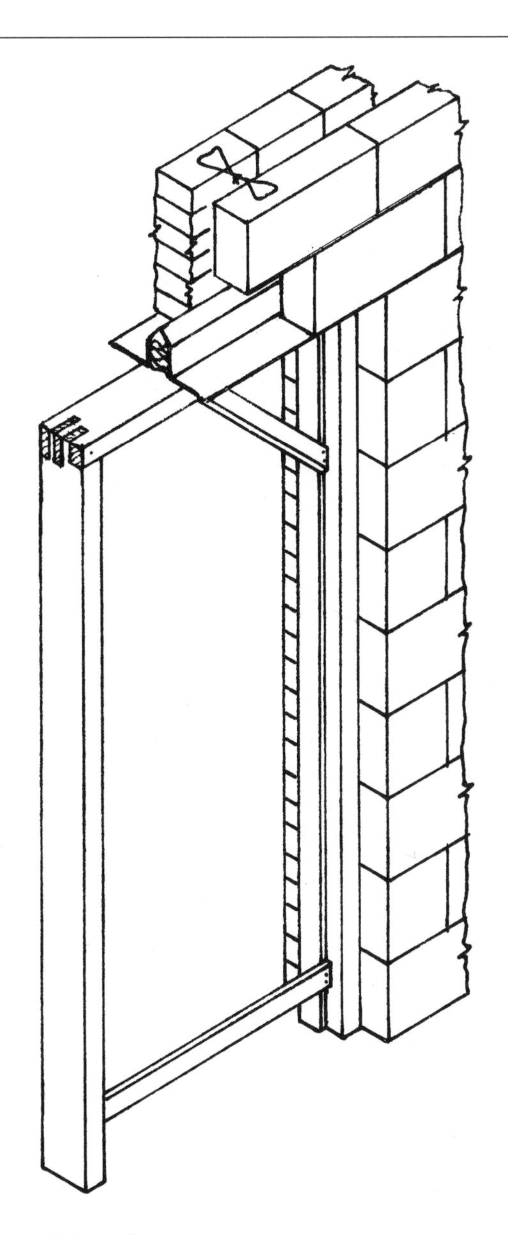

Figure 6.2 Door frame

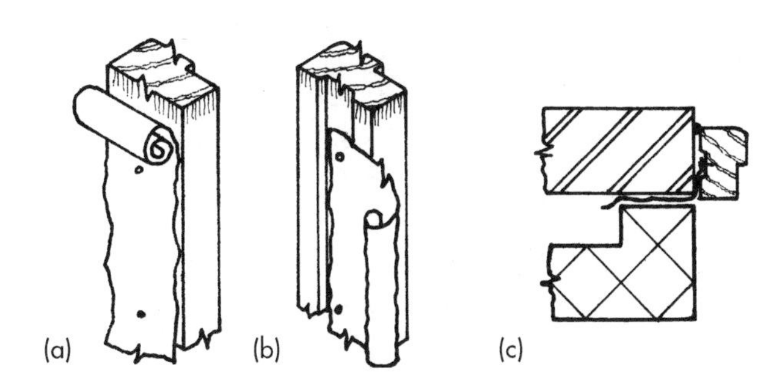

Figure 6.3 Frame protection material applied **(a)** full width, **(b)** and **(c)** to inner edges and cavity

of decay, built-in frames should be treated with preservative before being fixed, preferably at the manufacturing stage. Usually, at the time of being built in, the abutment-surfaces of the frame have a continuous strip of damp-proof material such as bituminous felt, plastic film (to BS 743) or waterproof building paper fixed to the full width, as in Figure 6.3(a), or fixed to the inner edges, as in (b) and (c). This shows the use of plastic film fixed to frame-jambs and sandwiched between brickwork and blockwork to stop any moisture in the outer wall from bridging the cavity.

6.1.5 Protection strips

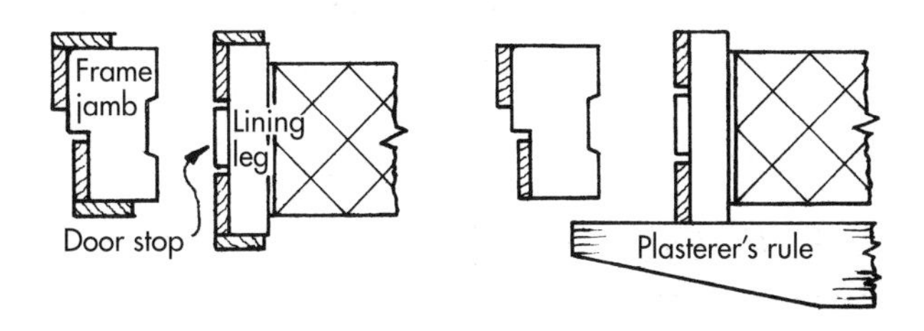

Figure 6.4 Protection strips

Figure 6.4: The inner edges of the jambs (sides) of frames and the legs (sides) of linings, especially at low level, should be protected from wheelbarrow damage and other careless movement of material and plant, by being covered with temporary wooden strips. These can be of whatever size; the illustration shows 38 × 12 mm strips fixed lightly with 38 mm oval nails to face and edges of the jambs and legs. Of course, the strips fixed to the edges would have to be removed if a wet plastering operation were to take place, as opposed to modern dry lining. This, as shown, would allow the plasterer's rule (straightedge) access to a guiding edge.

6.1.6 Setting up the Frame

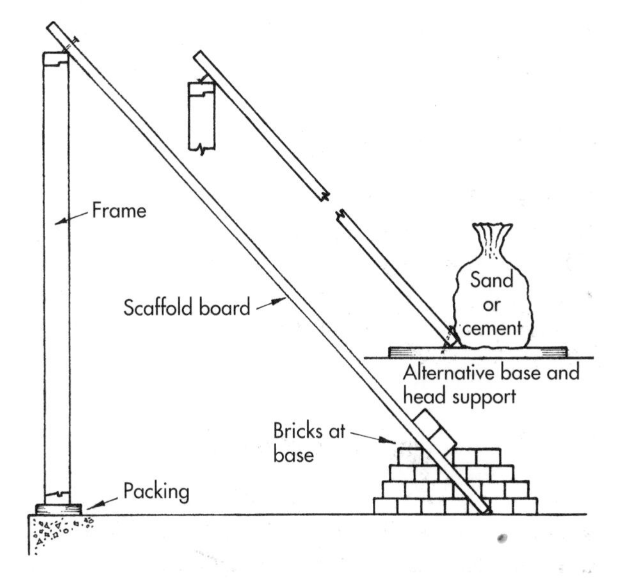

Figure 6.5 Setting up the frame

Figure 6.5: Built-in door frames are set up immediately before starting to build the walls, or sometimes after the first brick-course has been laid. The position of each door-opening reveal is set out and the frame is stood up in position with one or two scaffold boards supporting it at the head. Two boards (or two nails in one board) are better, especially if the frame is twisted. A 75 mm wire nail driven through the top of the board(s) holds the frame. The nail is driven through before lifting the board into position and the nail-point rests on the frame – it is not driven in.

Safe-nailing

Alternatively, as illustrated, the nail is only driven *into* the board – not through it – and the head of the nail is used to hold the frame. This is quite effective and a much safer practice if removal of the nail is eventually neglected.

The jambs are plumbed by spirit level, then bricks or blocks (or any heavy material) are piled at the foot of the scaffold boards to hold them in position. The head of the door frame must be checked for level and adjusted if necessary.

6.1.7 Vertical Adjustments

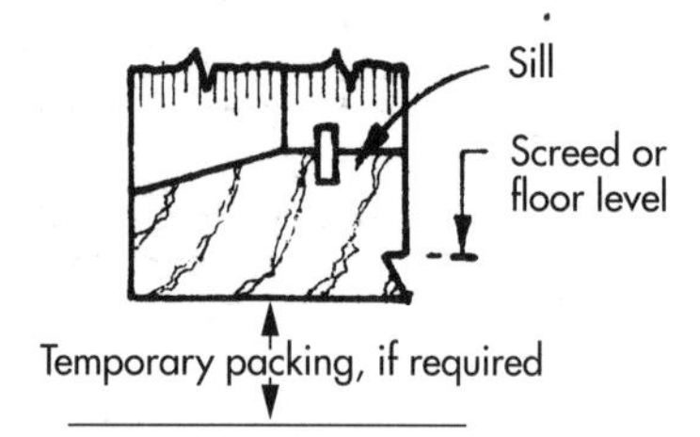

Figure 6.6 Vertical adjustments

Figure 6.6: These adjustments, under the sill, must also take into account the eventual finished floor level (ffl). It is also important to ensure that the head of the frame meets brick courses. If the brickwork is flush or slightly higher than the top of the frame, this will suit the seating of the open-back or classic profile shape cavity lintel. Any slight gap between frame-head and the underside of the lintel will eventually be sealed when the jambs and head are gunned around with a flexible frame sealant.

6.2 FIXING DOOR FRAMES

6.2.1 First Set

Figure 6.7: As the brickwork or blockwork proceeds, metal ties are usually fixed to the jambs and built into the bed joints of the mortar. The first set of ties (one to each jamb) must be fixed at low level, on the first course of blockwork, or on the second or third course of brickwork – no higher (Figure 6.7(a)).

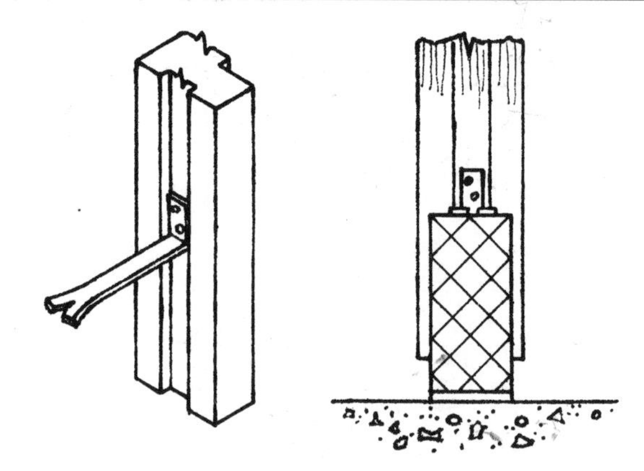

(a) Frame-cramp lug recessed ideally in grooved jambs

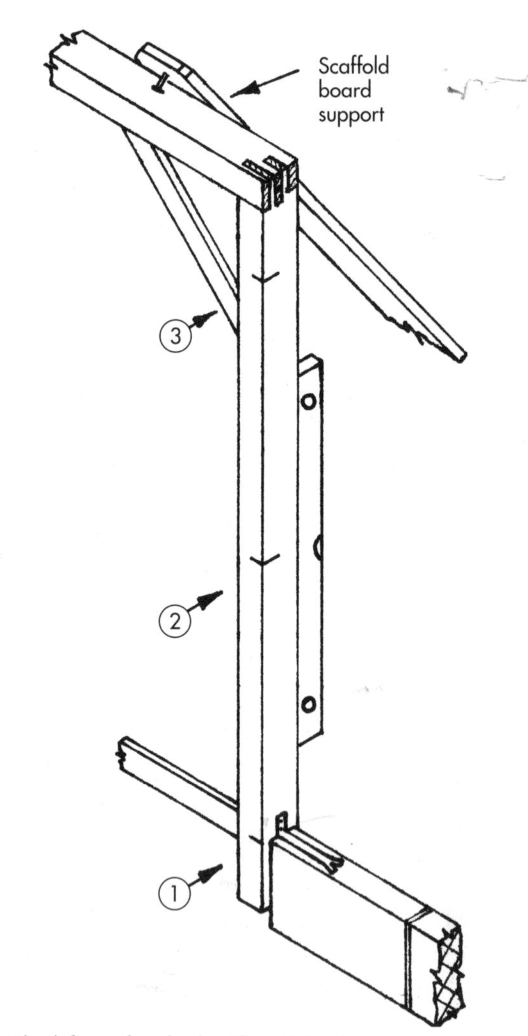

(b) Check frame for plumb with spirit level

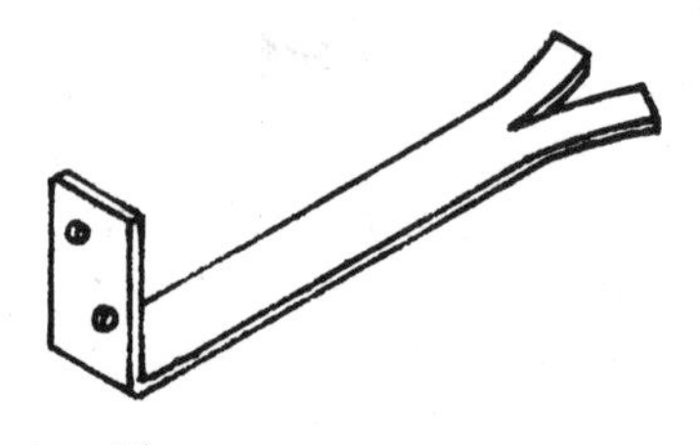

(c) Galvanized steel frame cramp

Figure 6.7 Fixing ties

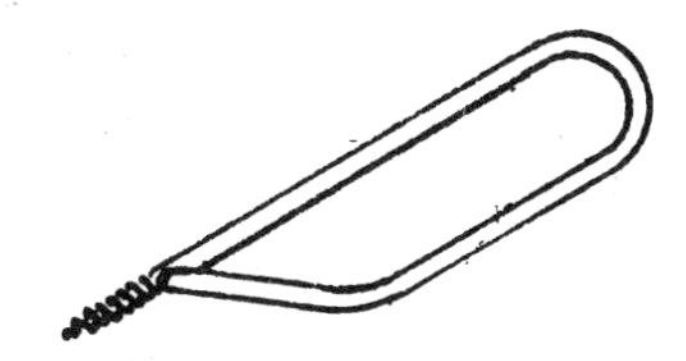

(d) Owlett's zinc-plated screw tie

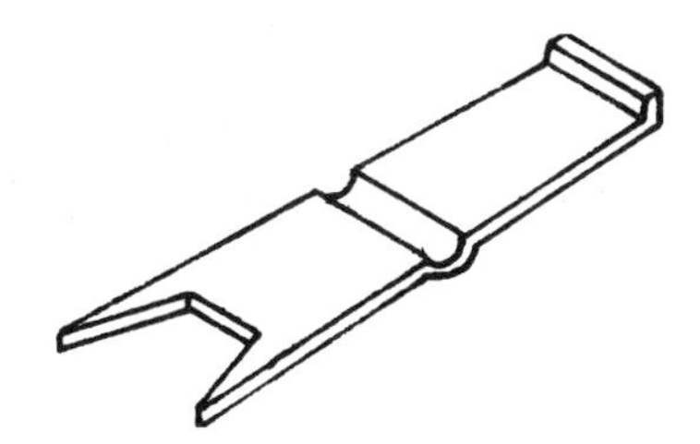

(e) Sherardized Holdfast

Figure 6.7 (continued) Fixing ties

6.2.2 Second and Third Sets

Figure 6.7(b): As the work progresses, the frame must be checked for plumbness from time to time in case the supporting scaffold boards have been accidentally knocked. At least two more sets of ties are fixed, at middle and near-top positions. For storey-height frames (from floor to ceiling, with fixed glass or fanlight above the door opening), a total of four sets of ties is advisable.

Different types of tie are available, each with points for and points against, as listed below.

Galvanized Steel Frame Cramp

Figure 6.7(c): These provide good fixings and, because they are screwed to the frame, any cramps (or ties) already fixed and bedded in mortar are not disturbed by hammering. Also, by resting on the last-laid brick, the next brick above the tie is easily bedded. The disadvantages are: handling small screws; requiring a screwdriver and bradawl (especially if no carpenters are on site yet, or they are sub-contract labour, not wanting to be involved); if no groove is in the frame, the upturned end of the cramp inhibits the next brick from touching the frame; doubts as to whether rust-proofed screws were used.

Owlett's Zinc-plated Screw Tie

Figure 6.7(d): These also provide good fixings and are screwed to the frame without vibration from hammering. They do not require screws, bradawl or screwdriver and can be offset or skewed to avoid the cavities in hollow blocks. On the debit side, the brickwork or blockwork has to be stopped one course below the required fixing to allow rotation of the loop when screwing in, then the brick or block beneath the tie is bedded – with some difficulty and loss of normal bedding-adhesion.

Sherardized Holdfast

Figure 6.7(e): Holdfasts are fixed quickly and easily, being driven in by hammer. The spiked ends spread outwards when driven into the wood, forming a fishtail with good holding power. The main disadvantage is that the hammering disturbs the frame and permanently loosens any Holdfasts already positioned in the still-green (unset) mortar.

Marking the Positions

Whatever type of frame cramps or ties are used, their intended positions, relative approximately to bed-joints, are best boldly marked on the jambs with a soft pencil (as seen at points (1), (2) and (3) of Figure 6.7(b)), to act as a reminder to the bricklayer as the work rises.

Braces and Stretcher

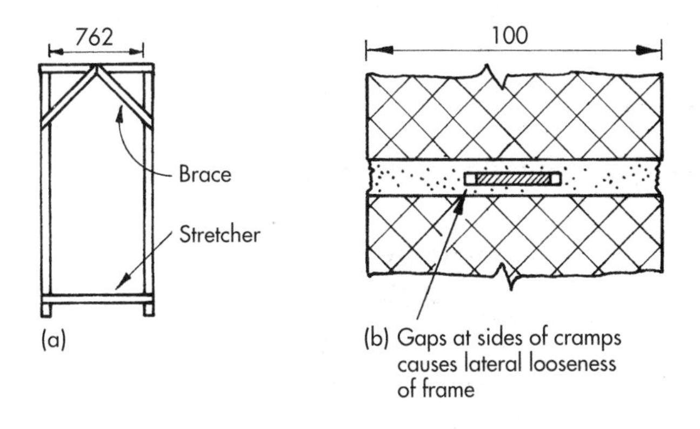

Figure 6.8(a) Braces and stretcher. (b) Effect of early removal

Figure 6.8: Usually, one or two strips of sawn timber, of about 50 × 18 mm section, acting as temporary diagonal *corner braces*, are nailed lightly to the frame to keep it square at the head. Another piece, called a *stretcher*, is fixed near the bottom to keep the jambs set apart at the correct width. Ready-made frames delivered to the site, already have braces and a stretcher – or a sill.

Early Removal

Figure 6.8(b): Although there is good reason on site to remove the braces and stretcher at the outset, because they obstruct easy passage through the opening, they should not be knocked or removed until the surrounding brickwork or blockwork is set. The illustration shows gaps at the sides of a frame cramp which will cause lateral looseness of the frame. This is due to the frame being accidentally knocked or the braces and/or the stretcher being removed (knocked off) too early, before the mortar was set.

6.2.3 Alternative Fixing Method

Figure 6.9 Frame-fix screw

Figure 6.9: A popular practice used nowadays for fixing door frames is to position and build them in without any fixings, care being taken with level and plumbness – especially lateral plumbness, affecting the brick reveals. Eventually, when the brickwork is completed and set, the frame is checked for plumbness on its face edges, minor adjustments made, then drilling and fixing to the brick reveals is carried out with nylon-sleeved Frame-fix or Hammer-fix screws. At this stage, rather than risk the quality of the fixing in the recently set mortar joints, more reliable fixings will be achieved by drilling into the bricks.

6.3 FRAME DETAIL

6.3.1 Weathering the Sill

Figure 6.10(a): For purposes of weathering and structural transition between exterior and interior levels and finish, external door frames usually have hardwood sills – sometimes referred to as thresholds. Traditionally, water bars, with a sectional size of 25 × 6 mm, ran along a groove in the top of the sill, protruding 12 mm to form a water check/draught excluder. These bars were made of brass or galvanized steel, but when this form of weathering is used nowadays, the bar is available in grey or brown-coloured rigid plastic with a flexible face-side strip which acts as a draught seal against the rebated edge of the door. In this arrangement, a weatherboard must be fitted to the bottom face of the door.

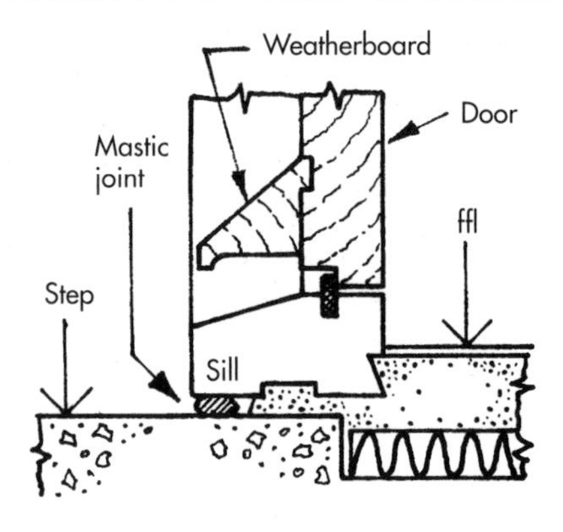

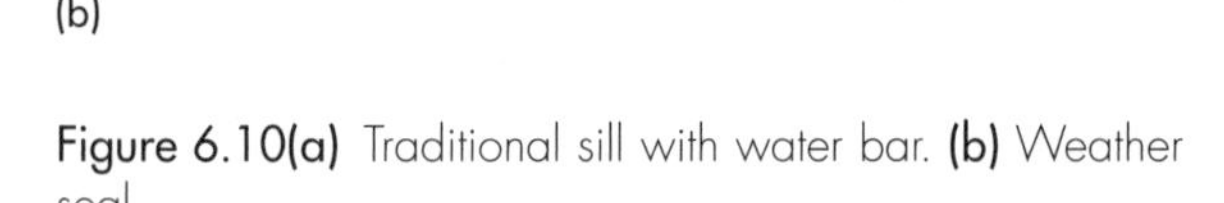

Figure 6.10(a) Traditional sill with water bar. (b) Weather seal

6.3.2 Modern Weathering Method

Figure 6.10(b): There are now very effective weather seals available for fixing to the threshold or sill of exterior frames. The wooden sill does not require the groove for a water bar, as before, but if a groove is present, it should be filled with a frame-sealant compound. These weather seals/draught excluders, resembling an open channel, are usually made of extruded aluminium in natural colour or with a brass effect finish.

Fitting and Fixing

When being fitted, the weather-seal channel is simply cut to length, fitted with rubber seals to the manufacturer's instructions and screwed in position to the sill or threshold. The door usually has to be reduced by about 25 mm at the bottom and the metal channel may require to be sealed at each end with silicone or a frame-sealant compound. This prevents moisture seepage from the channel and possible wet rot to frame or sill. Any rainwater that does enter the channel should drain out to the exterior, via weep holes in the inside front-edge of the channel.

6.3.2 Anchorage of Jambs

Figure 6.11: Sills also provide excellent anchorage for the feet of the jambs which are double-tenoned into them with comb joints (Figure 6.11(a)). Although usually through-jointed, the softwood jambs would be better protected if only stub-jointed into the hardwood sill. If the frame is of the type without a sill, 12 or 18 mm diameter metal dowels can be used to secure the jambs (Figure 6.11(b)). A hole is drilled up into the foot of each jamb with an electric or cordless drill and a combination auger bit, or with a brace and a Jennings' twist bit, and the dowels are hammered in to leave a

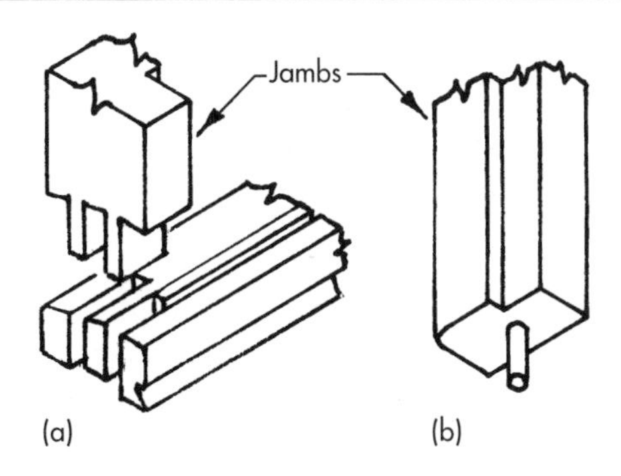

Figure 6.11 Anchorage of jambs. (a) Double tenon. (b) Dowel

40–50 mm protrusion. When the frame is being set up, the protruding dowels are either set into dowel holes or they are bedded in concrete or sand-and-cement floor screed, etc. Galvanized steel pipe, of a suitable diameter, can also be used for dowels.

6.3.3 Sunken Rebates or Planted Stops

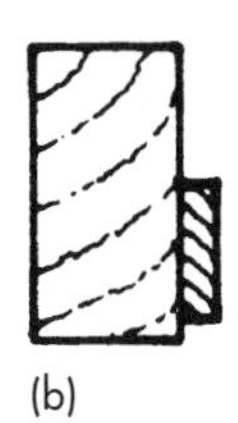

Figure 6.12 Rebates and stops

Figure 6.12: External frames should be rebated from solid wood to house the door (Figure 6: 12(a)), but internal types may have separate door stops fixed to the jambs and underside of the head to form a rebate (Figure 6.12(b)). The former are referred to as *sunken* or *stuck* rebates and the latter as *loose* or *planted* door stops.

6.3.4 Door-frame Joints

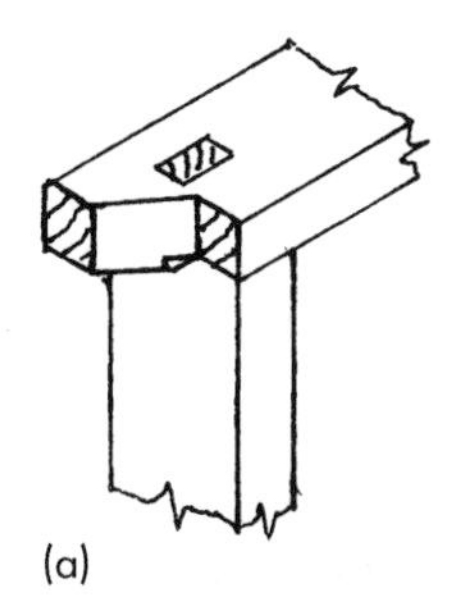

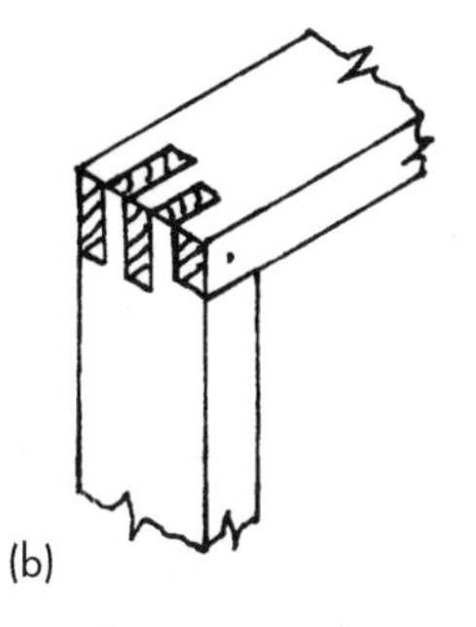

Figure 6.13 Door frame joints

Figure 6.13: Traditionally (Figure 6.13(a)), mortice and tenon joints were used to join the jambs to the head and sill. Projections of the head and sill, called *horns*, were left on to strengthen the joint and to anchor the frame when built into the brickwork. However, horns do not lend themselves so well to cavity-wall construction and bricklayers tend to cut them off anyway. In line with this practice (Figure 6.13(b)), the jambs of frames are now comb-jointed to the head and sill and pinned through the face, producing strong corner joints without the need for horns.

6.4 FIXING DOOR LININGS

6.4.1 Choice of Fixings

Linings may be fixed to timber stud partitions with 75 mm oval nails, brad-head or lost-head type or, traditionally, with cut clasp nails – which, when punched in, leave a larger hole to fill, but provide a really secure fixing. Alternatively, linings may be counterbored, screwed and pelleted. Screwing and pelleting is usually restricted to the fixing of hardwood linings, but can be justified on good-quality softwood jobs.

6.4.2 Fixing Problems

Although cut clasp nails may also be used when fixing to such lightweight partition blocks as Thermalite, other aerated building blocks, such as Celcon, do not hold these fixings very well. Therefore, when fixing linings to walls such as this, it is better to use mid-width positioned *through fixings* (drilled and fixed through timber and wall in one operation), such as nylon-sleeved Frame-fix screws.

6.4.3 Screwing and Pelleting

Figure 6.14(a)–(f) outlines the various steps involved with screwing and pelleting in relation to the following points:

(a) lining counterbored at the selected fixing points on each leg with a 12 mm diameter centre or twist bit, about 9 mm deep;
(b) shankholes drilled to suit gauge of screw;
(c) lining screwed into position, taking care not to damage the edges of the counterbored holes wih the revolving blade of the screwdriver;
(d) pellets then glued, entered lightly into holes, lined up with the grain direction and driven in carefully;
(e) bulk of pellet-surplus removed with chisel;
(f) remaining pellet-surplus cleaned off with a block plane or smoothing plane.

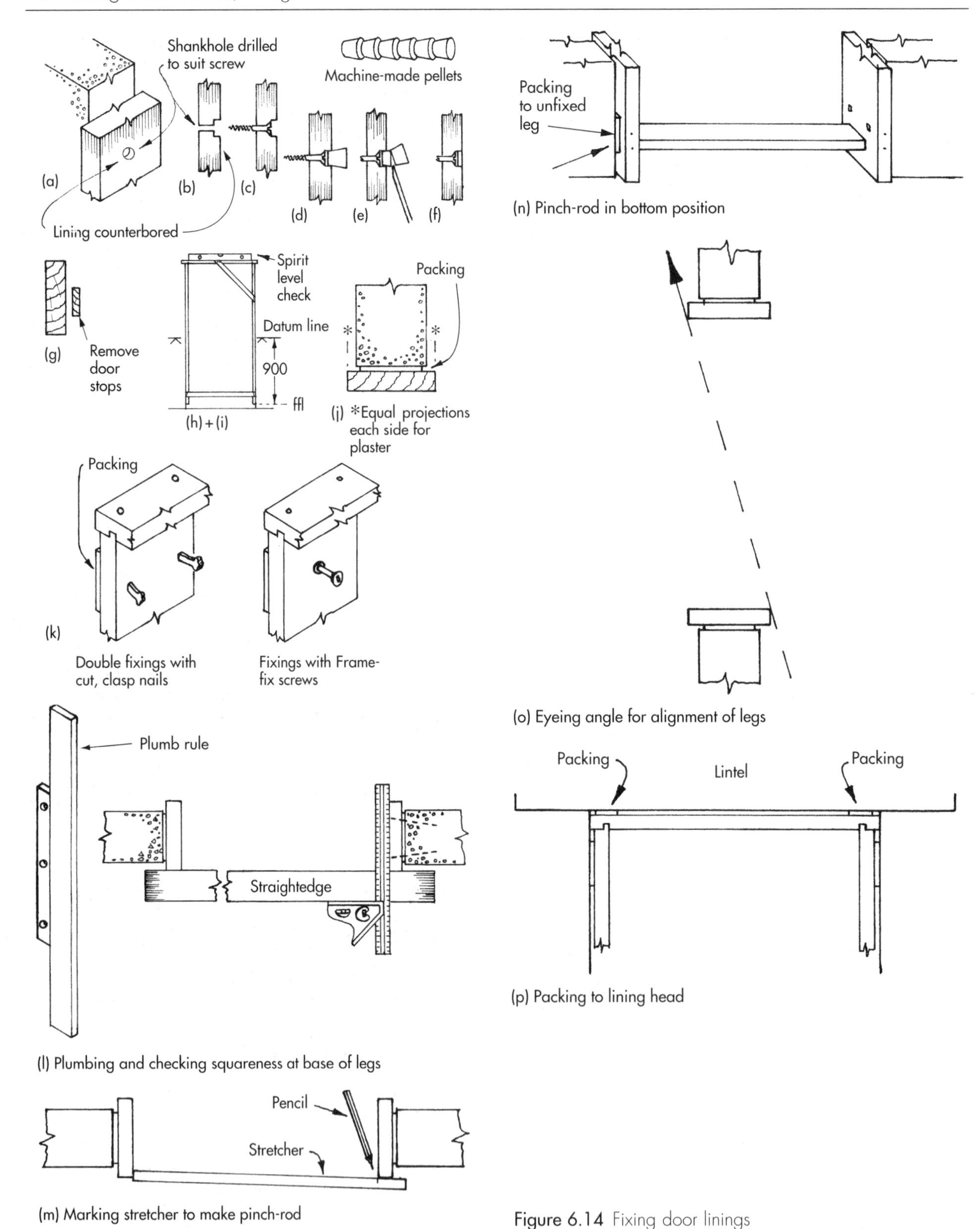

Figure 6.14 Fixing door linings

6.4.4 Fixing Technique

Owing to a lining's relative thinness and flexibility, the fixing operation can be problematic and unmanageable unless a set procedure is adopted. The following fixing technique, illustrated by Figures 6.14(g)–(p), is therefore recommended.

First Steps

(g) First remove the door stops which are usually nailed lightly in an approximate position on the lining legs and set aside.

(h) If working on unfinished concrete floors, check the finished floor level (ffl) in relation to the base of the lining. As illustrated, this is best done by measuring down from a pedetermined datum line set at 900 mm or 1 m above ffl. Place packing pieces under the lining's legs, as necessary.

Wedging and Packing

(i) Stabilize the lining in an approximate position by placing small wedges temporarily above each leg, in the gap between lintel and lining-head, then check the head with a spirit level and adjust at the base, if necessary. If gaps exist between the structural opening and the back of the lining's legs, as is usual, pack these out with plastic shims, obtainable in varying colour-coded thicknesses, or with pieces of non-splitting material such as hardboard or plywood, initially on each side of the top fixing positions only.

Adjusting and Initial Fixing

(j) Adjust the top of the lining to establish equal projections on each side of the opening for eventual plaster-thickness on the wall surfaces, or to form equal abutments for the edges of dry-lining boards.

(k) Now fix the lining near the top, *through* the packings on each leg, either with two nails per fixing or with mid-width screw fixings.

Plumbing, Squaring and Fixing

(l) Now plumb the lining on the face sides and edges with a long (1.8 m) spirit level, or alternatively, a 1.8 m straightedge with a short spirit level placed on it. Pack the bottom position each side as required and fix through one packing only. Check for squareness of the fixed leg at the base with a straightedge and try-square, as illustrated, then pack and complete the intermediate fixings on the same leg, checking before and after each fixing with the long level or straightedge. The amount of fixing points on each side should ideally be five, but not less then four. These should be placed at about 100 mm from the underside of the head, 100 mm from the ffl, and two or three intermediate points on each side.

Converting Stretcher to Pinch Rod

(m) Next, remove the stretcher from its position at the base of the lining, denail it, hold it up to a position just below the lining head, mark exact inside lining-width and cut to make a *pinch rod*.

(n) Fit the pinch rod in the bottom section of the lining-legs, as shown, and pack accordingly behind the lower fixing point on the unfixed leg.

Checking, Aligning and Final Fixing

(o) Check the plumbness of the unfixed leg for correct sideways position and check the alignment by sighting across the face edges as illustrated. Now fix the bottom, then the intermediate points, moving the pinch rod to each fixing area and packing out, if necessary, before fixing. The lining head is not normally fixed to the lintel unless the opening exceeds the normal width.

Packing to Lining Head

(p) However, the head does require to be held firmly by replacing the intitial temporary wedges mentioned in point (i) with packing or plastic shims driven into the gap between the lintel and head, at the two extremes only, immediately above each leg. Failure to complete this detail can result in the head becoming partially disjointed from the legs, at the tongued-housing joint, when final nailing of the head door-stop is completed at a later stage.

Finishing Touches

Remove the corner brace or braces (although this could have been done earlier, after points (g) or (h) to facilitate easier working). If nailed, punch in all nail fixings to about 3 mm below the surface. If screwed with nylon-sleeved Frame-fix screws, check that these are at least flush or slightly below the surface (having been fixed in the mid-width area, they should be covered by the door stops). Or, if counterbored for pellets, complete the pelleting operation as outlined in points (d) to (f). Finally, replace the door stops in their temporary position and fix protection strips, if considered necessary, as mentioned earlier in the text relating to Figure 6.4.

6.5 SETTING UP INTERNAL FRAMES PRIOR TO BUILDING BLOCK-PARTITIONS

6.5.1 Introduction

In these situations, the first step is to set out the wall positions on the concrete floor in accordance with the architect's drawing. Providing the structural walls are square to each other, the various internal partitions required need only be measured out at two extreme points from any wall to form parallels and squares.

6.5.2 Spotting

Figure 6.15: This setting out is best done with a steel tape rule and a method of marking known as *spotting*. A spot of mortar (about half a trowelful) is placed in position on the floor, then trowelled down to form a thin slither. The tape rule, peferably being held at one end, is

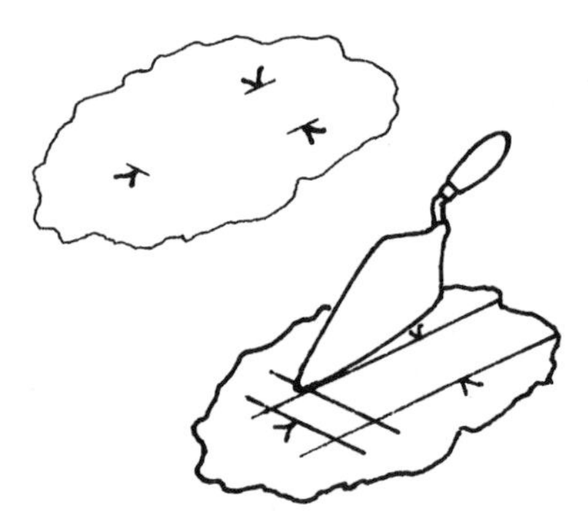

Figure 6.15 Spotting

pulled taut over the mortar spot and the trowel-tip is cut through it at the required measurement.

6.5.3 Use of Builder's Square

Short-length offset walls can be set out by using a builder's square (a wooden square, the making of which is covered in Chapter 4). Once the position of all the walls is determined, the next step is to set out the required door openings, staightedges can be used to join extreme marks and to allow further spots to be placed at intermediate positions to indicate the openings.

6.5.4 Positioning the Frame

Figure 16.6: The frames should be stood in position to relate to the setting out – preferably when the mortar spots are set – and some means of holding them at the foot and head must be devised. The following optional methods can be used.

Holding Frame at Foot

Figure 6.16(a): The first course of blocks can be laid and when set will act as a means of steadying the feet in one direction, while loose blocks on either side will hold the position in the other direction. Alternatively, notched pieces of wood can be placed against the feet or against the protruding metal dowels on the opening side and fixed to the concrete floor by means of a cartridge tool.

Holding the Frame at Head

Figure 6.16(b): The head of each frame can be held by a leaning scaffold board or boards – but a far better method is to use wall cleats and a system of top braces. The braces, of say 50 × 18 mm sawn timber, must be triangulated or placed in such a way as to create stability. The wall cleats, about 300 mm long, of say 100 × 25 mm timber, are normally fixed to the walls with 75 mm cut clasp nails, with heads left protruding for easy removal later.

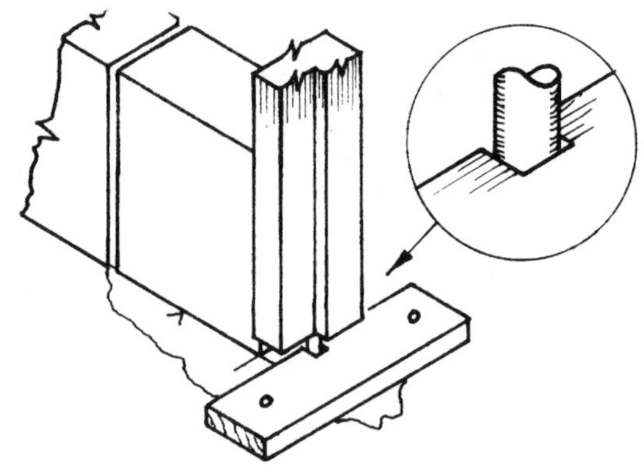

(a) Holding frame at foot

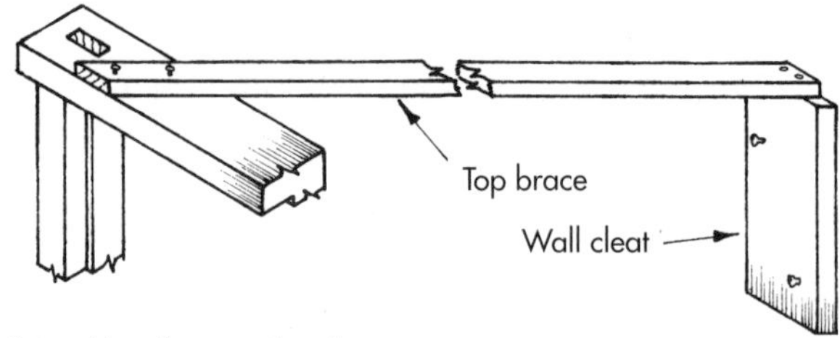

(b) Holding frame at head

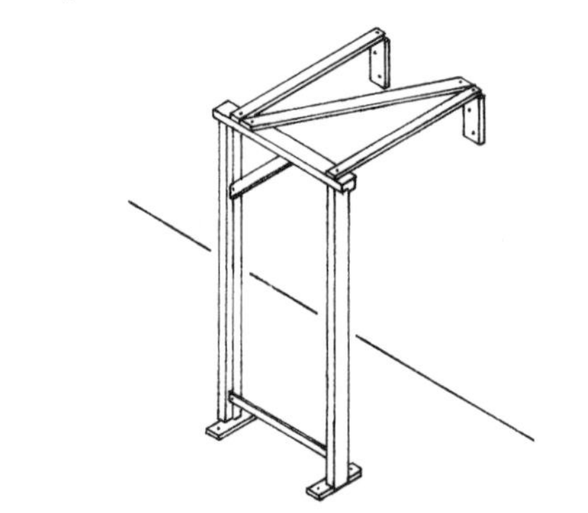

(c) Isometric view of braced frame

Figure 6.16 Positioning the frame

6.6 STOREY FRAMES

6.6.1 Internal and External with Fanlights

Figure 6.17: Storey frames may be internal or external and, as the name implies, fully occupy the vertical space between the floor and ceiling. The frame comprises two extended jambs, a head, a transom above the door and, usually, a hardwood sill on external types. The frame-space above the door can be (a) directly glazed, (b) contain louvres, (c) house a fixed sash, or (d) contain an opening sash (fanlight), opening outwards, or (e) inwards, for ventilation.

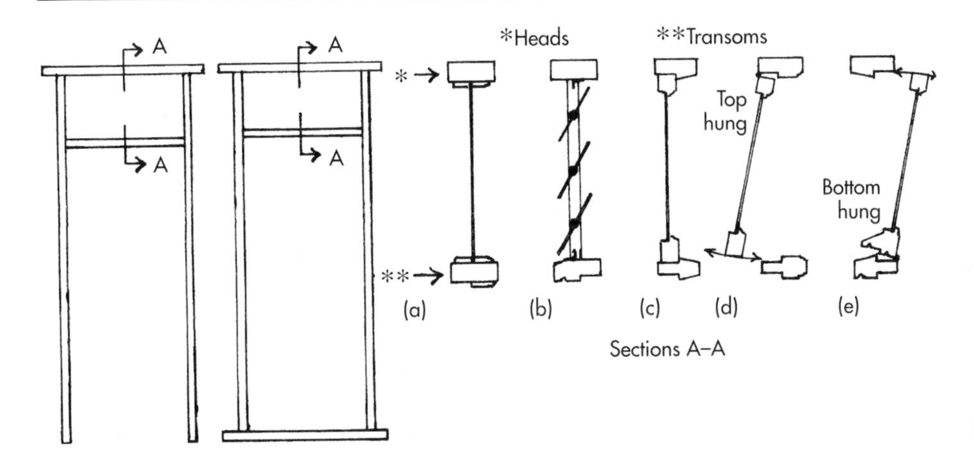

Figure 6.17 Storey frames

6.6.2 Extended Jambs Without Fanlight

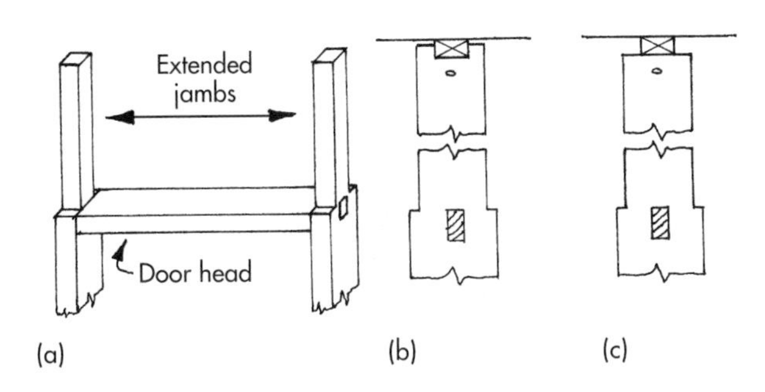

Figure 6.18(a) Extended jambs without fanlight. **(b)** Jambs notched. **(c)** Jambs butted

Figure 6.18(a): Another type of storey frame which can be used on block partitions of less than 100 mm thickness, would be minus the fanlight aperture, but have extended jambs to allow for some form of fixing to the ceiling. This would give greater rigidity to the thinner wall whose strength might otherwise be impaired by the introduction of an opening. The jambs protruding above head level should be reduced by the plaster thickness on each outer edge and, after the wall is built, these edges should be covered with a strip of expanded metal lath – unless dry-lining methods are to be used.

6.6.3 Fixing to Ceiling

Figures 6.18(b) and (c): When the ceiling above the storey frame is timber-joisted, the extended jambs are fixed to the sides of the joists or, more likely, to purpose-placed noggings between the joists. When the construction is of concrete, a timber batten or ground can be 'shot' onto the ceiling with a cartridge fixing tool and the jamb-ends fixed to this by (b) notching (cogging) or (c) butting and skew-nailing.

6.7 SUBFRAMES

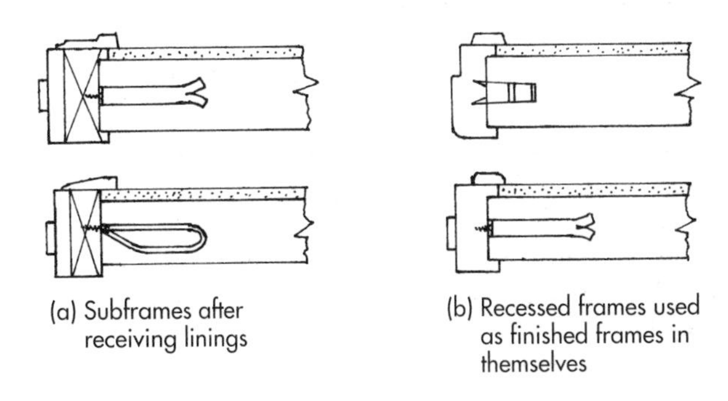

Figure 6.19 Subframes

Figure 6.19: If used, these can be recessed to take the thickness of the blocks. The recess helps to stabilize the block wall during construction and the subframe is ideal in providing an eventual means of fixing the lining and architraves – especially if these were in hardwood. Subframes, if built in as illustrated, also alleviate the fixing problems experienced with lightweight aerated blockwork. When being built in, the frames can receive metal ties or frame cramps as normal and may be used as (a) subframes to receive linings or (b) frames in themselves.

6.8 BUILT-UP LININGS

Figure 6.20: Although built-up linings, covering the full width of the reveals and soffit are not normally used nowadays, the subject is mentioned here in case it should be met on repair or conversion work. These linings, labelled (a) on the figure, are built up on grounds (b). Grounds are foundation battens which, if set up accurately, packed and fixed properly in the first-fixing operation, provide a good and true fixing base for the separate parts of the lining and/or architraves in the second-fixing operation.

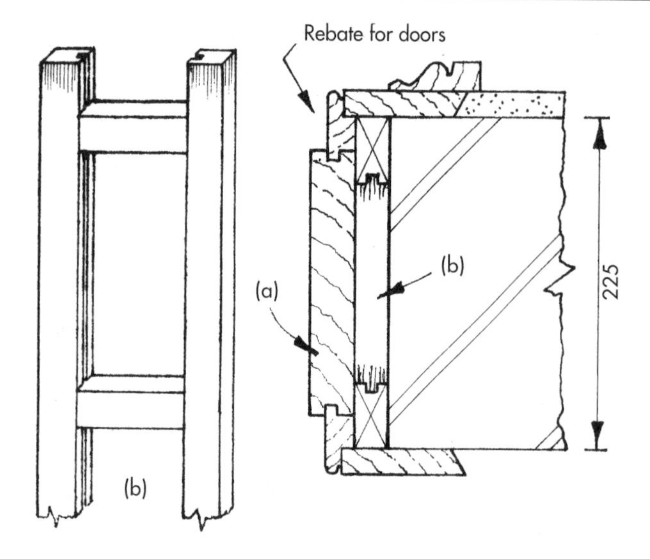

Figure 6.20 50 × 25 mm framed grounds for built-up linings

6.8.1 Framed Grounds

Figure 6.20: Framed grounds, illustrated at (b), consisting of two verticals and multi-spaced horizontal members morticed and tenoned together, looking like a ladder, were shop-made and fixed on site to suit built-up linings. These linings, used on walls of 225 mm thickness and above, were so constructed to minimize the effects of shrinkage across the face of the wider timber. The fixing technique for the three (two sides and a head) sections of framed grounds, would be similar to that used in fixing linings.

6.8.1 Separate Architrave-grounds

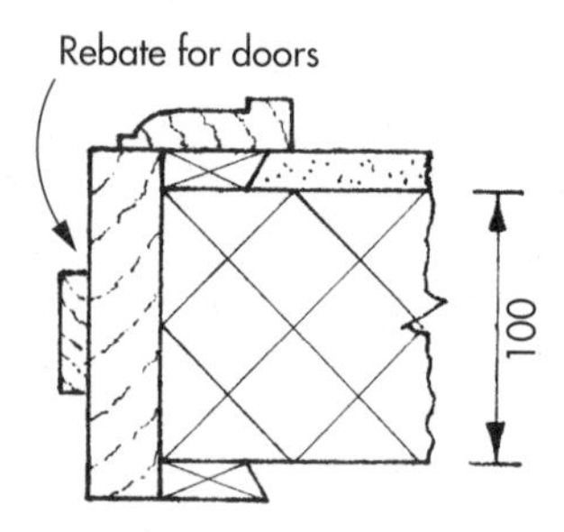

Figure 6.21 Grounds for architraves

Figure 6.21: Apart from the advantages of providing a wider fixing area for architraves, these grounds also protected lining-edges from becoming swollen by the wet plastering operation. As illustrated, these grounds were bevelled to retain the plaster on the outer edge. The width of the ground was such as to allow a minimum 6 mm overlap of the architrave on to the plaster surface. Cut clasp nails, in sizes of 50, 63 and 75 mm, were commonly used to fix the grounds to the mortar joints of the wall.

6.9 MOISTURE EFFECT FROM WET PLASTERING METHODS

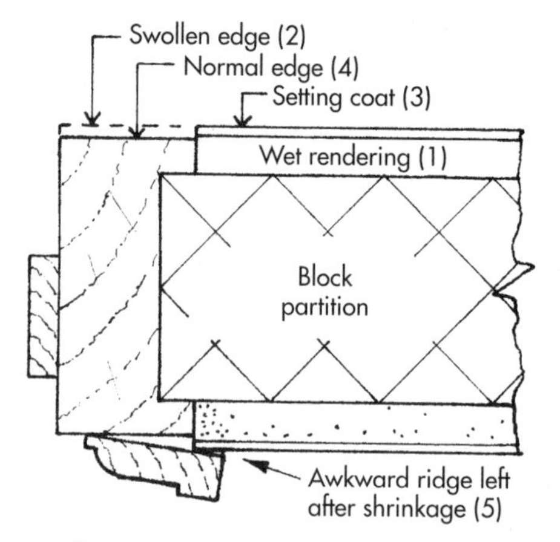

(a) Moisture effect

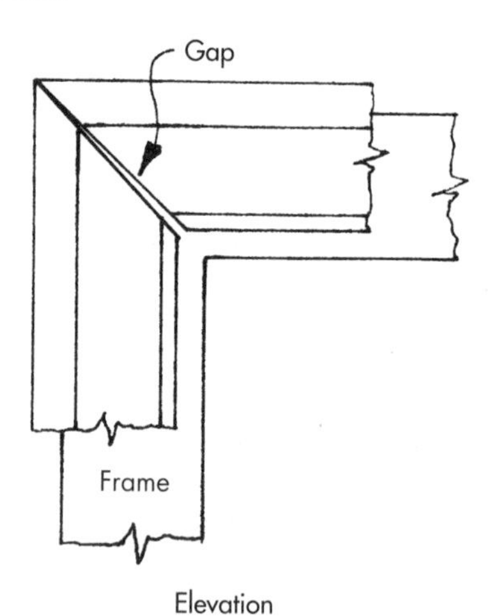

(b) Appearance of true-mitred architrave seated on raised plaster-edge

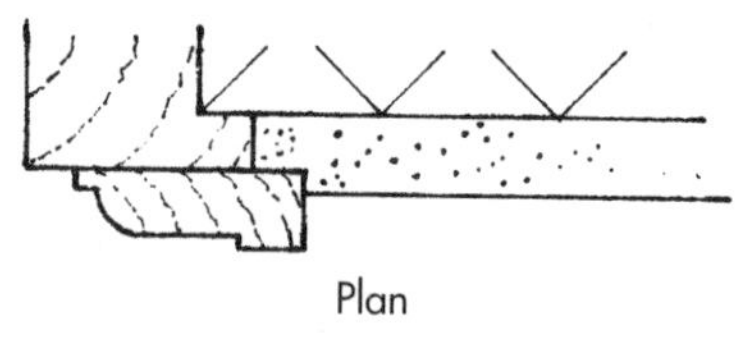

(c) Trimming plaster edges

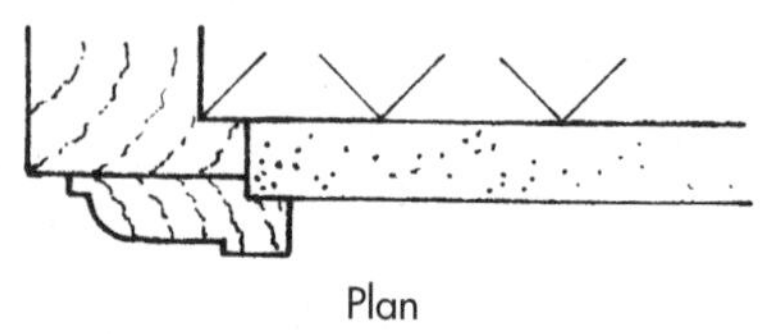

(d) Rebating architrave edges

Figure 6.22 Moisture effects and their solution

Figure 6.22(a): When frames, linings or grounds are set up and wet plastering methods are used, the effect of this should be realized as it is often detrimental to the finished work. In the first instance, excessive moisture from the wet rendering/floating coat, labelled (1) in the figure, against the timber lining, causes the timber to swell (2). While still in this state, the plasterer usually applies the setting/finishing coat of plaster (3), flush to the swollen edges. The timber eventually loses moisture and shrinks back to near normal (4), leaving an awkward ridge between the wall surface and lining (5), which upsets the seating of the architrave and the trueness of the mitres at the head.

6.9.1 Solving the mitre problem

Figure 6.22(b)–(d): Geometrically, when architraves are not seated properly, true mitre-cuts will appear to be out of true, touching on the outer (acute) points and open on the inner (obtuse) surfaces (Figure 6.22 (b)). If the plaster is only slightly proud of the lining edges, sometimes it can be tapered off with a scraping knife (or the edge of the claw hammer). This will seat the architrave better, but minor adjustments to the mitre-fit may still have to be made. If so, a sharp smoothing plane or block plane is used. Figure 6.22(c) shows how the problem can also be solved by trimming the plaster edges more drastically with a bolster chisel, which is not ideal and involves making good, or by rebating the edges of the architraves (Figure 6.22(d)).

6.10 DOORSETS

6.10.1 Introduction

Figure 6.23: Doorsets comprise linings or frames with pre-hung doors attached, locks or latches and one or two sets of architraves in position. These units are supplied by specialist firms producing doorsets as a factory operation. The main advantage of this modern practice is a reduction in time-consuming site work by eliminating conventional door-hanging, architrave and lock-fitting and fixing.

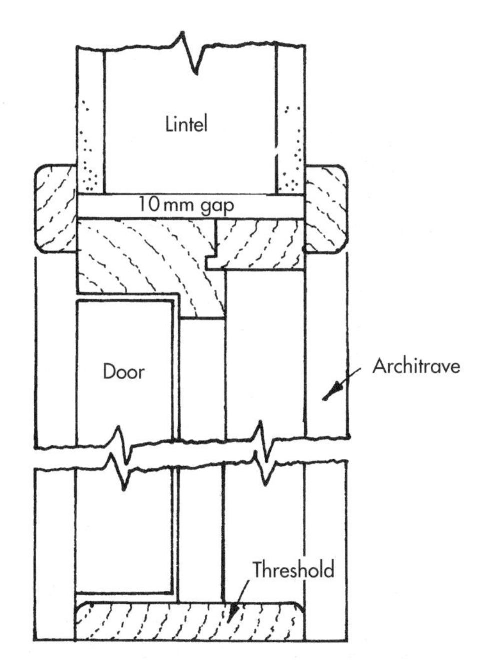

Figure 6.23 Vertical section through doorset

6.10.2 Suitability of doorsets

Because doorsets are complete units, they are not immediately suitable where conventional methods of construction, involving wet trades, are to be used. The issues against this practice include the protrusion of the architraves and hinges, which inhibits convential plastering methods, greater risk of damage to doors, and possible distortion of door-jointing tolerances due to moisture absorption by the lining from wet plaster.

6.10.3 Variations now Available

The possible variations now available are, therefore:

1. fixing coventional linings in situations where 'wet trades' are involved;
2. fixing conventional linings to openings in dry-lined walls;
3. using doorsets and fixing to openings in dry-lined walls; or
4. modifying the wet trade operation to enable the use of doorsets to openings in plastered walls.

6.10.4 Fixing Doorsets

Figure 6.24(a): The fixing of doorsets is somewhat similar to the fixing technique already covered for fixing conventional linings, except that doorsets are usually fixed after the dry-lining operation has been completed. This means that care must be taken around the opening to ensure that the finished wall thickness meets the exact width of the doorset lining. One way of doing this, is to fix temporary profile boards to the sides of the opening, similar to those shown in Figure 6.24(b), as a guide for the dry-lining fixer to work to. Other variations in technique include using the pre-hung door to check door-jointing tolerances as final confirmation of level and plumbness before completing the fixings.

6.10.5 Fixing Profiles

Figures 6.24(b)–(d): The modification mentioned in Section 6.10.3, variation 4, to enable the use of doorsets to openings in plastered walls, involves producing plywood profiles of minimum 12 mm thickness, cut to the finished wall thickness and fixed around the opening like a traditional lining. This can be done either temporarily as a guide for the wet plaster (Figure

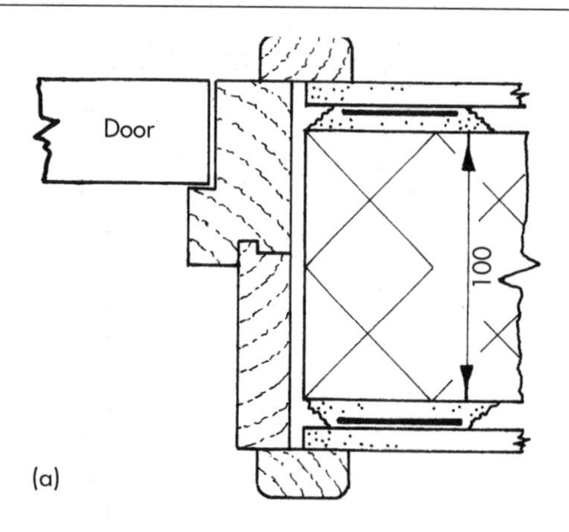

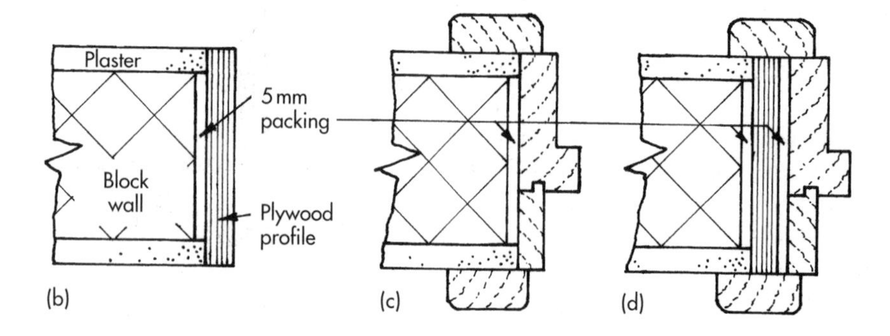

Figure 6.24(a) Horizontal section through doorset within dry-lined wall. **(b)** Profiles ensure correct wall thickness. **(c)** Receipt of doorset after profile removal. **(d)** Act as subframe for doorset.

6.24(b)), to receive doorsets after removal (Figure 6.24(c)), or permanently as an initial guide for plaster and subsequent subframe for the doorset fixing (Figure 6.24(d)).

6.11 FIRE-RESISTING DOORSETS

Figure 6.25: Doorsets with fire-resisting doors and frames are available. If they have been tested to the latest British Standards specification, instead of being referred to as ½-hour and 1-hour firecheck doors and frames, as in previous years, they should be referred to as 'fire-resisting doorsets' with a quoted stability/ integrity rating. This rating is expressed in minutes, such as 30/30 or 30/20, meaning 30 min stability/ 20 min integrity. *Stability* refers to the point of collapse, when the doorset becomes ineffective as a barrier to fire spread. *Integrity* refers to holes or gaps concealed in the construction when cold, or to cracks and fissures that develop under test.

6.11.1 Frame and Door Details

Figure 6.25: As illustrated, fire-resisting doorsets are identifiable by frames with sunken rebates of 25 mm depth and varying thicknesses of door, according to the amount of fire resistance required. Fire-resisting doors are usually:

1. made up of solid-core timber construction, clad with thin plywood, looking like thick blockboard;
2. built up of ply-clad framing with mineral infill; or
3. made up of timber frame, plasterboard, asbestos fibreboard and bonded plywood facings.

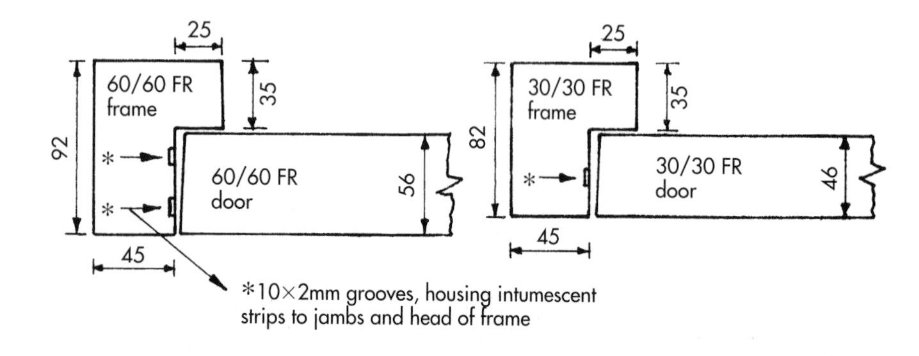

Figure 6.25 Fire-resisting doorsets

6.11.2 Intumescent strips

As illustrated, the gaps (joints) between door and frame usually contain intumescent strips which swell up when heated, thereby sealing the top and side edges of the door to increase the fire resistance. Intumescent strips give a fairly good seal to hot smoke, but as they do not become active until temperatures of 200–250°C are reached, they have no resistance to cold smoke.

6.11.3 Final Details

Figure 6.26: When fitting a fire-resisting doorset, before fixing the second set of architraves, pack the gaps between the frame and wall with mineral wool or similar fire-resisting material.

The latest TRADA (Timber Research and Development Association) Wood Information Sheet on fire-resisting doorsets recommends that narrow – not broadleaf – steel hinges should be used, to allow continuous intumescent strip to jamb edges. Slim locks, preferably painted with intumescent paint or paste, should be fitted; the thickness and thermal mass of these locks must be minimal. Over-morticing must be avoided, otherwise these hidden gaps will, in effect, reduce the *integrity* rating of the doorset.

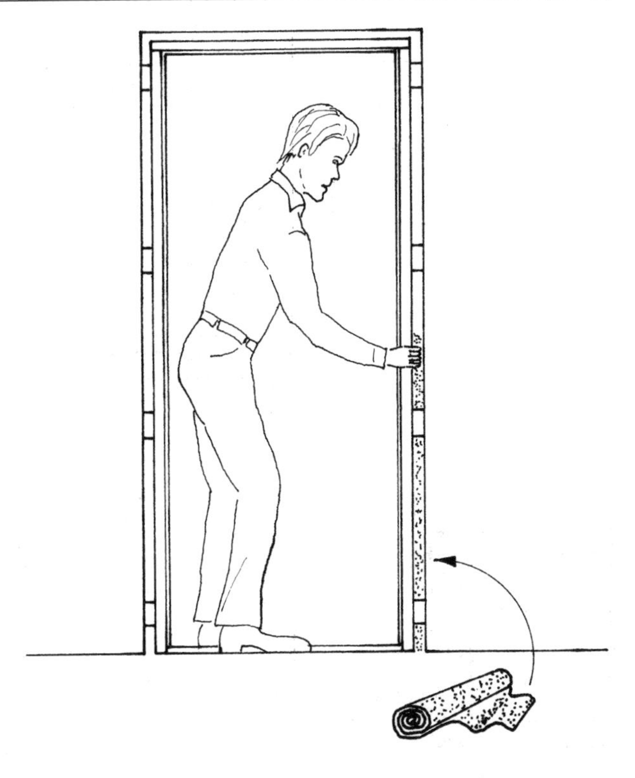

Figure 6.26 Mineral wool packing to gaps of fire-resisting frame

7

Fixing Wooden and uPVC Windows

7.1 INTRODUCTION

An important consideration which determines the method of fixing windows, is whether they are to be built-in as the brickwork proceeds, or fixed afterwards in the openings formed in the brickwork. This decision is related to the type of windows being installed and whether they are robust enough to withstand the ordeal of being used as profiles at the green brickwork stage.

7.2 CASEMENT WINDOWS

7.2.1 Wooden Casement Windows

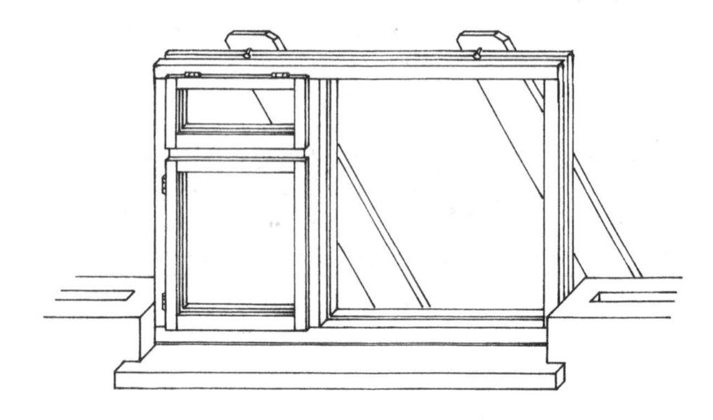

Figure 7.1 Built-in window frame

Figure 7.1: Casement windows made of wood are usually built in as the brickwork proceeds. They are secured with separate fixing devices, traditionally referred to as *frame cramps*, which are covered in detail in Chapter 6. Essentially, according to the type used, they are either screwed or hammered into the wooden side-jambs as the brickwork rises, to be built into the bed joints. Two or three cramps each side is usual. Like built-in door frames, these windows, after being positioned, are plumbed and supported at the head with one or two weighted scaffold boards pitched up from the oversite or floor. If the windows have a separate sill of stone or pre-cast concrete, usually these must be bedded first and protected with temporary boards on their outer face sides and edges. Projecting sills, formed with sloping bricks-on-edge, are usually built at a later stage, the windows having been packed up accordingly to allow for this.

7.2.2 uPVC Casement Windows

Windows made of uPVC are usually fixed after the opening is formed, by screw fixings drilled through the box-section jambs into the masonry reveal on each side, or – if these fixings clash with the cavity seal – by screwing into projecting side lugs which have been pre-cut from multi-holed galvanized strap and screwed to the sides of the window before insertion. To allow for expansion and fitting, either the window openings are built with a 6 mm tolerance added in height and width, or the windows are ordered with a 6 mm tolerance *deducted* in height and width. In either case, to ensure the correct size of opening is built, temporary wooden profile frames made of 50 × 50 mm or 75 × 50 mm prepared softwood, with 6 mm WBP plywood corner plates, as illustrated, are constructed and placed in position during the brickwork and blockwork operation.

Profile Frames

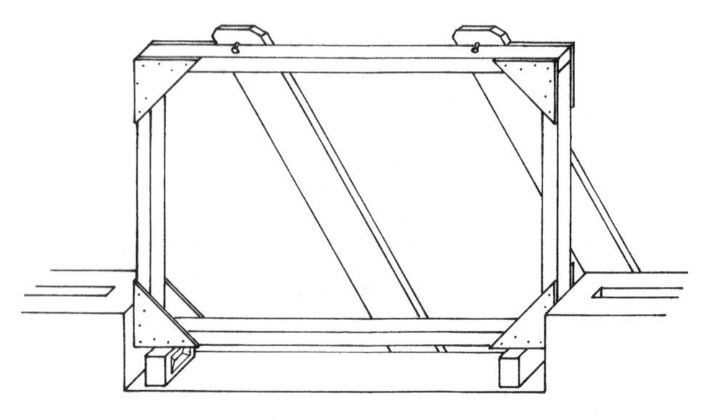

Figure 7.2 Temporary profile frame

Figure 7.2: The temporary profile frames may be made on site or in the workshop and can be removed soon after the brickwork/blockwork has set, or left in position until the windows are to be installed. After careful removal, the frames may be dismantled, stored or reused immediately. When the uPVC windows are eventually fitted, the expansion/fitting tolerance is taken up equally all round – or as equally as possible – with special plastic shims. On the sides of the window, these shims, which are 'U' shaped, are slid around each screw fixing before the screw is fully tightened. Upon completion, or when all other building work is complete, any projecting shim is trimmed off and the gap around the window is gunned around with a silicone sealant.

7.3 GLAZING

7.3.1 DIY Glazing

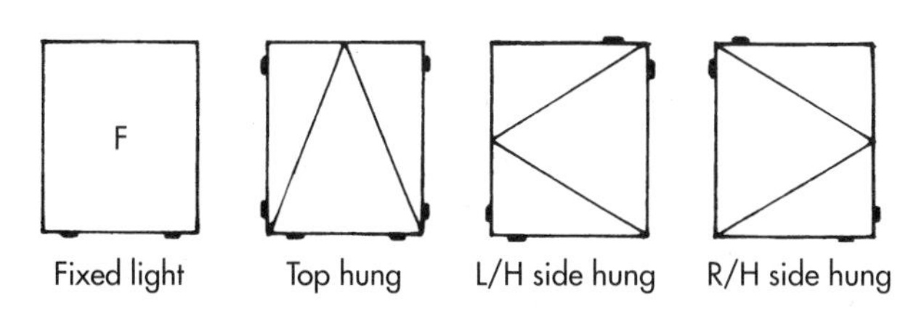

Figure 7.3 Position of setting and location blocks

Figure 7.3: If installing the double-glazed sealed units into the fixed uPVC windows yourself – which is easy enough once you know how – you must first understand that the units must be seated on plastic *setting blocks* and, if the window is an opening vent, have plastic *locating blocks* in various positions on the sides, as illustrated. Different thicknesses of plastic blocks may be required, but they are normally 3 mm thick, 25 mm long and should be as wide as the sealed unit's thickness (which is often 28 mm as standard nowadays).

7.3.2 Externally Beaded Glazing

Figure 7.4: With the setting and locating blocks fitted snugly in position between the glass and the uPVC casement, and the sealed unit pushed in to rest up against the upstanding inner edge, the prefitted shuffle beads, which were taken out to gain access for glazing, are now pivoted into the small front lip and snapped into position. Ideally, these should be pre-marked L/H, R/H, TOP and BOT, so that they relocate in their original positions.

Internal Access

On the other side of the glass now, the unit is pushed outwards, to be hard up against the external beads just fitted. This creates a small gap between the glazed unit and the grooved, upstanding edge of the window. The so-called *wedge gasket*, illustrated in position in Figure 7.4(a), is fed into this gap all round, forcing the unit forward and locking and sealing the external beading. The gasket, which has to be cut or partly-snipped on its concealed face at each corner (with at least a 25 mm length-allowance added each time to ensure a well-compacted fit), is not easy to push in. To assist with the task, it would be advisable to make a simple wooden caulking tool (Figure 7.4(b)) and, if still difficult, diluted washing-up liquid can be brushed into the gap prior to inserting the gasket.

7.3.3 Internally Beaded Glazing

Figure 7.5: uPVC casements with internally beaded sealed units are becoming quite popular nowadays, mainly for the following two reasons.

- They are thought to be more burglar-resistant.
- Being internally beaded means that the glazing operation is done from inside the building, which on certain jobs – especially those without scaffolding – can be an advantage.

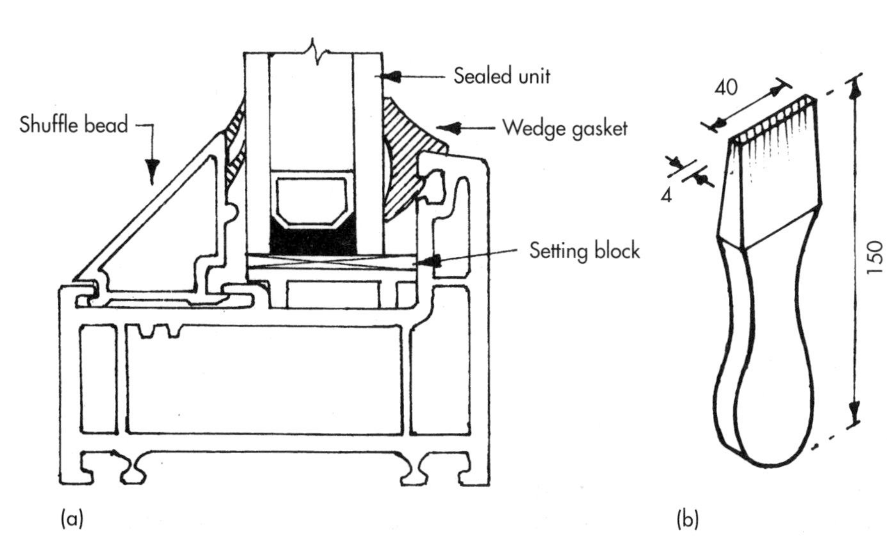

Figure 7.4(a) Externally beaded glazing. **(b)** Caulking tool

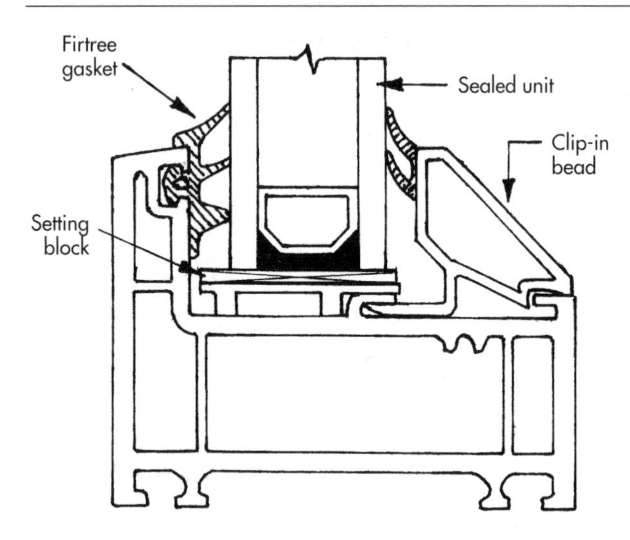

Figure 7.5 Internally beaded glazing

On this type of window, there is an *external* upstanding edge, grooved as before, but this time to receive a so-called *firtree gasket*, as illustrated, instead of the wedge gasket. The sealed unit, still on setting blocks and, if in an opening vent, secured with locating blocks, is pushed outwards to be hard up against the firtree-gasket edge, and the internal 'clip-in' beads are pushed and clipped fairly easily into place.

7.4 WINDOW BOARDS

Windows must be fixed into position before dry-lining or plastering takes place, to enable a satisfactory abutment of the lining against the window's head and sides. For the same reasons, window boards are also required to be fixed. These boards appear as a stepped extension of the sill and project beyond the plaster faces at the front and sides, like the projecting nosing of a stair tread. If made of MDF board or timber and fitting up against a wooden casement window, the back edge is usually tongued into a groove in the sill. If the window is of uPVC, the window board is only butted and may be held with an adhesive such as Gripfill – or the abutment may be covered with a small plastic cloaking-fillet or quadrant held with superglue.

7.4.1 Marking and Cutting the Window Board

Figure 7.6(a): As illustrated, a portion of board about 50 mm in from each end is marked and cut to fit the window reveals, and the machined nosing shape on the front edge is returned on the ends, by hand with a smoothing plane and finished with glasspaper.

7.4.2 Fitting and Fixing the Window Board

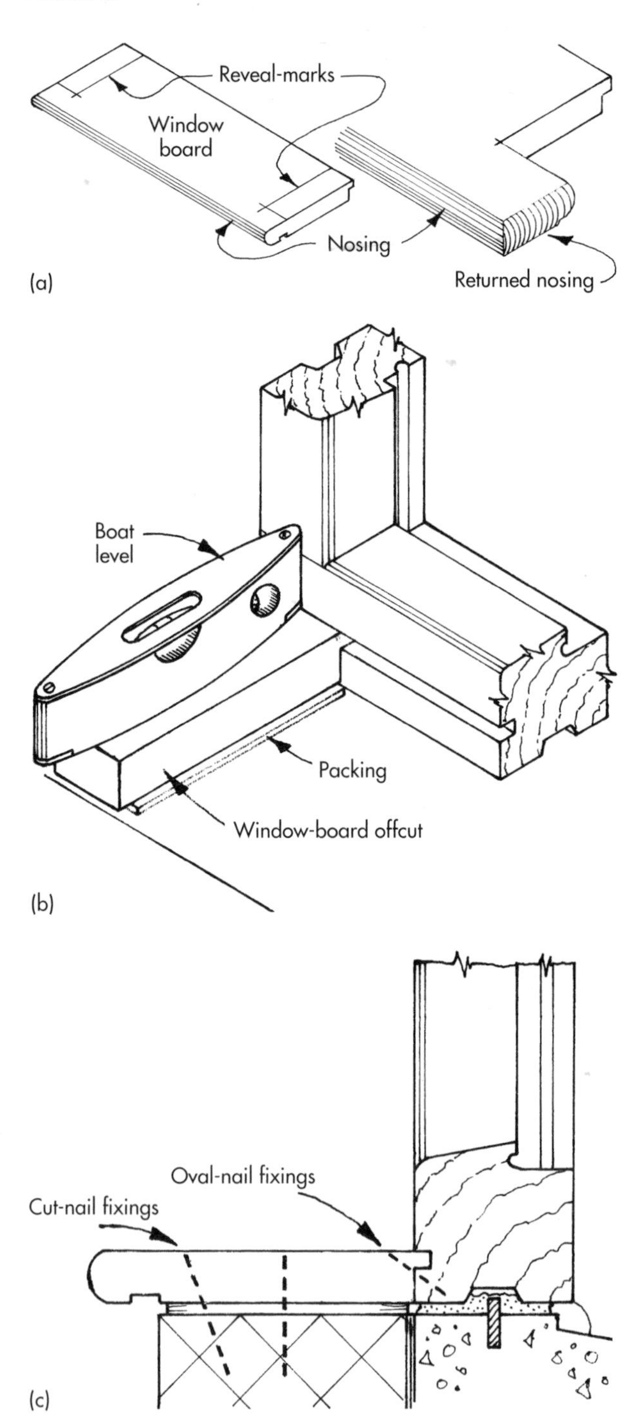

Figure 7.6 Marking and fitting a window board

Figures 7.6(b) and (c): Packing is usually required between the window board and the inner skin of blockwork, to level the board across its depth. Pieces of damp-proof course material, plastic shims or hardboard make ideal packings. If tongue-and-grooved, the boards are skew-nailed to the wooden sills with 38 mm oval nails or panel pins and fixed through the packings into the blockwork with either cut clasp nails or pelleted screw fixings. Packings may be established about every 450 mm prior to positioning the board, by using the end offcuts, as illustrated, with the tongue in the groove, a spirit level on top – or a try-square against the jambs – while trial packings are inserted.

8

Fixing Floor Joists and Flooring

8.1 INTRODUCTION

Although reinforced concrete floors of all kinds are used in large buildings such as blocks of flats or office blocks, timberfloors are still widely used in domestic dwelling houses, especially above ground-floor level. Floors are generally referred to according to their position in relation to the ground. These range upwards from ground-floor level, first floor, second floor and so on; they may also be classified technically as single or double floors, according to the cross-formation of the structural members.

8.1.1 Single and Double Floors

Figure 8.1(a): Suspended timber floors consist of board-on-edge like timbers known as *joists*, spaced parallel to each other at specified centres across the floor and, in the case of a *single floor*, resting between the extreme bearing points of the walls. In the case of a *double floor*, they rest on intermediate support(s) and the extreme bearing points of the walls. The top surface of the joists can be covered with various materials such as timber T&G floor boards, chipboard T&G flooring panels, plywood T&G flooring panels or Sterling OSB (oriented strand board) T&G flooring panels – and the underside of the joists covered with ceiling material such as Gyproc plasterboard.

8.1.2 Spacing of Joists

Figure 8.1(b): The spacing of the joists is related to the thickness of floor boarding or sheeting to be used; 400 mm centres (c/c) is required in domestic dwellings using ex. 22 mm T&G timber boarding, 18 mm chipboard T&G panels, 18 mm plywood T&G panels or 15 mm OSB T&G flooring panels. When the joists are spaced at 600 mm centres, three of these materials need to be thicker: the T&G timber boarding should be ex. 25 mm, the T&G chipboard should be 22 mm, and the OSB T&G panels should be 18 mm. Tongue and groove flooring panels are critically 2400 or 2440 mm long and 600 mm wide. The length needs to be considered when setting up the joists, because the staggered cross-joints of these panels must bear centrally on the joists. Therefore, if the panel is a metric modular length of 2400 mm, then the joist-spacings should be 2400 ÷ 6 = 400 mm c/c., or 2400 ÷ 4 = 600 mm c/c. However, if the panel is based on an imperial length of 8 ft, converted

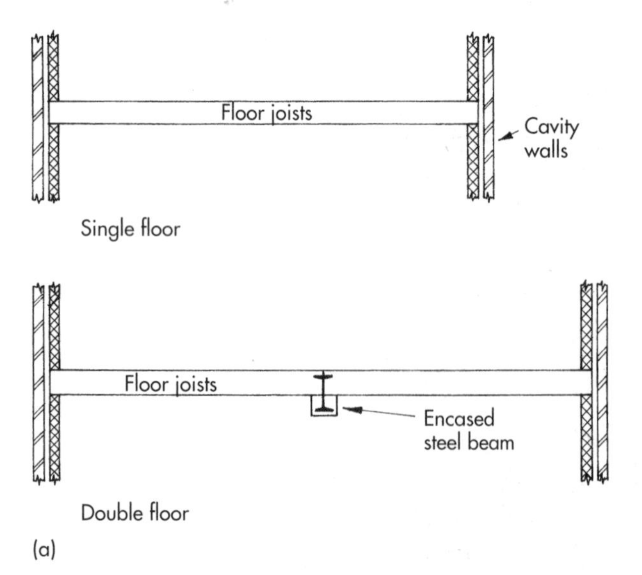

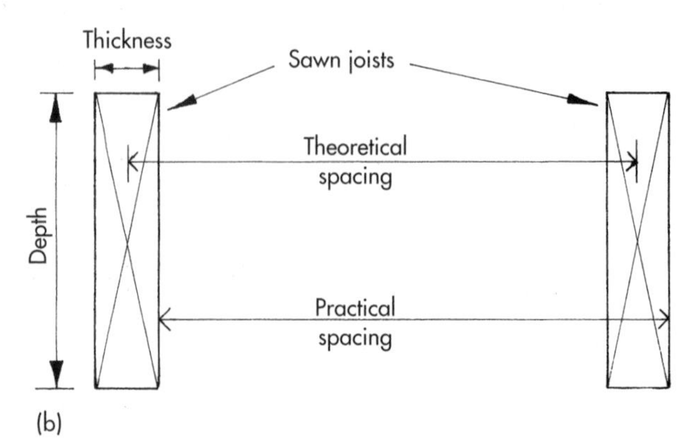

Figure 8.1(a) Suspended floor joists. (b) Spacing of joists

to 2440 mm, then the joist-spacings should be 2440 ÷ 6 = 406.6 mm c/c (16 in) or 2440 ÷ 4 = 610 mm c/c (24 in). On upper floor levels, these considerations also apply to the cross-joints of the plasterboard sheets to be used later on in the ceiling below.

8.1.3 Size of Joists

The sectional size of joists is always specified and need not concern the site carpenter or builder. The subject enters into the theory of structures and mechanics and is, therefore, a separate area of study. However, for domestic dwellings, a simple rule-of-thumb calculation has existed in the trade for many years, expressed as

$$\frac{\text{span}}{2} + 2$$

but this is only an approximate method, which errs on the side of safety. In Imperial measurement, this was expressed as

$$\text{depth of joist in inches} = \frac{\text{span in feet}}{2} + 2$$

For example, if the span of the joists is 14 ft 0 in

$$\text{depth of joist} = \frac{14}{2} + 2 = 9 \text{ in}$$

In metric measurement, the formula is converted to

$$\text{depth of joist in centimetres} = \frac{\text{span in decimetres}}{2} + 2$$

For example, for a joist span of 4 m

$$\text{depth of joist} = \frac{40}{2} + 2 = 22 \text{ cm} = 220 \text{ mm}$$

The thickness of joists, by this method, is usually standardized at 50 mm. The nearest commercial size, therefore, would be 225 × 50 mm.

8.1.4 Structurally Graded Timber

By comparison, table A1 for floor joists, given in The Building Regulations' AD (approved document) A1/2, specifies joists of 220 × 38 mm section at 400 mm centres, for a maximum span of 4.43 m. However, it should be noted that SC3 (strength class 3 – now referred to as C16) structurally graded timber is specified. Such timber is now commonly relied upon for structural uses and is covered by BS 4978 – and BS 5268: Part 2: 1991. Certain standards and criteria are laid down regarding the size and position of knots, the slope of grain, etc., and the assigning of species and grade combinations to strength classes SC3 (C16) and SC4 (C24).

8.2 GROUND FLOORS

8.2.1 Suspended Timber Floor

Figure 8.2(a): The first of the various types of ground floor that involves the carpenter, although rarely used nowadays, is the suspended timber floor. Joists of 100 × 50 mm section, are generally used, spaced at 400–600 mm centres. They rest on 100 × 50 mm timber wall-plates and are skew-nailed to these from each side with 100 mm round-head wire nails – or alternatively fixed with framing anchors such as MAFCO Trip-L-Grip, type AL and AR. The wall plates are bedded on half-brick-wide sleeper walls with a damp-proof-course material sandwiched in the mortar joint. The sleeper walls, which should be honeycombed for underfloor air circulation, were traditionally built at 1.8 m centres to support the 100 × 50 mm sawn joists. However, if using structurally-graded timber, designated as SC3 (structural class 3) – also known as C16 (class 16) – the walls may be built at 2.08 m centres to support the joists

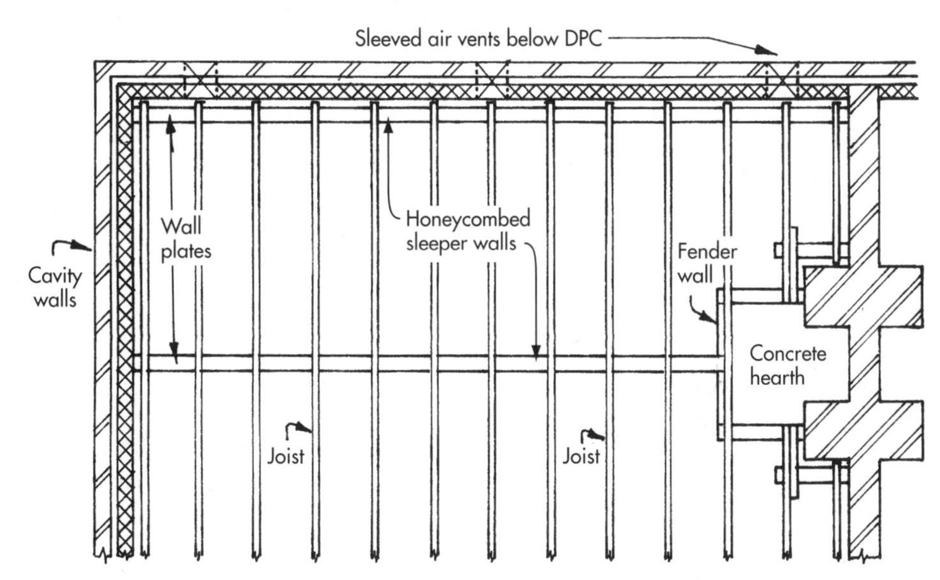

Figure 8.2(a) Part-plan view of exposed floor

at 400 mm c/c, and 1.67 m centres to support the joists at 600 mm c/c. If SC4 (structural class 4) - also known as C24 (class 24) - joists are used, the walls may be built at 2.2 m centres to support the joists at 400 mm c/c, and 1.82 m centres to support the joists at 600 mm c/c. The honeycombed sleeper walls are usually built onto the concrete oversite, rather than onto separate foundations.

8.2.2 Regulation Requirements

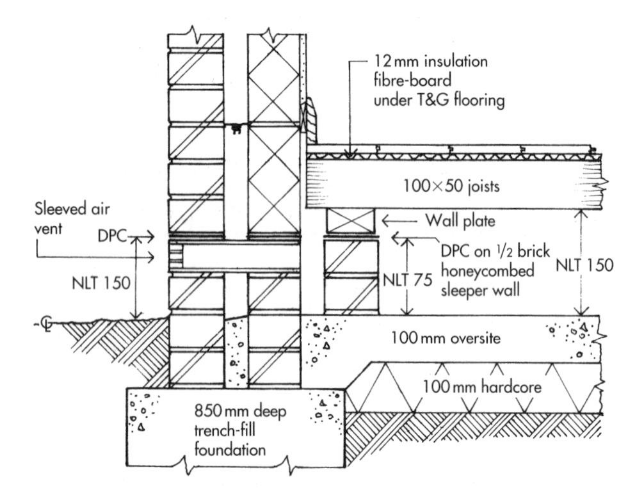

Figure 8.2(b) Sectional view through floor and wall (NLT = not less than)

Figure 8.2(b): Part C of the Building Regulations, which is concerned with protecting buildings from dampness, requires the site to be effectively cleared of turf and other vegetable matter and the concrete oversite to be of 100 mm minimum thickness and to a specified mix, laid on clean hardcore and finished with a trowel or spade finish. The top surface of the concrete oversite should not be lower than the highest level of the ground or paving adjoining the external walls of the building. The space above the concrete to the underside of the wall plates must not be less than 75 mm and not less than 150 mm to the underside of the joists. The space should be clear of debris (broken bricks, shavings, offcuts of timber, etc.) and be adequately through-ventilated with a ventilation area equivalent to 1500 mm² per metre run of wall. As illustrated, the damp-proof course in the cavity wall should be not less than 150 mm above the adjoining ground or paving – and the top of the cavity-fill should be not less than 150 mm below the level of the lowest damp-proof course (as indicated in Figure 8.9 for a *surface-battened floor*).

8.2.3 Bedding the Wall Plates

Figure 8.2(c): The first operation is to cut the wall plates to length, bearing in mind that the ends of these should be kept away from the walls by approximately 12 mm. After laying and spreading mortar on the sleeper walls,

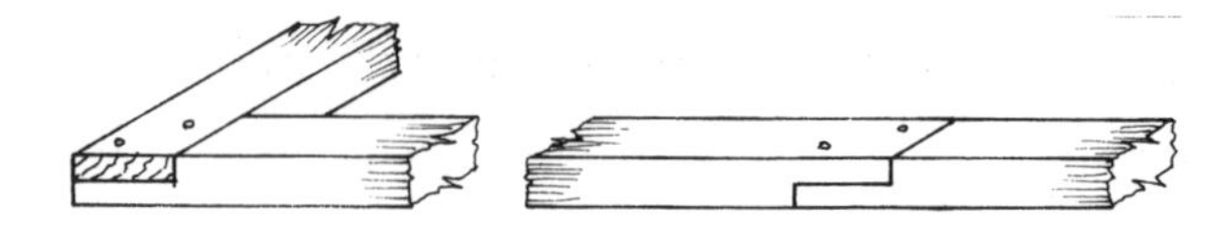

Figure 8.2(c) Wall-plate joints

rolling out and flattening the DPC material, more mortar is laid and the wall plates are bedded and levelled into position with a spirit level. Once the first plate has been bedded and levelled, the others, as well as being levelled in length, must also be checked for level cross-wise, using the first plate as a datum. If any wall plate cannot be laid in one piece, or changes direction, it should be jointed with a half-lap joint.

8.2.4 Laying and Fixing the Joists

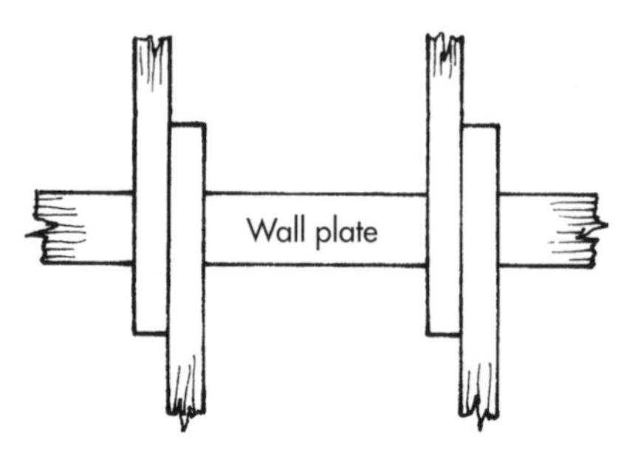

Figure 8.2(d) Joining joists on sleeper walls

Figure 8.2(d): When the wall plates are set, the joists can be cut to length and fixed in position by nailing or anchoring – bearing in mind that the ends of the joists should also be kept away from the walls by approximately 12 mm. The first joist is fixed parallel to the wall, with a 50 mm gap running along its wall-side face, to create more reliable edge-bearings and to facilitate easier board fixings. The second joist can be fixed at 400 mm centres from the first if timber boarding is to be used, but if edge-finished flooring panels are to be used, then the second joist should be fixed at 400 or 406 mm + 12 mm expansion gap from the wall – *not* the first joist. Subsequent joists are fixed at the required spacing until the opposite wall is reached. The last spacing is usually under or slightly over size. Joists joined on sleeper walls are usually overlapped, as illustrated, and side-nailed. To improve the insulation qualities of this type of floor, insulation such as 50 mm thick Rockwool slabs could be preset between the joists (Figure 8.2(e)), in addition to the overlaid insulation fibreboard suggested in Figures 8.2(b) and (f).

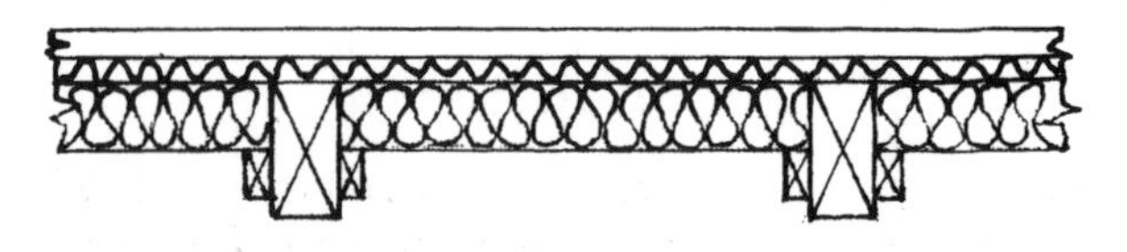

Figure 8.2(e) Additional insulation

8.2.5 Providing a Fireplace and Hearth

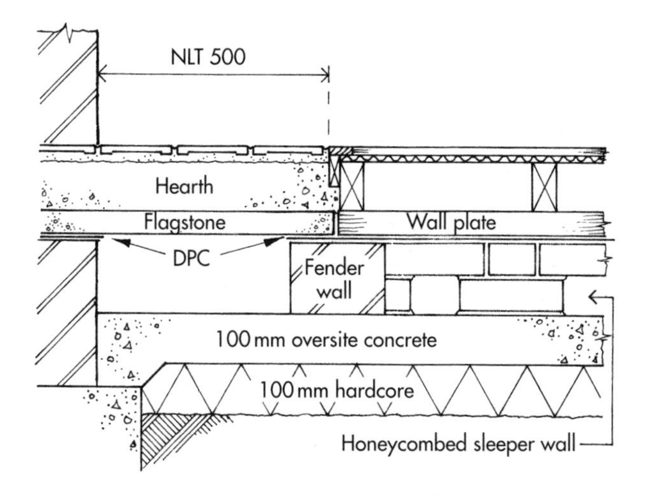

Figure 8.2(f) Secional view through hearth

Figure 8.2(f): Although omitted in recent years, there seems to be a demand for – and some return to – traditional fireplaces able to take a gas or electric fire in the lounge or living room. Therefore, if a fireplace and the required concrete hearth are to protrude into the floor area, the hearth can be contained below floor level within a one-brick-thick *fender wall* built around the fireplace. The ends of the joists rest on wall plates supported by half the thickness of the fender wall. The other half supports the concrete hearth. Part J of the Building Regulations requires that no timber should come nearer to the fire opening than 500 mm from the front and 150 mm from each side (Figures 8.2(f) and (g)). Although not demanded by the Regulations, ideally the timbers should be pre-treated or treated on site with a preservative.

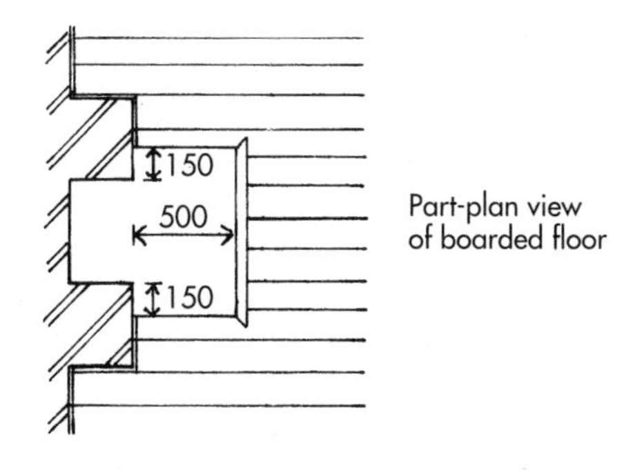

Figure 8.2(g) Regulation size of hearth

8.2.6 Flooring Materials

The flooring material can be tongue and groove (T&G) timber boarding, flooring-quality chipboard T&G flooring panels, plywood T&G flooring panels or OSB (oriented strand board) T&G flooring panels. Their thicknesses must suit the joist-spacing, as described above.

8.3 LAYING T&G TIMBER BOARDING

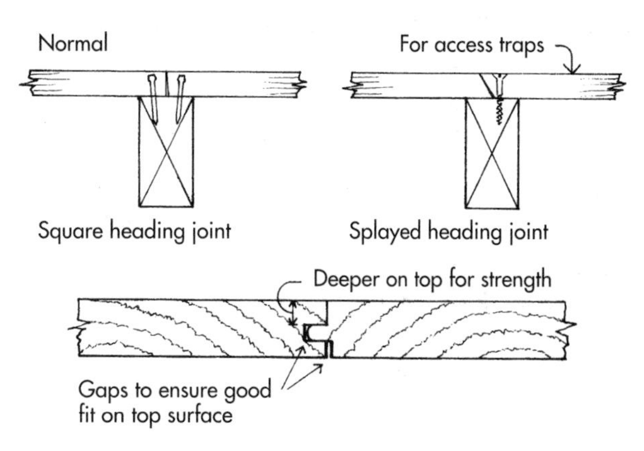

Figure 8.3 The right way up for T&G boards

Figure 8.3: When laying boarded floors, cross-joints (end grain or heading joints) should be kept to a minimum, if possible, and widely scattered. No two heading joints should line up on consecutive boards. On all sides, boards should be kept away from the walls by approximately 12 mm. This is to reduce the risk of picking up dampness from the walls and to allow for any movement across the boards due to expansion. All nails should be punched in 2–3 mm below the surface. Boards should be cramped up and fixed progressively in batches of five to six at a time. Tongues and grooves should be protected during cramping by placing offcuts of boarding between the cramps and the floor's edge. Fixings – cut floor-brads or lost-head wire nails – should be about $2\frac{1}{2}$ times the thickness of the board, i.e. for 20 mm boards use 50 mm nails. There should be two nails to each board fixing, about 16 mm in from the edges. Just prior to fixing, boards must be sorted and turned up the right way, as illustrated.

8.3.1 Fixing Procedure

Figure 8.4: Cramping can be done with patent metal floor cramps which saddle and grip the joists when wound up to exert pressure on the boards, or by using sets of folding wedges cut from tongued-and-grooved offcuts, 200–300 mm long. The first board is nailed down about 12 mm away from the wall, with small wedges inserted to retain the gap during cramping. Five or six more boards are cut to length and laid. When using wedges as cramps, a seventh board is cut and partially nailed, set away from the laid boards at a distance

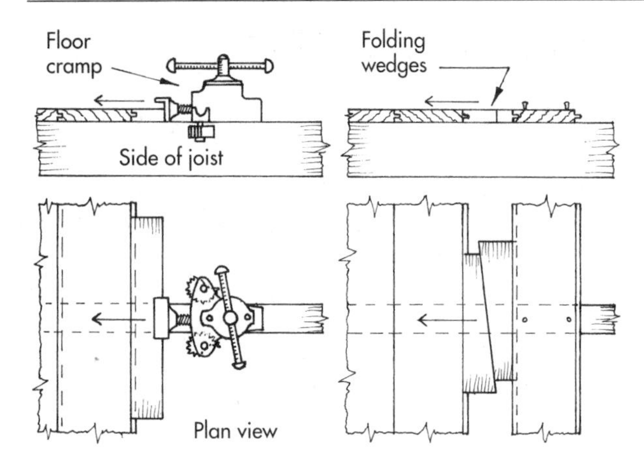

Figure 8.4 Cramping methods

equal to the least width of the pre-cut folding wedges. The wedges are inserted at about 1–1.5 m centres and driven in to a tight fit. The boards, having been marked over the centre of the joists, are then nailed. When complete, the wedges and the seventh board are released. This board becomes the first of another batch of boards to be laid and the sequence is repeated until the other wall is reached. The final batch of boards are levered and wedged from the wall with a wrecking bar or flooring chisel, then nailed; the last board having been checked and ripped to width to ensure a 12 mm gap from the wall on completion.

8.4 FLOATING FLOOR (WITH CONTINUOUS SUPPORT)

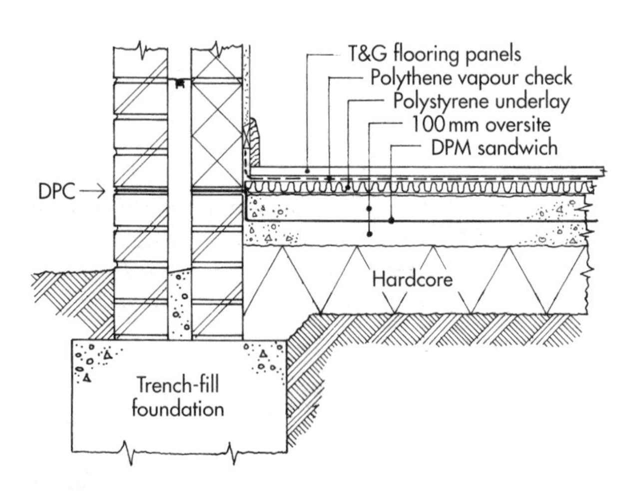

Figure 8.5 Floating floor

Figure 8.5: This type of floor, consisting of manufactured T&G flooring panels, laid and floating on close-butted 25 mm thick rigid polystyrene sheets, such as Jablite, and held down by its own weight and the perimeter skirting, turns a cold unyielding concrete base into a warm resilient floor at relatively low cost. The flooring panels used may be 18 mm T&G chipboard (preferably moisture-resistant grade), or 18 mm T&G plywood, with laid-sizes of 2400 × 600 mm or 2440 × 600 mm. Sterling OSB T&G flooring panels are not recommended for use on floating floors.

8.4.1 Laying Procedure

After the closely butted sheets of Jablite insulation have been laid over the concrete floor, a roll of 1000 gauge minimum thickness polythene sheeting is rolled out and laid over it to act as a vapour check. This should be turned up at least the flooring-thickness at the wall edges and all joints should be lapped by 300 mm and sealed with water-proof adhesive tape, such as Sellotape 1408. Next, the tongued-and-grooved panels are laid, taking care not to damage the polythene and leaving an expansion gap of 10–12 mm around all walls and other abutments. The cross-joints must be staggered to form a stretcher-bond pattern and all joints should be glued with polyvinyl acetate adhesive, such as Febond. Laying is started against the wall from one corner and when the other corner is reached, any reasonably-sized offcut can be returned to start the next row. Otherwise half a sheet should be cut to do this. Temporary wedges should be inserted around perimeter gaps until the glued joints set. A protective batten or flooring-offcut should be held against the flooring's edge when hammering panels into position. Finally, before fixing the skirting, the perimeter gap should be checked and cleaned out, if necessary.

8.4.2 Other Points to Note

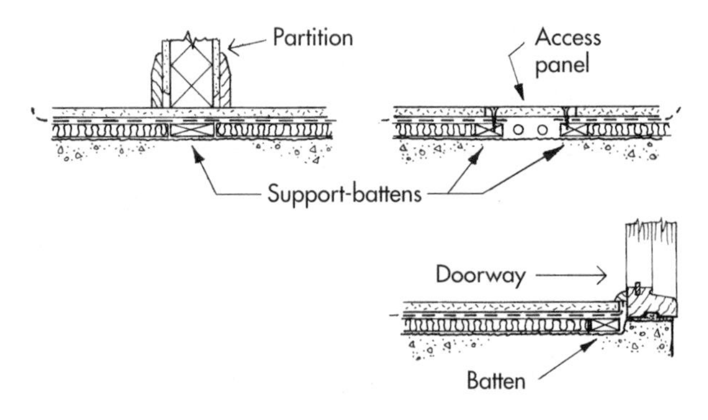

Figure 8.6 Battens required within the insulation

Figure 8.6: Any pipes or conduits to be accommodated within the insulation material must be securely fixed to the oversite concrete and the thickness of insulation material may need increasing to exceed the diameter of the largest pipe. Hot-water pipes should not come in direct contact with any polystyrene underlay. Preservative-treated battens should be fixed to the slab to give support to any concentrated load on the floor, such as a partition or the foot of a staircase, etc. Such battens are also required where an access panel is to be formed and where the floor abuts a doorway, as illustrated. As well as being stored carefully on site, flooring panels should be *conditioned* by being laid loosely in the area to be floored for at least 24 hours before fixing.

8.5 FLOATING FLOOR (WITH DISCONTINUOUS SUPPORT)

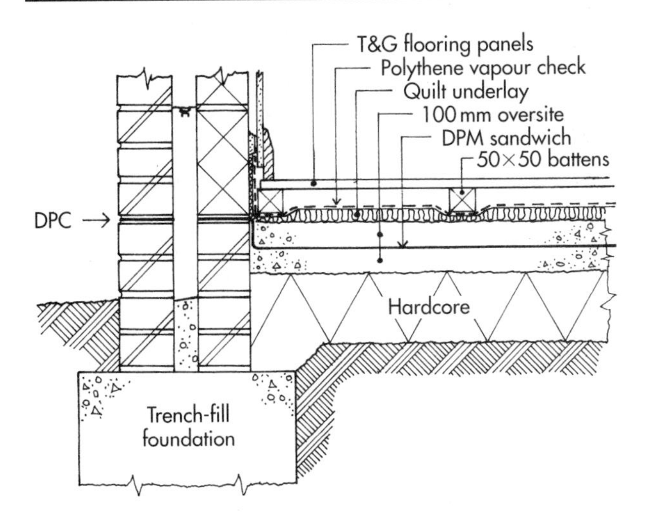

Figure 8.7 Battened floating floor

Figure 8.7: This type of floor is referred to as a *battened floating floor*. Staggered flooring panels are laid and fixed to a framework of 50 × 50 mm battens (the final notes on upper floors (page 83) give fixing details of panels on joists, which also apply here in fixing panels to battens). The battens are laid unfixed on a resilient insulating quilt of glass fibre or mineral wool of minimum 13 mm uncompressed thickness, resting on a dry and damp-proofed concrete oversite and covered with a vapour barrier as described previously. When fixed to the battens, the floor is floating on the insulation quilt and is held down by its own weight and the perimeter skirting.

8.6 FILLET OR BATTENED FLOORS

Another type of floor to be considered, consists of 50 × 50 mm sawn fillets or battens, spaced at 400 mm or 600 mm centres and either embedded in the concrete oversite or fixed on top. Both of these applications are covered by Part C of the Building Regulations' Approved Document Regulation 7, and the detailed examples given here are aimed at meeting the requirements for timber in *floors supported directly by the ground*.

8.6.1 Embedded-fillet Floor

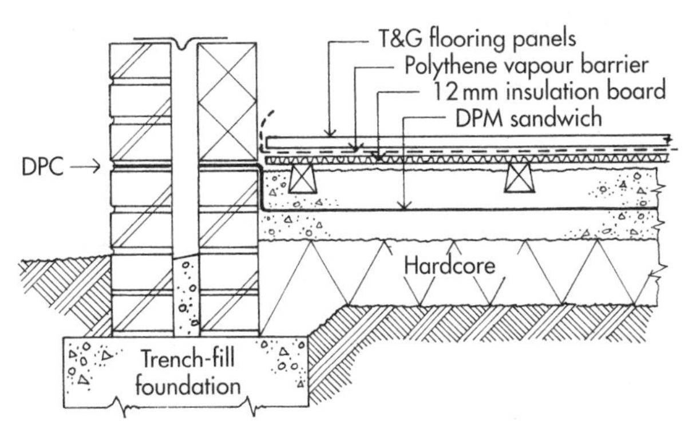

Figure 8.8 Embedded-fillet floor

Figure 8.8: The timber fillets are splayed to a dovetail shape and must be pressure-treated with preservative in accordance with BS 1282: 1975, *Guide to the choice, use and application of wood preservatives*, prior to being inserted in the floor. The concrete oversite must incorporate a damp-proof sandwich membrane consisting of a continuous layer of hot-applied soft bitumen of coal tar pitch not less than 3 mm thick, or at least three coats of bitumen solution, bitumen/rubber emulsion or tar/rubber emulsion. After the DPM has been applied and has set, the splayed fillets, having been cut to length and the cut-ends resealed with preservative, are bedded in position at the required centres and levelled. This can be done by placing small deposits of concrete in which the fillets are laid and tamped to level positions. When set, the top half of the concrete sandwich is laid, using the fillets as screeding rules.

8.6.2 Surface-battened Floor

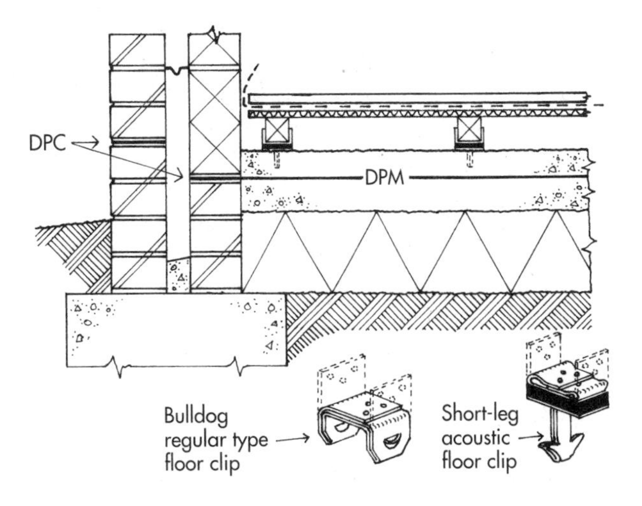

Figure 8.9 Surface-battened floor

Figure 8.9: A damp-proof sandwich membrane, as described above, is required and must be joined to the damp-proof course in the walls. Standard or acoustic

Bulldog floor clips can be used to hold the battens in position at the required spacing. The clips are pushed into the plastic concrete at 600 mm centres within 30 minutes of laying and levelling. A raised plank is placed across the concrete to support the operative and a batten marked with the clip-centres is used to act as a guide for spacing and aligning the clips. When laying the battened floor, after the concrete is set and thoroughly dry, the ears of the clips are raised with the claw hammer, the battens are inserted and fixed with special friction-tight nails supplied with the clips. As illustrated, both floors may be insulated with 12 mm thick bitumen-impregnated fibreboard, covered with a polythene vapour barrier prior to being floored with T&G chipboard, plywood or OSB flooring panels.

8.7 BEAM-AND-BLOCK FLOOR

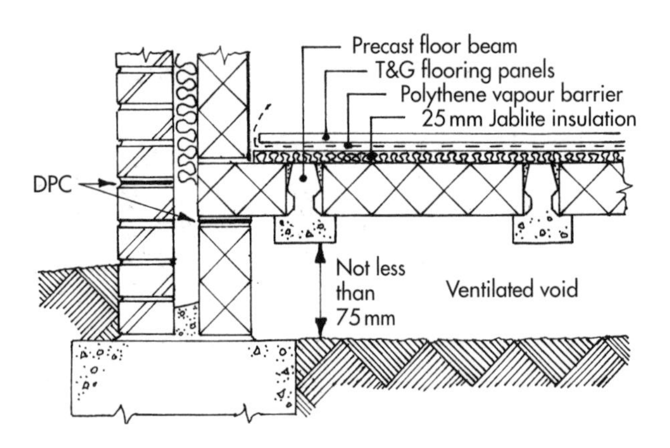

Figure 8.10 Floating floor on beam-and-block construction

Figure 8.10: This is the most modern construction used at ground floor level on dwelling houses. The precast reinforced concrete floor-beams have to be ordered to the lengths and number required. They take their bearings on the inner-skin of blockwork and, as illustrated, must be at least 75 mm above ground level to the underside of the beams. They are spaced out to take the length of 440 × 215 × 100 mm standard wall blocks, laid edge to edge and resting on the beams' protruding bottom-edges. A row of cut blocks is inevitably required along one side, as illustrated. The gaps between blocks and beams are filled with a brushed-over sand-and-cement grout. The void beneath the precast beams should be provided with through-draught ventilation, continuous through any intermediate cross sleeper-walls, with an actual ventilation area equivalent to 600 mm^2 per metre run of wall. The floating floor laid on this sub-floor construction will be as described previously under 'Laying Procedure'.

8.8 UPPER FLOORS

8.8.1 Timber Joists

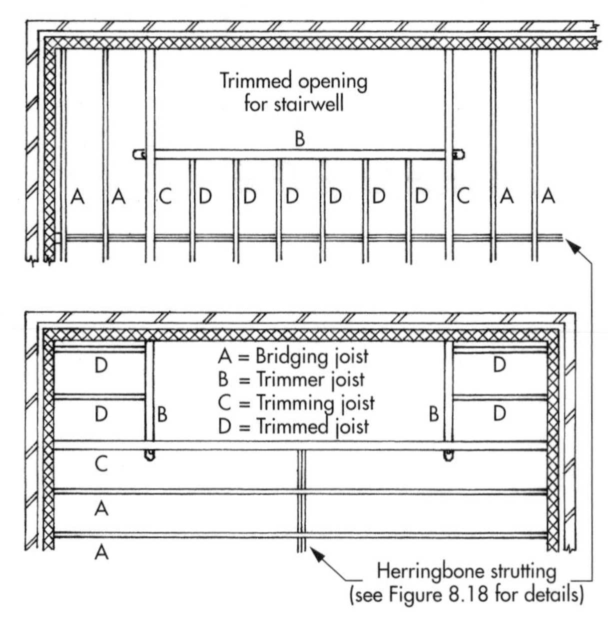

Figure 8.11(a) Part-plan views of alternative joist-arrangements around trimmed opening

Figure 8.11(a): In dwelling houses, suspended timber floors at first-floor level and above are usually single floors comprising a series of joists supported only by the extreme bearing points of the structural walls. These joists are called *bridging joists*, but any joists that are affected by an opening in the floor, such as for a stairwell or a concrete hearth in front of a chimney-breast opening are called *trimmer, trimming* and *trimmed joists*, as illustrated. Because the trimmer carries the trimmed joists and transfers this load to the trimming joist(s), both the trimmer and the trimming joists are made thicker than the bridging joists by 12.5–25 mm, or double joists, bolted together, are used. The depth of the joists, as mentioned in the opening pages of this chapter, does not usually concern the site carpenter or builder, such structural detail being the responsibility of the architectural team. However, if ever needed, the size of joists relevant to the span and joist-spacing, can be gained easily enough by reference to Tables A1 and A2 for floor joists in *The Building Regulations' Approved Document A* from HMSO Publications Centre or bookshops.

8.8.2 Framing Joints

Figure 8.11(b): Traditionally, a tusk tenon joint was used between the trimming joist and the trimmer – and is given here for reference. This joint was proportioned as shown and was set out and cut on site with the aid of a

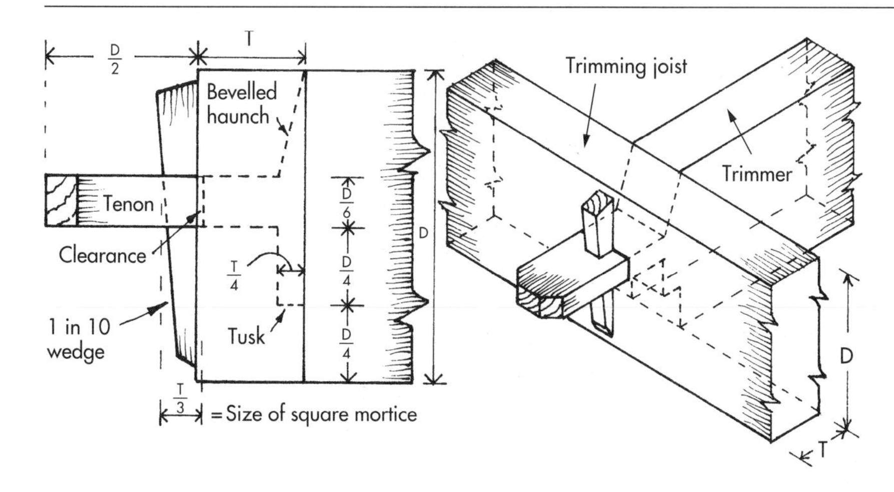

Figure 8.11(b) Traditional tusk tenon joint (T = thickness; D = depth)

crosscut saw, brace and twist bits (for the mortice and wedge hole) and firmer chisels, etc. The wedge was cut to a shallow angle of about 1 in 10 ratio to inhibit rejection, made as long as possible and, upon assembly, was driven into a offset draw-bore mortice in the tenon. The offset clearance that was needed to effect the drawn-tight fit between the two structural members, is indicated in the illustration. The slope on the bottom of the wedge was to facilitate entry and the top slope lent itself to the angle of the hammer blow with less risk of shearing the short grain. When jointing, particular care was taken to ensure that the bearing surfaces of the tusk and tenon were not slack against the stopped housing and the mortice.

8.8.3 Blind Tenon and Housing Joints

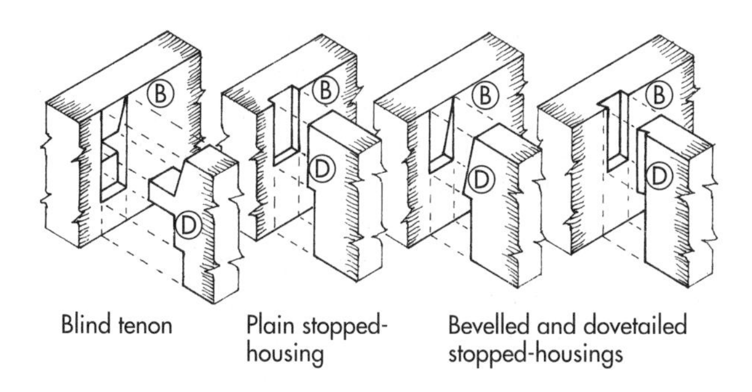

Figure 8.11(c) Traditional framing joints

Figure 8.11(c): Traditional joints used between trimmed joists (D) and trimmer joist (B), varied between a *blind tenon* and a *plain stopped-housing*. Other joints, seen more in textbooks than in practice, included a *bevelled stopped-housing* and a *dovetailed stopped-housing*. The blind tenon joint was made to the same proportions as the tusk tenon, but did not have a wedge or projecting tenon. The plain stopped-housing joint was set out and gauged to cut into the trimmer 12.5 mm on the top edge and half the joist-depth on the side. It was quickly formed by making three diagonal saw cuts across the grain (two on the waste side of the lines, one in the mid-area), chopping a relief slot at the bottom of the housing and chisel-paring from above.

8.8.4 Modern Framing Anchors

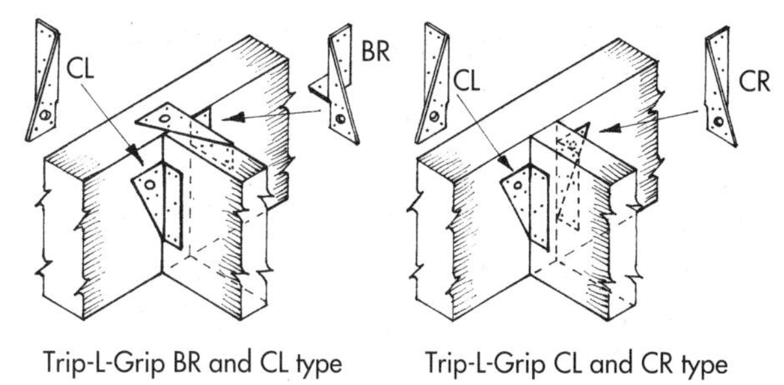

Figure 8.12 Modern framing anchors

Figure 8.12: Metal timber-connectors are now extensively used to replace the above-mentioned joints, in the form of metal framing anchors and, more commonly, timber-to-timber joist hangers. The advantages to be gained in using these connectors are a saving of labour hours and, in the case of the hangers, more effective support of the trimmer or trimmed joists, by the bearing being at the bottom of the load. However, it must be mentioned that traditional framing joints have held up to the test of time in houses of several hundred years of age. When using sherardized framing anchors, such as MAFCO Trip-L-Grip, for floor joists, the loads to be carried are such that each trimmed joint should comprise both a B type and a C or two C type anchors. When using two C types (CL and CR), one on each side of the joist, they should be slightly staggered to avoid nail-lines clashing. The anchors are recommended to be fixed with 3 mm diameter by 30 mm sherardized clout nails.

8.8.5 Timber-to-timber Joist Hangers

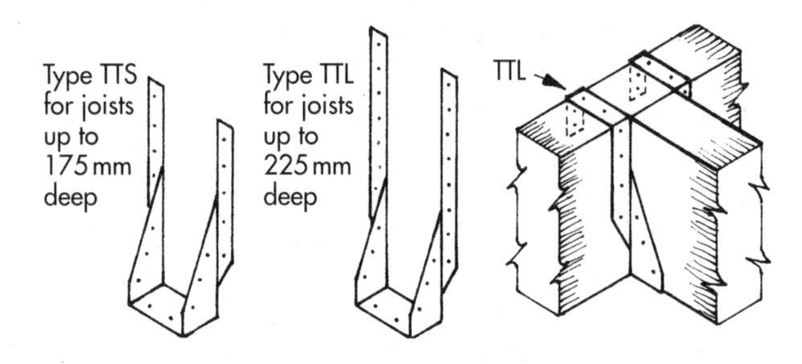

Figure 8.13 Timber-to-timber joist hangers

Figure 8.13: Steel joist hangers, such as those manufactured by Catnic–Holstran, type TT (timber to timber), S and L (short and long), are made from 1 mm galvanized steel with pre-punched nail holes. With the aid of a hammer, the straps are easily bent over the joists, as illustrated, and fixed with 32 mm galvanized plasterboard nails. Another advantage of the thin-gauge metal is that hangers do not require housing into the top or bottom edges of the joists. Although perhaps not structurally necessary, it is advisable to place nail-fixings in all the available fixing holes.

8.8.6 Double Floors

Figure 8.14(a): The sensible structural rule of timber joisted floors, is that the joists should always bridge across the shortest span of an area, unless a double floor is required, whereby a steel beam (or beams) bridges the shortest span and the timber joists run the longest span, bearing on the intermediate beams(s), as illustrated. The protruding beam at ceiling level is encased in several ways to achieve fire-resistance and a visual finish. One way of doing this is to make a quantity of U-shaped frames – known as *cradles* or *cradling* – using 50 × 50 mm or 38 × 38 mm timber, with lapped or half-lapped, clench-nailed joints at each corner, and fix them, one to each joist-side, as indicated at Figure 14(b), close to the beam's bottom and flange-edges. These frames are then clad with 12.5 mm plasterboard or other such non-combustible material.

8.8.7 Solid-wall Bearings

Figure 8.15: The old practice of building the ends of joists into solid (non-cavity) walls is now frowned upon, because of the increased risk of timber decay through lateral damp-penetration. As illustrated, the modern practice is to use steel joist hangers, such as those manufactured by Catnic–Holstran, type TW (timber to wall), made from 2.5 mm galvanized steel. When fixing, 32 mm galvanized plasterboard nails are recommended. Owing to a double metal-flange on the bottom, equalling a thickness of 5 mm, the bottom edge of the joists require notching out to achieve a flat surface for the pasterboard ceiling.

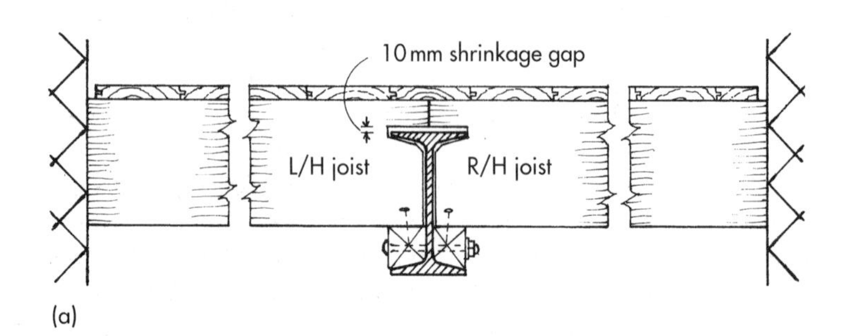

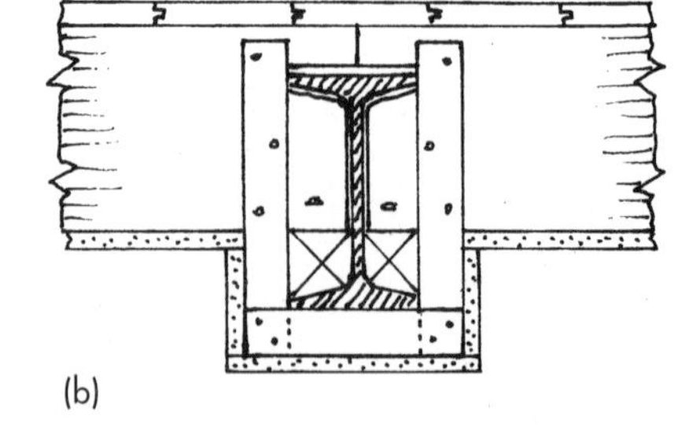

Figure 8.14(a) Double floor. (b) Cradling

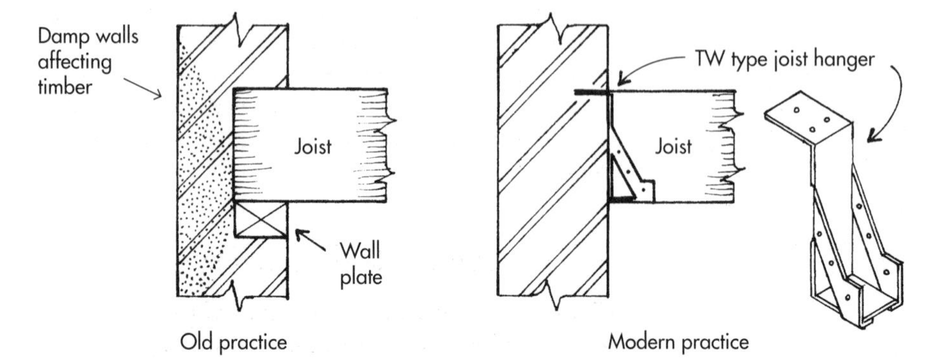

Figure 8.15 Solid-wall bearings

8.8.8 Cavity-wall Bearings

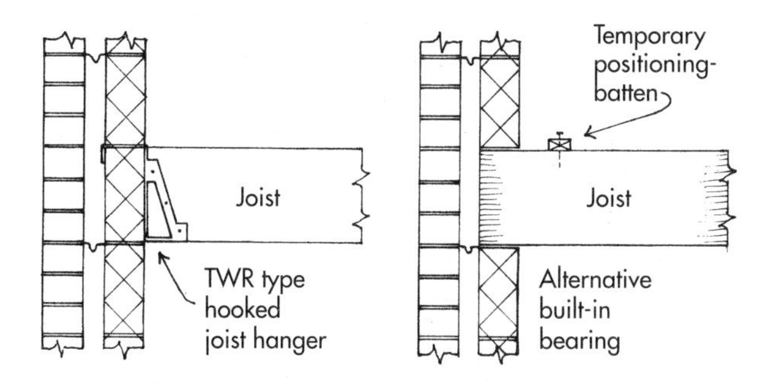

Figure 8.16 Cavity-wall bearings

Figure 8.16: TW (timber to wall), or, as illustrated, TWR (timber to wall return) joist hangers with a turn-down top flange to ensure correct and safe anchorage, especially when there is insufficient weight above, may be used for cavity walls. Alternatively, the ends of joists, which should be treated with timber preservative, are positioned, levelled up and built into the inner skin of the cavity wall. Care must be taken to ensure that the joists do not protrude past the inner face of the bearing-wall, into the cavity. The temporary positioning-batten, illustrated, should be attached to the scaffold or return wall at its end(s) to create stability and to stop the joists toppling sideways until they are built in.

8.9 STRUTTING

8.9.1 Introduction

Strutting in suspended timber floors is used to give additional strength by interconnection between joists. This removes the unwanted individuality of each joist and effects equal distribution of the weight and prevents joists bending sideways. Struts should be used where spans exceed 50 times the joist thickness. Therefore, with 50 mm thick joists, a single row of central struts should be used when the span exceeds 2.5 m and two rows are required for spans over 5 m and up to 7.5 m.

8.9.2 Solid Strutting

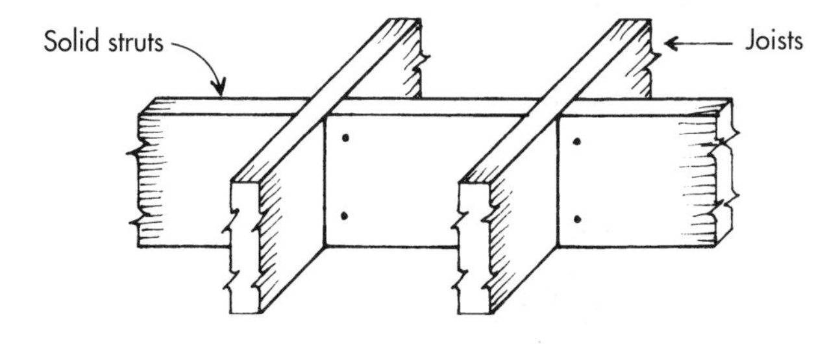

Figure 8.17 Old practice of strutting

Figure 8.17: The old practice of strutting the floor with solid noggings is now frowned upon as being costly in material, adding unnecessary weight and creating an inflexible floor. However, Section 5 of the *New Build Policy* Technical Manuals recommends that solid strutting should be used instead of herringbone strutting where the distance between joists is greater than three times the depth of the joists. (Note that the manuals mentioned here are registered-builders' guidance notes supplied by Zurich Municipal Insurance, whose surveyors monitor the building work during construction and issue a certificate on completion, guaranteeing the building's fitness for a period of ten years. This scheme is similar to the one run by the NHBC.)

8.9.3 Herringbone Strutting

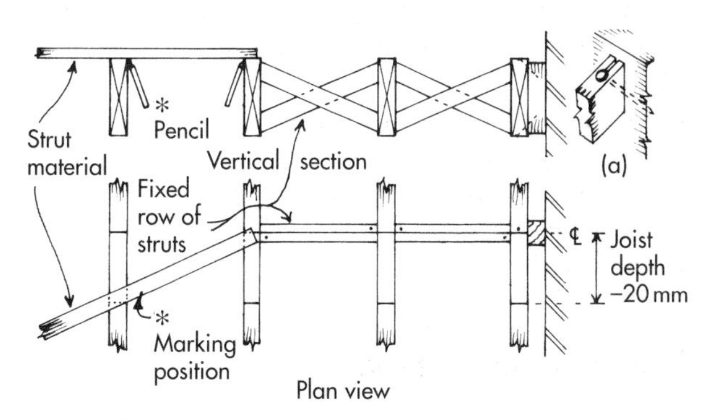

Figure 8.18 Plan and sectional view of herringbone strutting

Figure 8.18: This traditional method of strutting, using 38 × 38 mm or 50 × 38 mm sawn timber, although still effective and commonly used, nowadays has to compete with struts made of steel. The method of fixing involves marking a chalk line across the joists, usually in the centre of the floor as stipulated in the introductory notes. From this, marks are squared down the sides of the joists and – in the case of timber strutting – another line is struck on top with the chalk line, parallel to the first and set apart by the joist-depth minus 20–25 mm. As illustrated, the strutting material is laid diagonally within these lines and marked from below to produce the required plumb-cuts (vertical faces of an angle). Cutting and fixing the struts is done in a kneeling position from above, using 50–63 mm round-head wire nails. Prior to fixing the struts, the joists running along each opposite wall should be packed – technically wedged – and nailed immediately behind the line of struts. As indicated at (a), a sawcut can be made at the end of each strut to receive the nail fixing and eliminate splitting.

8.9.4 Catnic Steel Joist-struts

Figure 8.19: As illustrated, two types of galvanized steel herringbone struts are produced to compete with

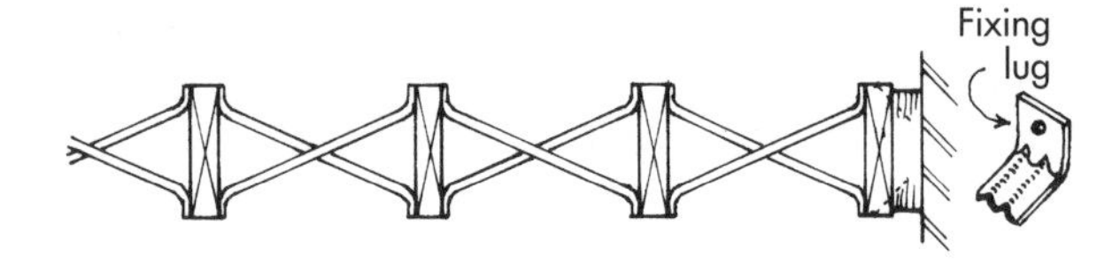

Figure 8.19 Catnic steel joist-struts

traditional wooden strutting. The first type, by Catnic–Holstran, have up-turned and down-turned lugs for fixings with minimum 38 mm round-head wire nails. As before, fixing is done from above.

8.9.5 Batjam Steel Joist-struts

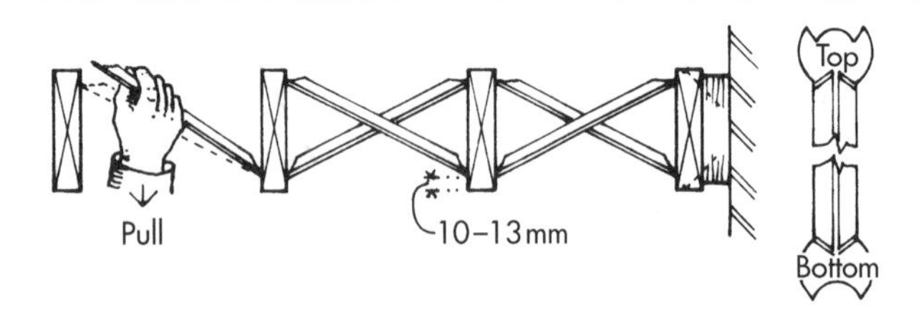

Figure 8.20 Batjam steel joist-struts

Figure 8.20: The second type, with the well-know 'BAT' trademark, has forked ends which simply bed themselves into the joists when forced in at the bottom and pulled down firmly at the top. This time, fixing is done from below. One minor disadvantage with steel strutting, which is made to suit joist centres of 400, 450 and 600 mm, is that there are always one or two places in most floors that do not conform to size and require reduced-size struts. When this occurs, it is necessary to use a few wooden struts in these areas.

8.9.6 Horizontal Restraint Straps

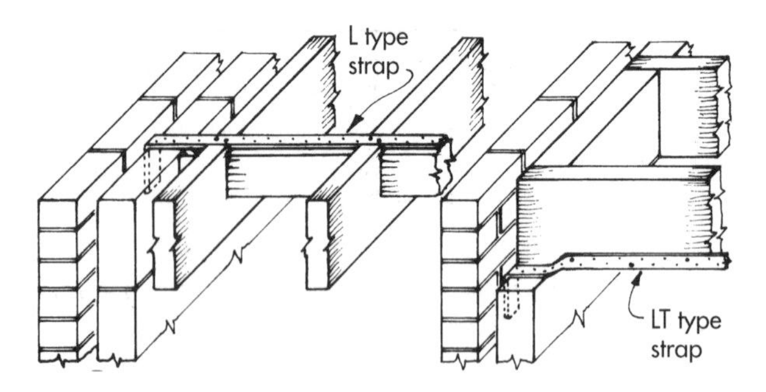

Figure 8.21 30 × 5 mm restraint straps for joists parallel or at right angles to wall

Figure 8.21: Modern construction methods involving lighter-weight materials in roofs and walls, has led to the need for anchoring straps, referred to in *The Building Regulations' Approved Document A*, to restrict the possible movement of roofs and walls likely to be affected by wind pressure. Such straps are made from galvanized mild-steel strip, 5 mm thick for horizontal restraint and 2.5 mm thick for vertical restraint. The straps are 30 mm wide and up to 1.6 m in length. Holes are punched along the length at 15 mm staggered centres.

8.9.7 Joists Parallel or at Right Angles to Wall

As illustrated, the straps, which may be on top or on the underside, require notching-in when the floor joists run parallel to the supported wall, so as not to clash with the seating of the flooring or ceiling materials – but only require surface-fixing on the sides when the joists are at right-angles to the supported wall. The walls should be anchored to the floor joists at centres of not more than 2 m on any wall exceeding 3 m in length, including internal load-bearing walls, irrespective of length. When joists run parallel to the wall to be supported, the straps should bridge across at least three joists and have noggings and end-packing tightly fitted and fixed between the bridged spaces. The noggings should be at least 38 mm wide by half the depth of the joists. If the straps are fixed on the underside, the noggings and packing should be equal to the depth of the joists. There should be at least four 38 mm × 8 gauge screw-fixings in each strap.

8.10 FITTING AND FIXING JOISTS

8.10.1 Procedure – Joists Built-in to Cavity Wall

When the load-bearing walls have been built up to storey height and allowed to set, the joists may be fitted. After cutting to length and sealing or re-sealing the ends with preservative, the joists – with any cambered edges turned upwards – are then spaced out to form the skeleton floor and temporary battens are fixed near each end to hold the joists securely in position. The side-fixed restraint straps are then fixed and the joists are built-in by one course of blocks being laid all round. When set, the notches may be cut for those restraint straps that run across the joist tops, the 50 mm packing and the noggings cut and fixed, then the straps are screwed into position – and the blockwork may proceed.

8.10.2 Sequence of Fixing Joists

Figure 8.22: Normally, the first consideration is to position the trimming joists and trimmer of any intended opening, then, from this formation, the trimmed and

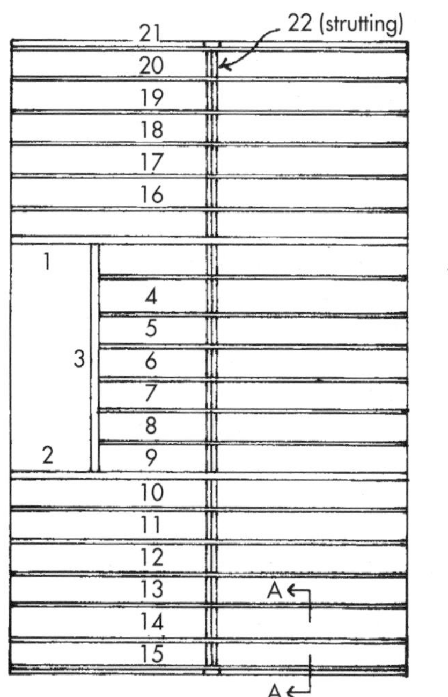

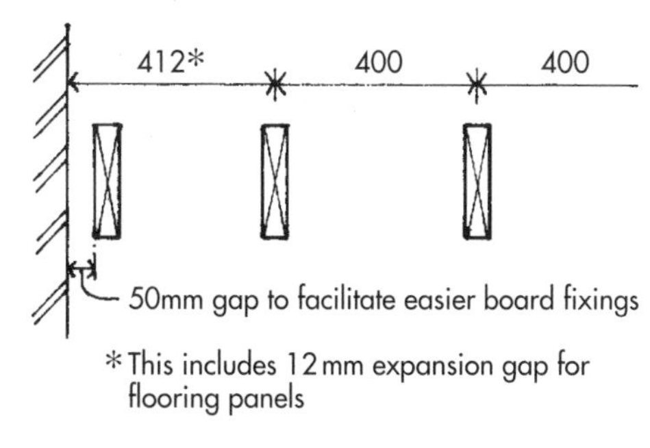

Figure 8.22 Sequence of fixing joists

bridging joists are spaced out. Joists should be checked for alignment with a straightedge or line and, if necessary, packed up with offcuts of thin material such as felt DPC or oil-tempered hardboard – or lowered by minimal paring of the joist-bearing area. However, if regularized joists are used, the need for vertical adjustments should be eliminated. Herringbone strutting is fixed later, after the bricklayers have finished their work.

8.10.3 Procedure – Joists Bearing on Joist Hangers

Figure 8.23: When joist hangers are to be used as an alternative to the joists being built-into the inner skin of blockwork, the blockwork is built to the top of the floor-joist level and the joists are cut to length and positioned at the same time as placing the joist hangers. The joists must be cut carefully to the correct length, to hold the hangers snugly against the walls, yet not push the blockwork from its set position. As illustrated, the restraint straps are fixed to the upper edge on the side, when the joists are at right-angles to the supported wall, or, alternatively, restraint-type joist hangers can be used (Figure 8.23(b)) where required.

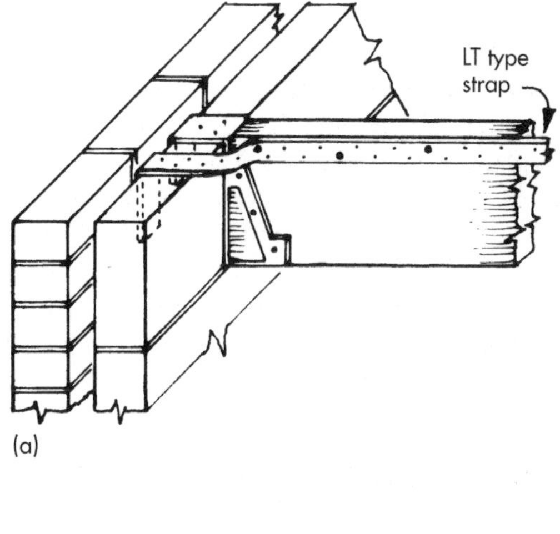

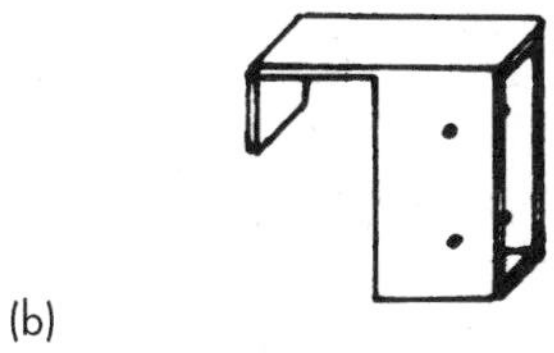

Figure 8.23(a) Restraint strap in relation to joist hanger. (b) Restraint-type joist hanger

8.11 FIXING FLOORING PANELS ON JOISTS

8.11.1 Introduction

Tongue and groove (T&G) flooring panels of chipboard, plywood or Sterling OSB, machined on all four edges with compatible tongue and groove profiles and with laid-measure dimensions of 2400 or 2440 mm × 600 mm, can be laid on suspended timber floors in the following thicknesses related to joist-spacings: 18 mm chipboard, 18 mm plywood and 15 mm Sterling OSB for joists at 400 mm centres – and 22 mm chipboard, 18 mm plywood (as before) and 18 mm Sterling OSB for joists at 600 mm centres.

8.11.2 Fixing Procedure for T&G Panels

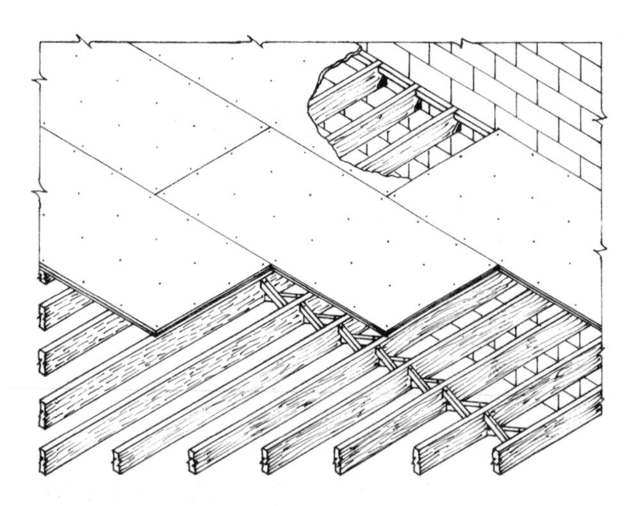

Figure 8.24 Fixing T&G flooring panels

Figure 8.24: The T&G boards, as illustrated, are laid with the long edges across the joists. The short edges bear centrally on the joists and *only* the long edges against the walls must be supported by noggings of at least 38 mm width – but preferably of 50 mm width and 75 mm depth. The boards should be nailed with three or four nails to each joist, two at about 25 mm from each edge and one or two nails equidistant between. The nails should be 45 mm × 10 gauge annular-ring shank type for floor thicknesses up to 18 mm and 56 mm × 10 gauge for floors of 22 mm thickness. All joints should be glued with polyvinyl acetate (PVA) adhesive, preferably the waterproof type. Gluing of joints, which is often skimped, is important to eliminate an aggravating squeaky floor.

8.11.3 Fixing Square-edged Panels

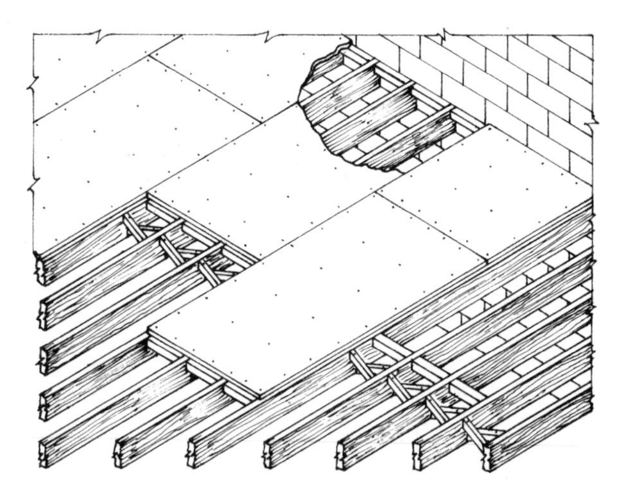

Figure 8.25 Fixing square-edged flooring panels

Figure 8.25: These boards, as illustrated, are laid with the long edges bearing centrally on the joists. *All* short edges, including the edges against the walls, must be supported with, preferably, 75 × 50 mm noggings tightly fitted and fixed between the joists. As before, the boards should be fixed with 45 mm or 56 mm × 10 gauge annular-ring shank nails at 300 mm centres around the edges and at 400–500 mm centres on intermediate joists. Nail fixings should be at least 9 mm in from the edge of the boards.

Cross joints on both types of board must be staggered and expansion gaps of 10–12 mm allowed around the perimeter of walls and any abutment. Sterling OSB square-edged panels are recommended to have a 3 mm expansion gap between boards in addition to the perimeter gap. Traps in the floor must be supported on all four edges and fixed with 45 mm × 8 gauge countersunk screws. As stated on floating floors, all panels should be *conditioned* by laying loosely in the area to be floored for 24 hours before being fixed.

9

Fixing Interior and Exterior Timber Grounds

9.1 INTRODUCTION

Timber grounds are either sawn or prepared battens, fixed to walls or steel sections, to create a true and/or receptive fixing surface. Depending on the material the grounds are to be attached to, they may be *fired* onto steel flanges or webs, brickwork and concrete with a cartridge powered tool, fixed to lightweight aerated blocks with cut clasp nails, or fixed to blockwork and brickwork with Fischer type Hammer-fix screws.

9.2 SKIRTING GROUNDS

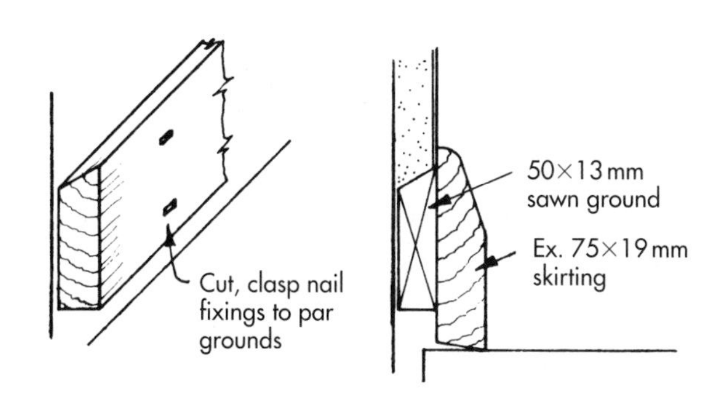

Figure 9.1(a) Skirting grounds

Figure 9.1(a): Although grounds were traditionally only used on good class work to promote truer plastered surfaces and provide a means of fixing for the skirting, they are included here because research has shown that a small percentage of dwellings being built still use wet-plastering techniques and therefore may use grounds. The grounds are bevelled on their top edge to retain the bottom edge of the plaster and must, of course, be equal to the required plaster thickness. As packing pieces (offcuts of damp-proof course material and plastic shims are ideal) are often required on uneven walls, the grounds should be slightly less thick than the required plaster thickness.

9.2.1 Fixing Technique

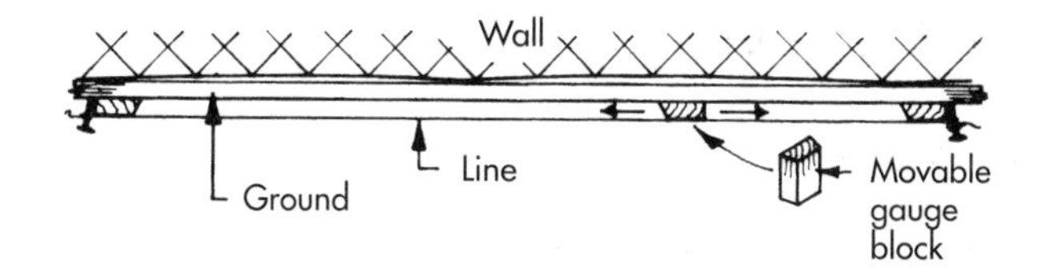

Figure 9.1(b) Checking straightness of ground with string-line and gauge blocks

Figure 9.1(b): The top of the grounds should be levelled (or parallel to the finished floor) and set up to finish between 6 and 10 mm below the anticipated skirting height. Long grounds should be fixed at each end and have a string line pulled taut along the face. Two pieces of offcut ground, one at each end, are pushed in between the line and ground, while a third piece of offcut ground is tried between the taut line and the unfixed ground at 600–900 mm intervals, packed if necessary and fixed. Shorter grounds may be checked for straightness with a timber or aluminium straightedge. Internal and external angles are butt jointed – not mitred. On external angles, run the first ground about 50 mm past the corner, butt the end of the second ground up to this and when fixed, cut off the first projection flush to the second ground's face.

9.2.2 Deep and Built-up Skirtings

Figure 9.1(c): Grounds for deep or built-up skirtings may be required on refurbishment, maintenance and repair work. Such grounds, as illustrated, have a longitudinal top ground and vertical, face-plumbed soldier pieces of ground fixed at 600–900 mm centres. Depending upon the particular skirting design and height, additional stepped soldiers may be required to be fixed onto the first row.

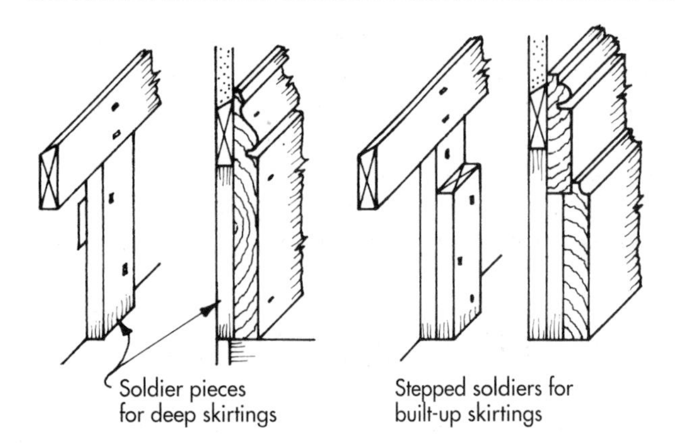

Figure 9.1(c) Soldier pieces for deep skirtings and stepped soldiers for built-up skirtings

9.3 ARCHITRAVE GROUNDS

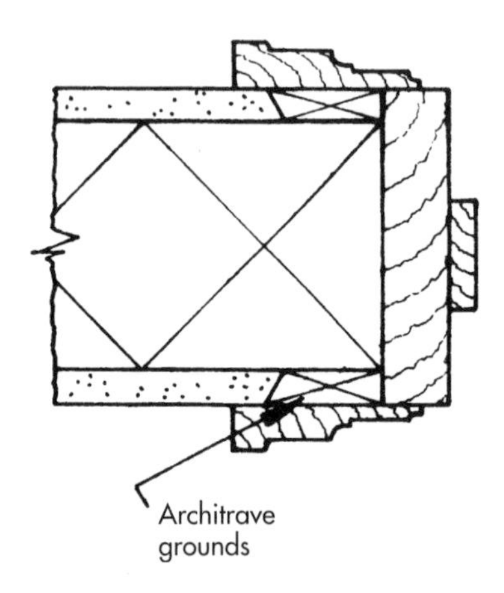

Figure 9.2 Architrave grounds

Figure 9.2: Because the modern architrave section is relatively narrow, the traditional use of grounds in these situations is rarely required nowadays. Similar to skirting grounds, the edge against the plaster was bevelled and had to be concealed under the outer-architrave edge by about 6–10 mm. These grounds helped to keep the wet plaster away from the lining's edges and provided a true and receptive fixing surface for the outer-architrave edges.

9.4 APRON-LINING GROUNDS

Figure 9.3: Grounds are often required behind the apron lining around the edge of a trimmed stairwell. This is to bring the face of the lining to a position equal to the centre of the newel post and to further support the projecting landing-nosing. The grounds may be longitudinal if being faced with an MDF or plywood apron

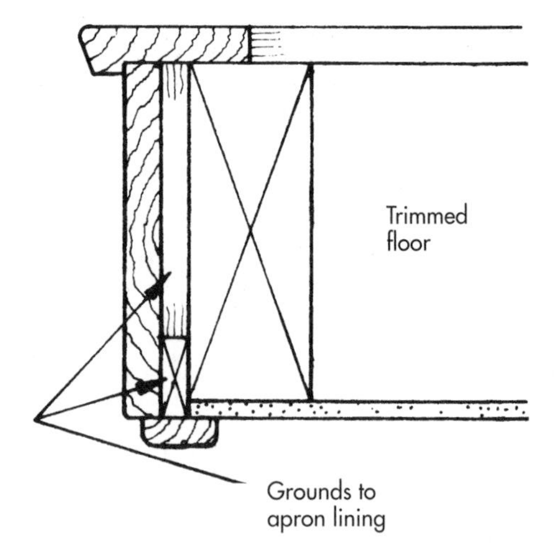

Figure 9.3 Grounds to apron lining

lining, or in the form of vertical soldier pieces if a timber lining is being used. The soldier pieces give better support by being across the grain to offset any cupping of the lining.

9.5 WALL-PANELLING GROUNDS

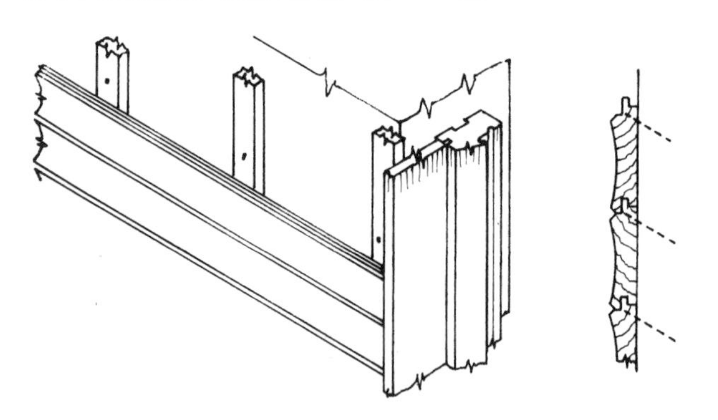

Figure 9.4 Horizontal cladding secret-nailed to vertical grounds

Figure 9.4: Grounds – without bevelled edges and sized about 50 × 25 mm – may be fixed horizontally across a wall and spaced at 600 mm centres from floor to ceiling, acting as a straight and plumb fixing medium for vertical boarding. Alternatively they may be fixed vertically at similar centres across the wall, acting again as a true surface and fixing medium for horizontal boarding. The technique in these situations is, having fixed two extreme grounds straight and true (one horizontally near the floor, one near the ceiling – or one vertically on the extreme left, one on the extreme right), they are used as a fixed datum for all the in between grounds to relate to, by means of a straightedge or string line.

9.6 FRAMED GROUNDS

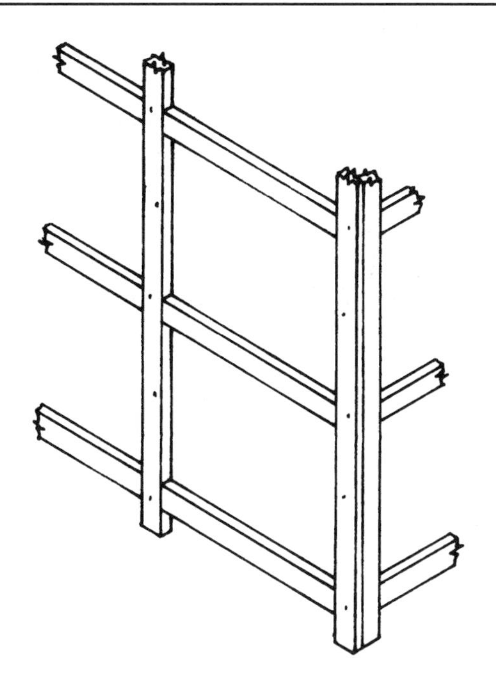

Figure 9.5 Framed grounds

Figure 9.5: Occasionally, on traditional forms of wall panelling, framed grounds are still used. Basically, these consist of either an arrangement of grooved uprights and tongued cross-rails of about 50 × 25 mm section, or morticed uprights and stub-tenoned cross-rails of a similar section. The fixing technique for framed grounds is as explained above for wall-panelling grounds.

9.7 EXTERNAL GROUNDS

Fixing timber grounds for external work such as for timber or plastic cladding will again involve a fixing technique similar to that described above for wall-panelling grounds. The main difference will be that the grounds should be tanalized or protimized, or protected with a similar acceptable preservative treatment.

10

Fixing Stairs and Balustrades

10.1 INTRODUCTION

Figure 10.1: Traditionally, a series or flight of steps, rising from one level to another, whether it be a floor to a landing or vice versa, was known as a *stair*, but is now more commonly referred to as *stairs* or a *staircase*. Originally, *stairs* was the plural of stair, meaning more than one flight of steps and the word *staircase* meant the space within which a stair was built. This space is now called a *stair well*. These more-recent, modern terms are used here.

10.1.1 Manoeuvrability

For reasons of easier transportation, manoeuvrability through doorways, and practical issues involved in the fitting and fixing, staircases usually arrive on site separated from the newel posts and balustrade, the bottom step (if such protrudes beyond the newel post, as with a bullnose step), the top riser board, the landing nosings and the apron linings.

10.1.2 When to Fix

Fixing is best done before dry lining or traditional plastering takes place, soon after the shell of the building is formed and the roof completed. This sequence allows the staircase to be fixed to the bare wall, so ensuring a better finish by the plasterer or dry lining being seated on the edges of the wall-string board, sealing any gaps that otherwise would appear if the stair-string were fixed to the dry-lined or plastered surface.

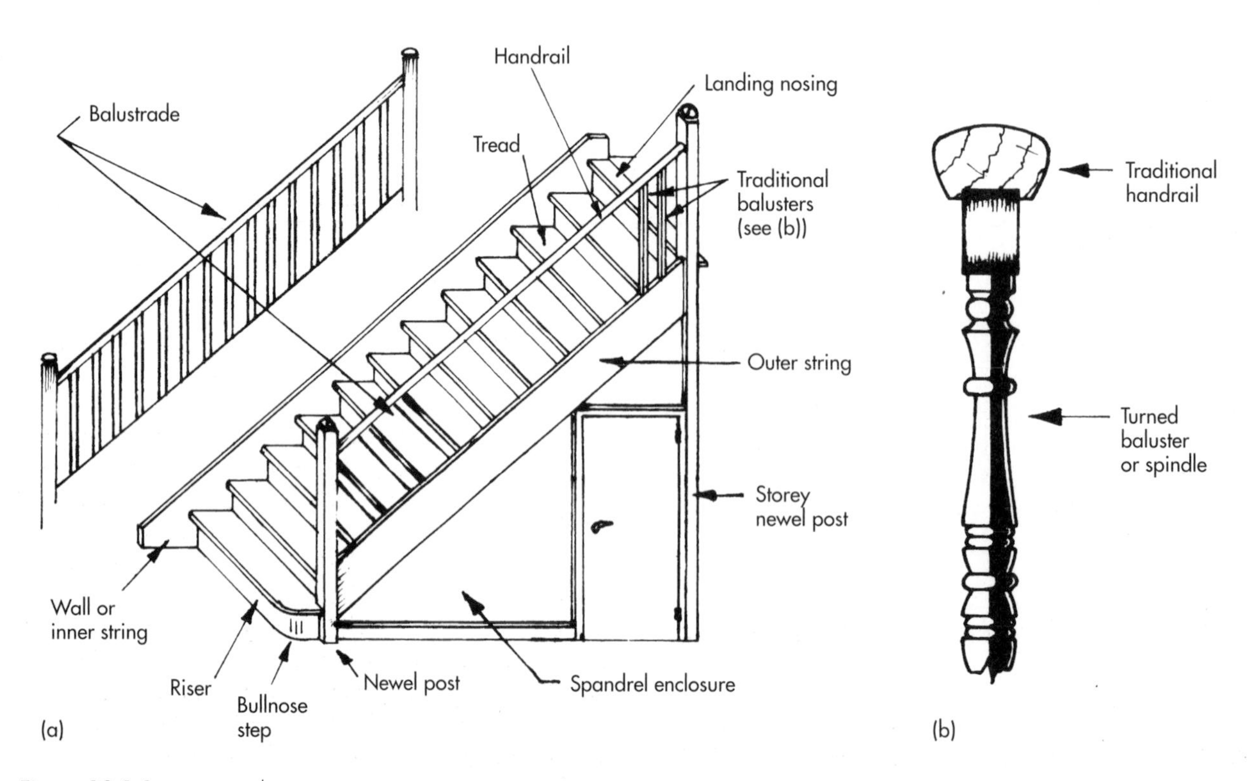

Figure 10.1 Stair terminology

10.1.3 Other Considerations

Fixing at this stage also effectively reduces the disproportionately thick-edge appearance of the wall string and, if worked out, perhaps by packing the wall string when fixing, it can be gauged so that the remaining thickness of wall string equals the thickness of the skirting board that will eventually abut its ends at the top and bottom of the staircase. This is an important point on good quality work, because abutting skirting boards ought to be flush with the string-face.

Another reason for installation at this stage is to allow building operatives quick and easy access to the upper floor(s). The following steps outline the operations involved in fitting and fixing a straight flight of stairs.

10.2 INSTALLATION PROCEDURE

10.2.1 Checking Floor Levels

Figure 10.2(a): Check whether the existing floors (upper and lower) are finished levels. In the case of a boarded or ply/chipboard/OSB sheeted floor, these are usually the levels to work to – as any additional floor covering can be assumed to cover the steps as well, thereby retaining equal rises to all steps. If, however, the ground floor is of concrete (slab form or beams and blocks) and has yet to receive a finishing material such as a 50 mm sand-and-cement screed, or a floating floor of polystyrene sheets (Jablite) and tongued and grooved chipboard panels, or sand/cement screed and wood parquet blocks, then the finished floor level (ffl) must be known or found out and established – and packing blocks prepared to fit under the bottom step.

10.2.2 Establishing the Finished Floor Level

Figure 10.2(b): At this stage of the job, the ffl has usually been established and may be found, ready to transfer from the bottom of door linings or the sills of external door frames. Alternatively, a bench mark above the site datum can be levelled across to the stair area, marked on the wall and measured down the set amount to the ffl.

10.2.3 Cutting for Floor and Skirting Abutment

Figure 10.2(c): Next, cut the wall string at the bottom to fit the ffl (even if the finished floor is yet to be laid). If not already cut or marked during manufacture, then

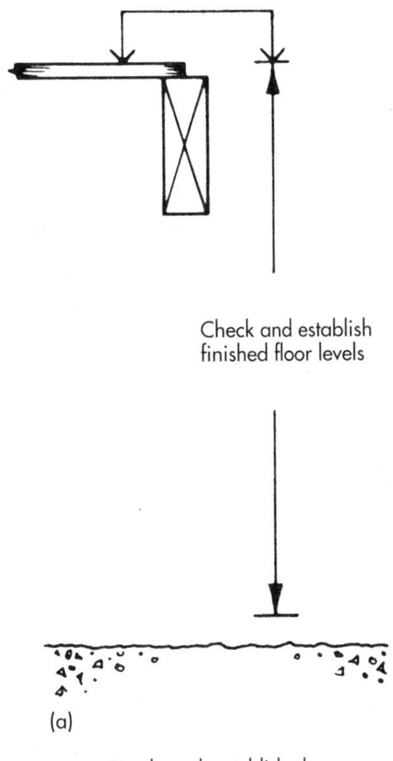

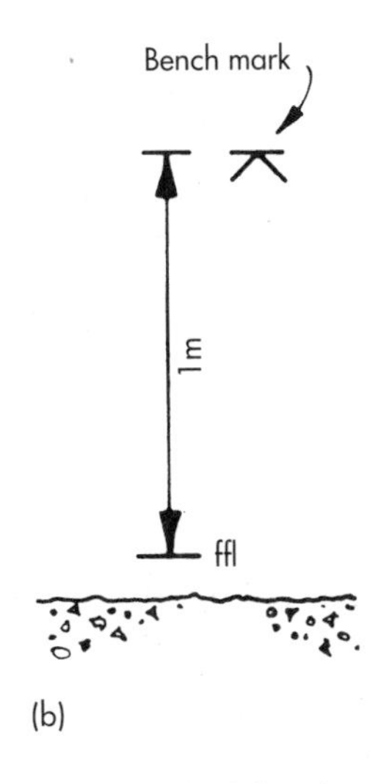

Figure 10.2(a) Establish finished floor levels. **(b)** Establish bench mark

simply measure down the depth-of-rise from the top of the first tread-housing (if such exists, as in the case of a bottom step left out for site-fixing), or measure down from the tread and mark a line through this point at right-angles to the face of the first riser-housing or riser board. Cut carefully on the waste side with a sharp saw (to produce a clean cut). Then mark the plumb cut to form the abutment joint between the string and skirting board, as indicated. To do this, set the skirting height, say 95 mm, on the blade of the combination mitre-square and square-up from the ffl cut, sliding along until the corner of the blade touches the edge of the

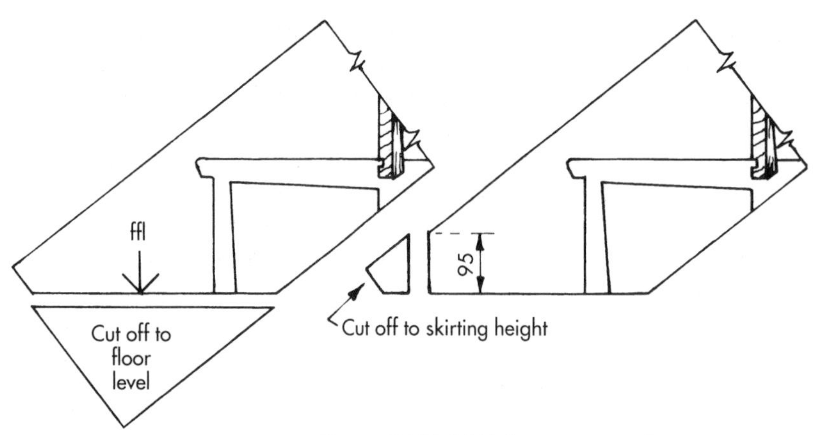

Figure 10.2(c) Wall-string cuts at bottom

string. At this point, mark the plumb line and cut with a panel or fine hardpoint saw.

10.2.4 Cutting to Fit Trimmer and Skirting Abutment

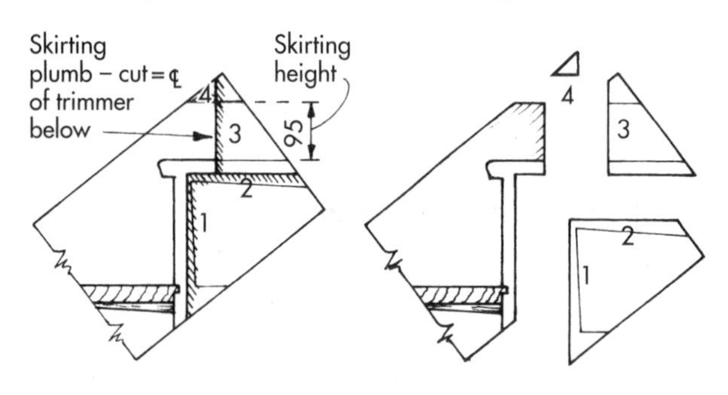

Figure 10.2(d) Wall-string cuts at top

Figure 10.2(d): At the top end of the wall string, more elaborate marking out and cutting is required to enable the staircase to fit against the landing/floor trimmer or trimming joist. This also includes preparation of the string to meet the skirting. As indicated, the cuts are made in four places:

1. within the riser housing, on a line equal to the back of the riser;
2. within the tread housing, on a line equal to the underside of the flooring;
3. on a plumb line equal approximately to the centre of the landing trimmer (this is for the skirting abutment);
4. a horizontal cut at the very top of the string, equal to the skirting height.

This last cut should be planed to a smooth finish, as it becomes a visual edge of the string margin.

10.2.5 Offering Up and Checking

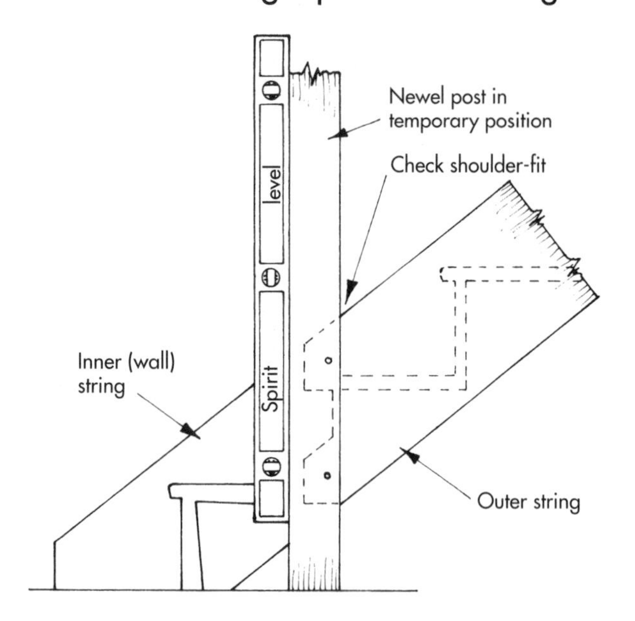

Figure 10.3 Checking for error in total rise

Figure 10.3: Now offer the staircase up into position, resting against the landing and packed up at the bottom, if necessary. Check the treads across the width and depth with a spirit level. Any inaccuracies registering in the depth of the tread will infer that either a fundamental error has been made in the mathematical division of the total rise of the staircase, or that the floor-to-floor storey height is not what it should be. A more positive way of confirming this will be to position the bottom newel post temporarily onto the outer-string tenons, making sure that the shoulder of the bare-faced tenon fits snugly against the newel, and checking for plumb with the spirit level, as illustrated.

10.2.6 Dealing with Inaccuracies

If inaccuracies are confirmed and they are only minor, they may have to be suffered, as very little can be done –

short of shoddy tactics such as adjusting the shoulder of the string-tenons to improve the plumb appearance of the newel posts. If inaccuracies in level and plumb are more serious, then measure the rise of one step carefully, multiply it by the total number of steps in the staircase and compare this figure with the actual measurement of the storey height from ffl below to ffl above. Armed with this information, it would be wise to confer with the site foreman or builder's agent before proceeding.

10.2.7 Fixing the Wall String

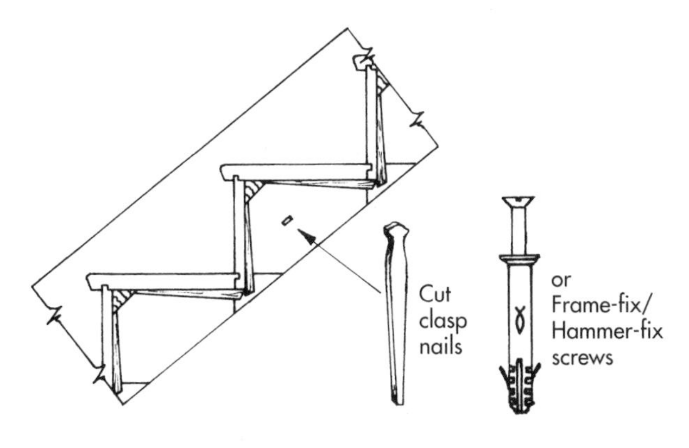

Figure 10.4 Fixing the wall string

Figure 10.4: After minor adjustments, if any, to the normal correctly fitting staircase, the next operation to consider is the fixing of the inner string to the wall. If the wall, being the inner-skin of cavity construction, is built of aerated lightweight material such as Celcon or Thermalite blocks, then nailing with 100 mm cut clasp nails will be satisfactory. These fixings are driven through the string on the underside of the treads, within the triangular area of every third or fourth step. However, if the wall is built with bricks or concrete blocks, and is not receptive to direct nailing, the wall string/wall will have to be drilled through in one operation to receive Fischer-type nylon-sleeved Frame-fix or Hammer-fix screws of 100 mm length. These may also be used, of course, if the first-mentioned aerated lightweight blocks do not prove to be dense enough to grip the cut clasp nail.

10.2.8 Preparing the Bottom Newel Post

Figure 10.5: Having decided on the method best suited to fixing the wall string, the next job is to prepare the bottom newel post to meet the floor level. The post is usually left longer at its lower end to allow for site treatment in various ways, according to the construction of the floor. Unless specified, the carpenter will decide exactly which way is suitable for a particular floor. The various methods of treatment at floor level are now described.

Newel Rests on Concrete

Figure 10.5(a): On concrete floors (slab form or beams and blocks), the newel post can be cut to rest on the concrete – although the end should be sealed with preservative and/or wrapped with a piece of polythene

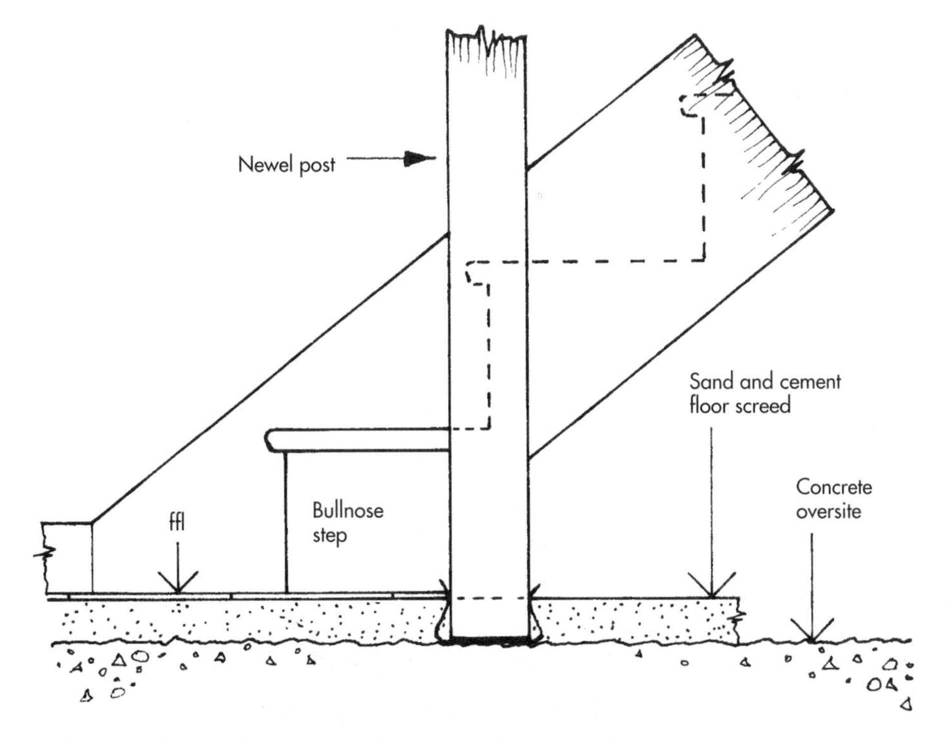

Figure 10.5(a) Screed bedded around newel

sheet. When, after installation of the staircase, the sand-and-cement floor screed is bedded and set around the post, a further degree of rigidity is achieved.

Secure with Metal Dowel

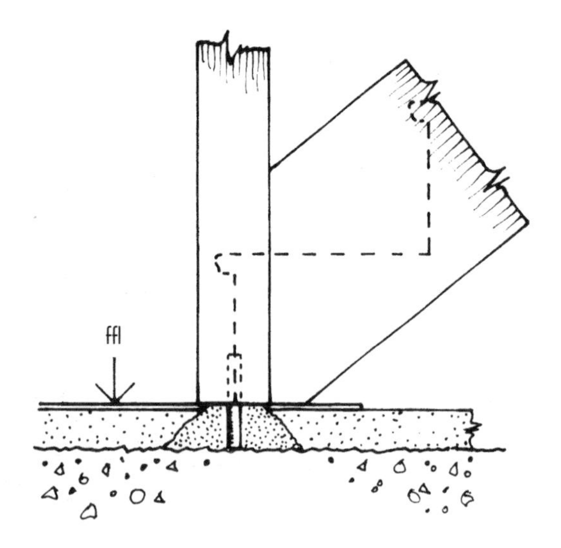

Figure 10.5(b) Screed bedded around metal dowel

Figure 10.5(b): Alternatively, as illustrated, the newel post can be cut off at floor level, be drilled up into the end grain and have a metal dowel inserted. The dowel, which can be cut from 18 mm diameter galvanized pipe, should be inserted for at least half its length and protrude to rest on the concrete. Separate, localized bedding, with a strong mix of sand and cement around the dowel, is recommended before the main floor screed is laid.

Housed into Floor

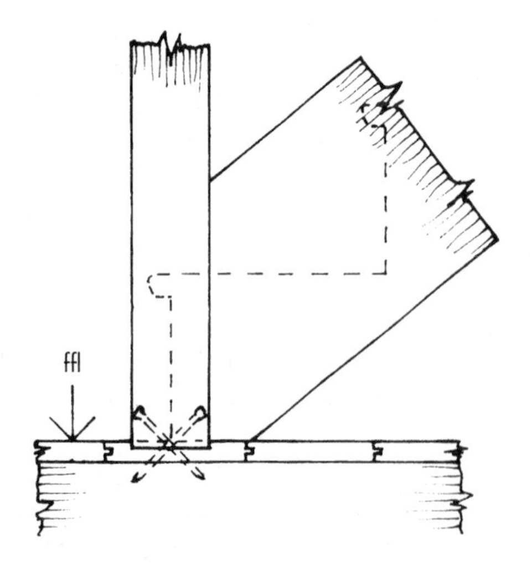

Figure 10.5(c) Newel post housed into floor

Figure 10.5(c): The position of the newel post is marked on the wooden or chipboard floor and chopped out to form a shallow housing, equal to about one-third of the floor thickness. The post should fit this snugly and be skew-nailed into position.

Skew-nailed

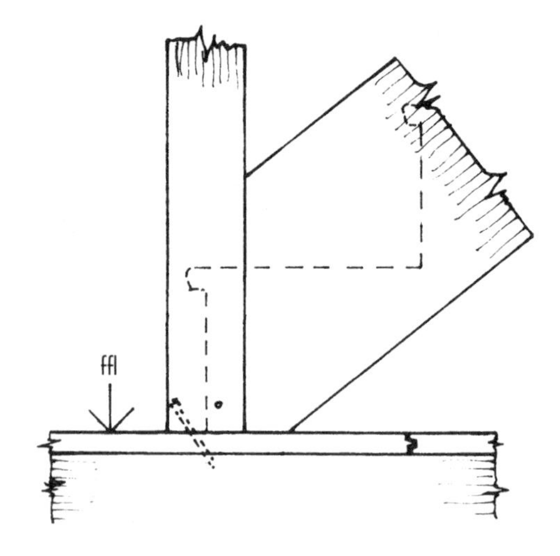

Figure 10.5(d) Newel post skew-nailed into floor

Figure 10.5(d): Alternatively, on wooden or chipboard floors, bottom newel posts are quite commonly cut off at floor level, seated without any jointing and skew-nailed into the floor material with 50 or 75 mm oval nails, punched under the surface. The degree of rigidity achieved by this is minimal – and the newel post's stability depends mainly on the jointed connection to the string and lower step(s); therefore, the gluing, pinning and screwing of these parts (see Figure 10.6) should not be skimped.

Bolted to Joist or Nogging

Figure 10.5(e): Finally, on suspended wooden-joisted floors, although more time-consuming, tedious and rarely done in practice, the newel post achieves a far greater degree of rigidity if it is taken through the floor in its full sectional size and coach-bolted to a joist or – more likely in practice – to a solid nogging. According to the precise position of the newel post, the nogging would be trimmed between nearby joists. If not accessible below, pieces of flooring would have to be left out to facilitate the insertion of the bolt.

Note that the type of staircase indicated in Figures 10.5(a)–(e) is nowadays quite common and has the face of its first riser board central to the newel, without any protruding step. Although aesthetically less attractive, it involves less expense.

10.2.9 Fixing the Newel

The bottom newel post, which will have been morticed and fitted during manufacture, is now ready to be

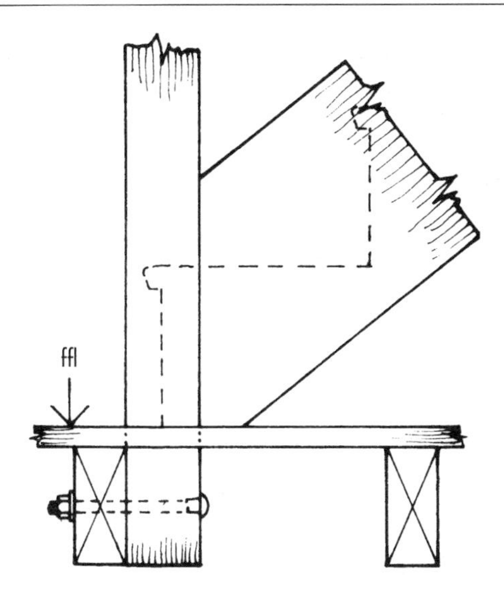

Figure 10.5(e) Newel post bolted to joist or nogging

permanently fixed to the outer string. This can be done with the staircase lying on its side or resting up against the landing above and propped up on saw stools, or similar, at the bottom. The mortice and tenon joint should already be drilled to receive 12 mm diameter wooden dowels (pins). The holes should be slightly off-set to enable the tapered pins to effect a wedging action when driven in, so drawing up the shoulders of the oblique (uncrampable by normal means) tenons to a good fit against the post.

Procedure

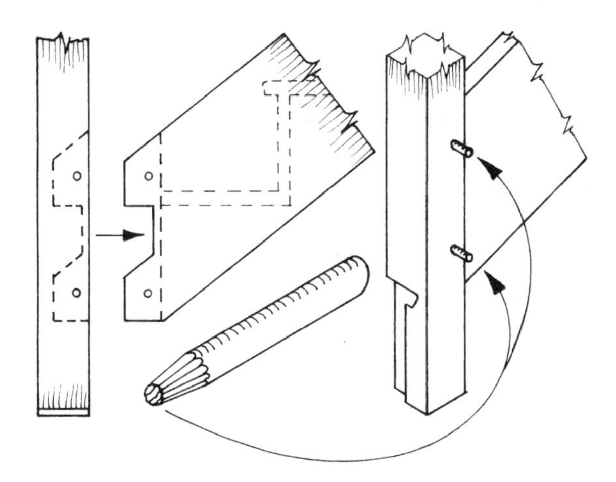

Figure 10.6 Gluing and pinning bottom newel

Figure 10.6: If pins are not supplied, cut off pieces of 12 mm diameter dowel rod, about 50 mm longer than the newel thickness and chisel the ends to a shallow taper of about 25 mm length. After trying the newel post into position, coat the joint with PVA (polyvinyl acetate) glue and reposition the newel. This is best done using a claw hammer onto a spare block of wood held against the lower face of the newel. When a reasonable fit has been achieved, a touch of glue is placed into the draw-bore holes and the tapered pins are driven in until no part of the taper remains within the newel – bearing in mind, however, that the lower dowel usually clashes with the step on the other side. Clean off excess glue with a damp rag or paper and then cut off the surplus dowel ends with a fine saw, near the newel's surface. Clean off the remainder with a block plane or smoothing plane.

10.2.10 Fitting a Protruding Step

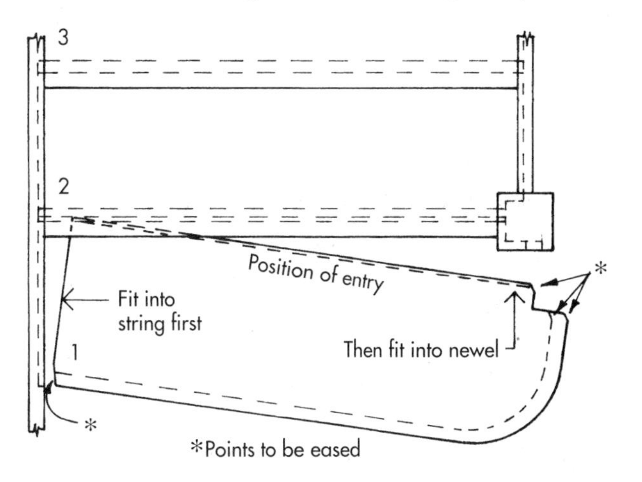

Figure 10.7 Fitting protruding bottom step

Figure 10.7: If the staircase has a bullnose (as illustrated) or splay-ended bottom step, which protrudes beyond the newel post, this is the next to be fixed. It should be realized that such steps cannot be attached during manufacture without the newel being permanently in position. The step may have to be fitted and, as shown, this usually involves slight easings to the front end of the tread entering the string housing, and the rear end of the tread and face-edge of the bullnosed riser entering the newel post housings. After a successful dry fit, glue the step into position and drive in the glued string wedges, screw the lower face of the second riser to the back-edge of the tread and, finally, screw the ends of the bottom two risers into the housings of the newel post.

10.2.11 Positioning the Staircase

Figure 10.8(a): Set the staircase back into its ultimate position, ready for the next operation of fitting and fixing the handrail and top newel post. As with the fixing of the bottom newel and step, the ideal position for the staircase is on its side, but available space rarely permits this, so methods of working *in situ* have to be devised. One method, as illustrated, is to push the staircase forward until enough height has been gained above the landing or upper floor to allow access to complete the work from that level. To make this arrangement safe, a

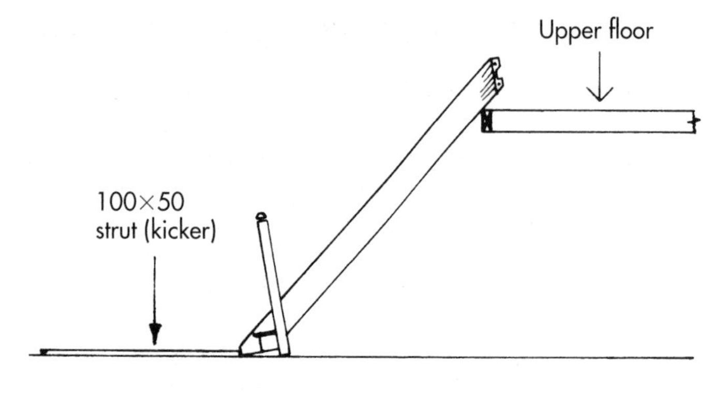

Figure 10.8(a) Positioning staircase to facilitate fixing of top newel and handrail

temporary kicker strut or struts of 100 × 50 mm section, should be lodged against the nearest cross-wall and extend to support the base of the staircase.

10.2.12 Notching the Top Newel

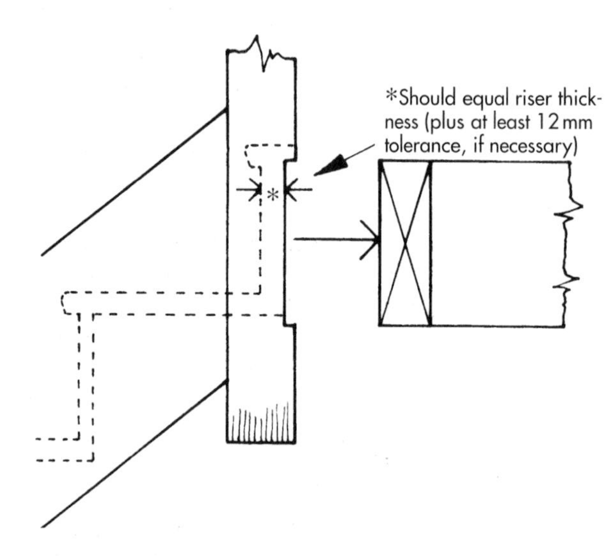

Figure 10.8(b) Newel notched to fit trimmer

Figure 10.8(b): Next, the newel may need to be notched-out (housed) to fit over the face of the landing trimmer. If so, this has a certain advantage of providing a good anchorage of the newel – and thereby that side of the staircase – to the top landing. However, whether this needs to be done or not depends on the newel's thickness, the thickness of the riser and whether a tolerance gap is to be allowed between the trimmer and the riser board (Figure 10.8(b)). The main reason for a tolerance gap is to overcome possible problems of the landing being out of square with the staircase. However, this is usually discovered in the early stages of offering up the staircase and may be proved to be unnecessary.

10.2.13 Fixing the Top Newel and Handrail

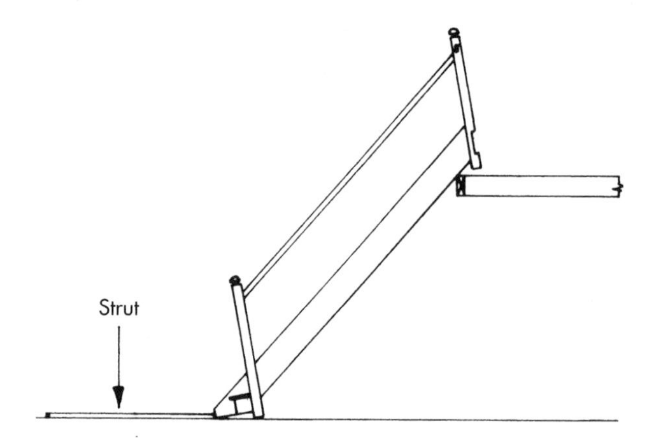

Figure 10.8(c) Fixing top newel and handrail

Figure 10.8(c): After checking the dry assembly of the newel post and handrail in relation to both newels, bearing in mind that all three would have been fitted together previously by the manufacturer and draw-bored, glue may be applied to the joints, the handrail located in the lower-newel mortice and held suspended while the top newel is fitted to the string and handrail tenons. Moving speedily, as with all gluing operations, the joints are knocked up and the glue-coated draw-bore pins are driven in to complete the assembly of the skeleton balustrade. Finally, remove the surplus dowels and clean up as described previously.

10.2.14 Preparing Top Riser and Nosing

Figure 10.9: Before the staircase can be set back into position, the top riser and the landing-nosing have to be fitted and fixed to each other, to the string housings, the newel-post housings and to the adjacent tread. This operation is often skimped, resulting in a loose top riser and a squeaky top step. To avoid this, attend to all the following points.

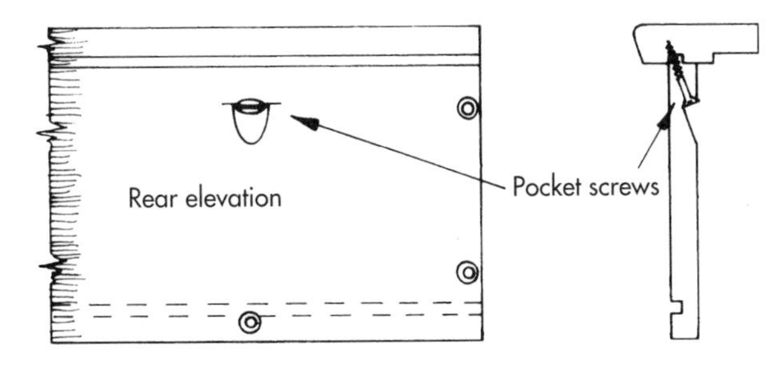

Figure 10.9 Preparing top riser and landing-nosing

Detailed Procedure

Check that the rebated side of the nosing is equal to the flooring thickness and, if thicker, ease with a rebate or shoulder plane. Then measure between the housings and cut the nosing and riser to the correct length. Now, because this particular step cannot have glue blocks set into its inner angle like the other steps (as they would clash with the trimmer), the best way to strengthen the joint is by *pocket screwing*. This is achieved by gouging or drilling shallow niches into the upper back-face of the riser board and by drilling oblique shank holes through these to create at least three fixings to the nosing piece. The riser is also drilled to receive two screws at each end and three or four along the bottom edge for the adjacent tread fixing.

Gluing Up

Next, glue the tongue-and-groove joint between the riser and nosing piece and insert the pocket screws. Set the step-shaped riser/nosing into the glued housings of the newel and string, up against the glued back-edge of the adjacent tread and insert the two screws at each end, followed by the three or four screws along the bottom edge. Clean off any excess glue on the face side.

10.2.15 Final Positioning and Fixing

Figure 10.10: The staircase is now ready for fixing. Remove the struts and lower carefully into position. Re-check the newel posts for plumb and check that the staircase is seated properly at top and bottom levels. Fix the bottom newel to the floor; fix the top newel to the trimmer by skew-nailing through the side with two 75 mm or 100 mm oval nails or – better still – by pocket-screwing (Figure 10.10(b)); nail the nosing to the landing trimmer with 50 mm or 56 mm floor brads or lost-head nails; then, finally, as described earlier, fix the inner string to the wall.

10.2.16 String-to-skirting Abutment

Figure 10.11: Where wet plastering techniques are being used (as they still are to some extent) and as opposed to dry-lining methods, timber skirting grounds should be fixed at least to the wall-string wall beyond the two extremes of the wall string. This ensures a flush abutment of the skirting where it meets the string. In practice, it is wise to set the grounds back 2 mm more than the given skirting thickness from the face of the string. This combats the effect of the timber ground swelling after gaining an excess of moisture from the rendering/floating coat of plaster. Note that the ground remains swollen, but the subsequent setting coat of finishing plaster, which is usually applied whilst the timber is swollen, remains about 2 mm proud when the ground eventually loses moisture and shrinks back to near normal.

10.2.17 String Mouldings

Figure 10.12: Finally, on the subject of wall strings, it must be mentioned that they might be moulded on their top edge to match the moulded edge – if any – of the skirting member. This entails extra work in the

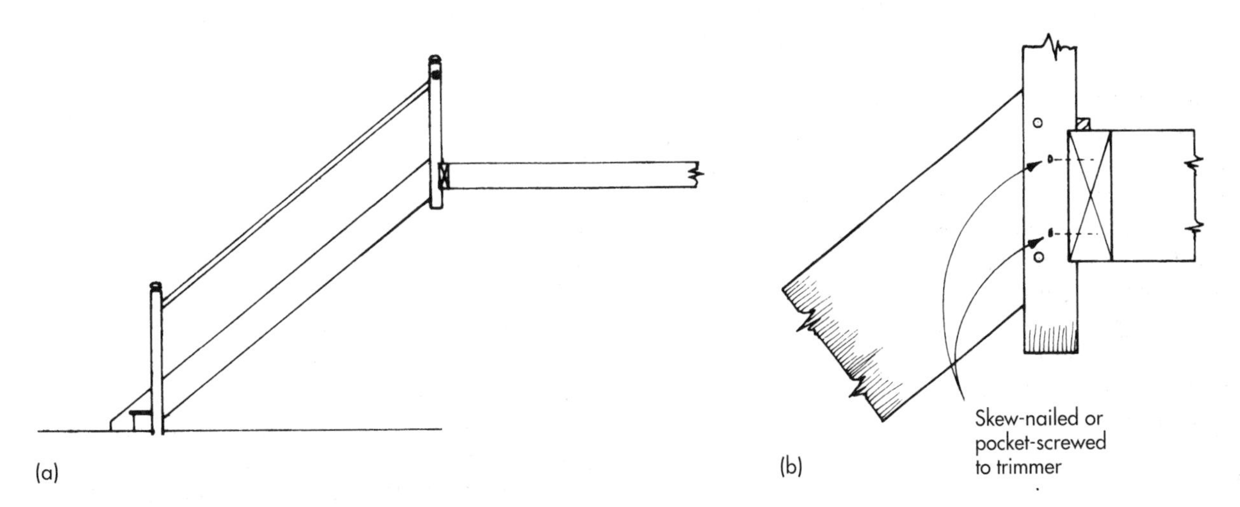

Figure 10.10(a) Staircase finally in position. (b) Skew-nailed or pocket-screwed to trimmer

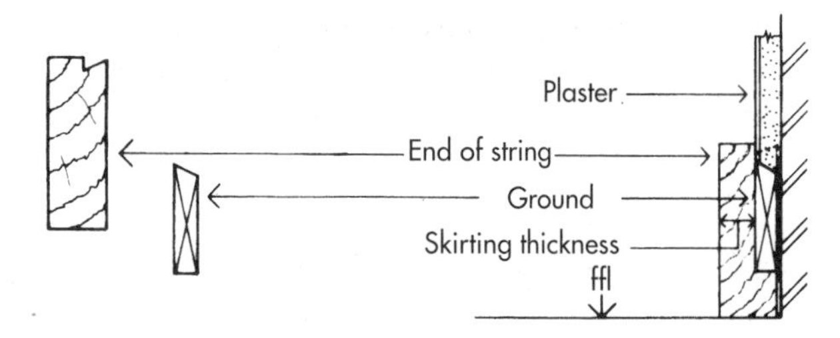

Figure 10.11 String-to-skirting abutment

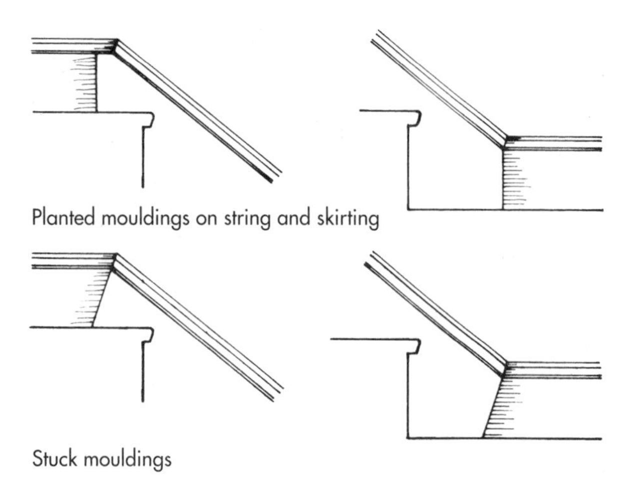

Figure 10.12 Mouldings

manufacture and/or on site, according to whether the shaped edge is a *stuck* moulding (machined out of the solid timber of the string and skirting), or a *planted* moulding (a separate moulding fixed by nails or pins to the plain, square edges of the string and skirting). As illustrated, only the moulding is bisected at the angle when planted, whereas with the stuck-moulded string and skirting, it will be easier and acceptable to let the bisected angle form a complete cut across the timber.

10.2.18 String-Easings

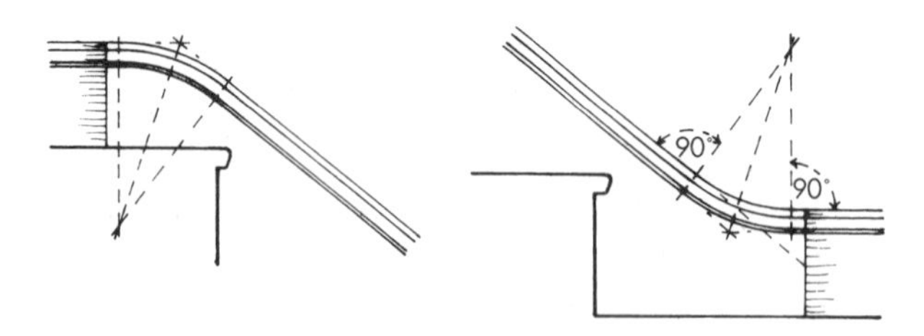

Figure 10.13 String-easings

Figure 10.13: As well as being moulded and forming obtuse and reflex angles, these string/skirting junctions might be required to be swept into a concave shape at the bottom and a convex shape at the top. This shaping is known as *easing* and, according to the moulding being stuck or planted, is either formed during manufacture, or fixed on site. It must be said that such work is uncommon nowadays because of the cost and the disinterest generally in moulded work, but could be met on repair, maintenance or refurbishment work.

10.2.19 Protection of Handrails, Newels and Steps

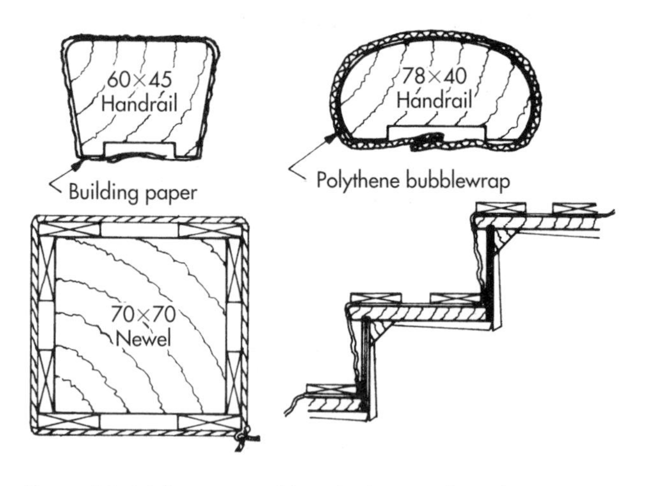

Figure 10.14 Protection of handrails, newels and steps

Figure 10.14: The remaining work on the staircase is best left until the second fixing stage, after the plasterer or dry-liner and other trades are finished. In the meantime, it is good practice to protect the handrail, newels and steps just after a staircase is installed, by wrapping building paper, heavy-gauge polythene or polythene bubblewrap around the handrails and newel posts and securing it with lashings of strong adhesive tape. Newels, which are usually more vulnerable, can be protected with additional wooden corner strips and tied or taped. If the handrail – or the staircase – is of hardwood, it should be sealed with diluted varnish and allowed to dry before being covered. The treads and risers should also be protected for as long as possible, by being covered with building paper or heavy-gauge polythene sheet, held into the shape of the steps by lightly nailed tread boards.

10.3 FIXING TAPERED STEPS

Traditionally, tapered steps – as they are now called – were referred to as *winders* or *winding steps* and they were incorporated into a variety of stair designs, used to change the direction of flight either at the bottom, halfway up, or at the top of the staircase. If the change in direction was 180°, there would be six winding steps, known as a half-space (half-turn) of winders. If turning through 90° – which was more common – there would be three winding steps (the square winder, the kite winder and the skew winder), known as a quarter-space (quarter-turn) of winders.

This terminology equates to landings, identified as quarter and half-space landings. Tapered steps usually replace landings to improve the headroom and when the 'going' of the staircase is greatly restricted. In certain cramped positions, there is often no alternative to them being used at both the top and bottom of the flight. However, tapered steps at middle and high levels of a flight, although not against the current regulations, are generally considered to be potentially dangerous and are usually avoided.

10.3.1 Tapered Steps at Bottom of Flight

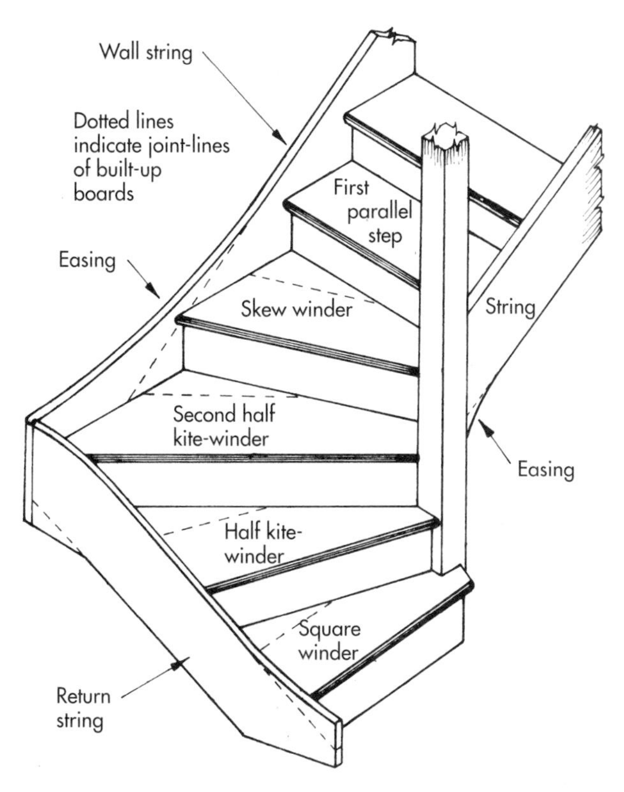

Figure 10.15 Tapered steps at bottom of flight

Figure 10.15: Tapered steps in non-geometrical staircases (those without wreathed strings and wreathed handrails, but with newel posts) are now mainly used at the bottom of the flight, in the form of four steps, as illustrated, to effect a quarter-turn. Although it is possible for some tapered-step arrangements to be completely formed and assembled in the shop, it is more common that they be formed and only partly assembled, then delivered to the site for fitting and fixing. The reasons for this, as with straight flights, are for easier transportation and manoeuvrability through doorways, etc., and for practical issues involved in the fitting and fixing operation. Such a flight would arrive on site separated from the newel posts and balustrade, the top riser board, the landing nosings and apron linings, the return string, the tapered treads and their corresponding riser boards, etc. The operations involved in fitting and fixing this type of staircase are generally the same as already described for straight flights, with certain obvious additions, as follows.

10.3.2 Fitting the Main Flight

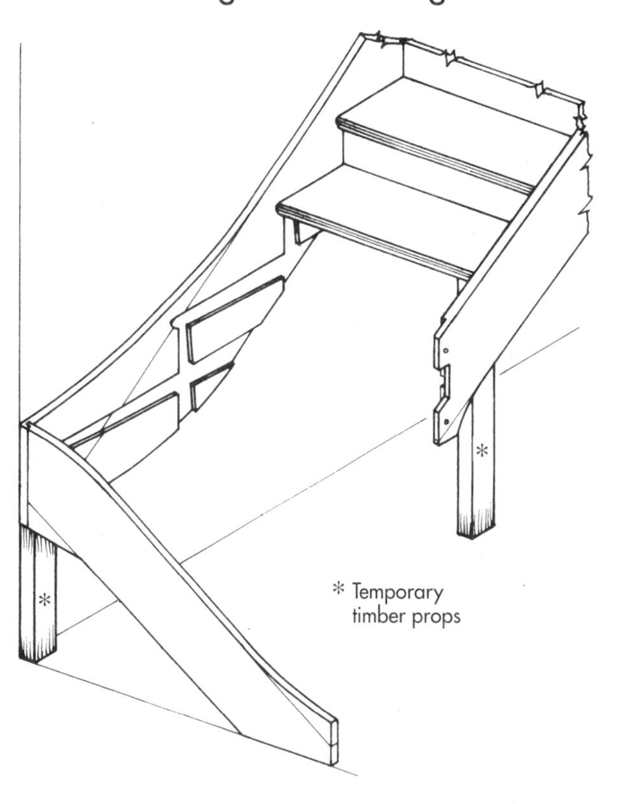

Figure 10.16 Fitting and checking the staircase

Figure 10.16: First, the main flight is offered up and fitted to the landing above and checked for level and plumb. To achieve this, built-up packing will be required at the bottom to compensate for the four missing tapered steps. Alternatively, two short timber props can be used, one under the bottom edge of the extended wall string, the other, as illustrated, up against the outer string, propping up the first available tread board. If supported like this, the staircase should remain firm, because, although not shown in the illustration, the end of the extended wall string butts up to the return wall.

10.3.3 Fitting the Return String

The return string, which connects to the main wall string with a tongued housing joint, is fitted and tried into position, the tops of the long tread-housings then being checked for level. The two strings ought to be at right-angles to each other, but this will depend largely on whether the return wall is truly square or not (which is another good reason for assembling and fitting these steps *in situ*).

10.3.4 Fixing the Main Flight

After these initial operations, the staircase will require repositioning to allow for the fitting and fixing of the newels and handrail. As outlined previously, this can be done by pushing the staircase up onto the landing and supporting the bottom end with packing and struts. The fitting of the skeletal balustrade then follows the sequence:

1. fix bottom newel post by gluing and pinning to string tenons;
2. glue handrail tenon and insert into bottom newel;
3. glue and fit top newel to handrail and outer string;
4. quickly complete the pinning of the unpinned joints;
5. fix top riser and landing nosing – after joining same together.

Back into position again, the main flight is re-checked and fixed as previously described. The tongued housing joint of the return wall-string is glued and fitted and the string fixed to the return wall.

10.3.5 Starting on the Tapered Steps

Finally, starting from the bottom, the tapered steps have to be fitted and fixed. This is the most difficult part of the whole operation and requires great care in checking and transferring details of the tread's shape and length from the housings of the strings and newel, to the separate treads.

10.3.6 Using a Pinch Rod

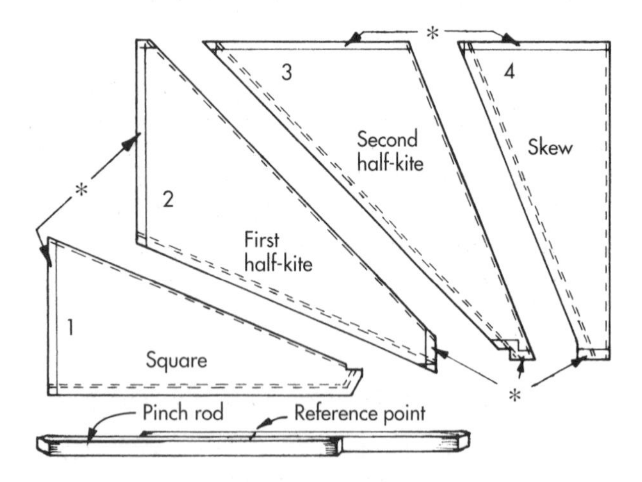

Figure 10.17(a) Pinch rod and tapered treads. Tolerances in length are indicated by*

Figure 10.17(a): The treads will be already marked and cut to a tapered shape, usually with tolerances of about 25 mm left on in length to offset any problems that may arise if the return wall (and thereby the return string) should be out of square. A common method for checking lengths in this situation, is to use a pinch rod formed by two overlapping laths (timber of small sectional size). The laths are held together tightly, expanded out to touch the two extremes, then marked across the two laths with a pencil line as a reference point, so that they can be released and put back together when marking the tread and/or riser.

10.3.7 Checking, Cutting, Forming and Fixing

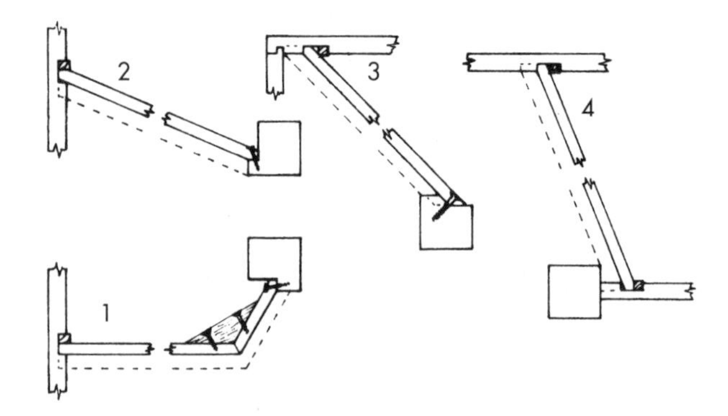

Figure 10.17(b) Riser-end shapes into string and newel. The dotted lines indicate nosings

Figure 10.17(b): Using such a method, the bottom riser is checked and cut to length. Then, with the aid of a carpenter's bevel and the pinch rod, the first tapered tread is checked, marked and cut to shape. After being tried into position (which often involves easing protruding corners of the tread and/or housings), the tread is fixed to the riser by gluing the joint between the two boards and gluing, rubbing and pinning (with panel pins) glue blocks on the inside angle. The housings are then glued, the step inserted and glued-wedges driven into the string housings and screws driven into the newel (as illustrated). On the bottom riser, at least, the wedges cannot normally be driven-in on the string side and will have to be tapered in thickness and driven-in sideways from the face.

10.3.8 Repetition and Completion

This technique of checking and cutting, forming and fixing, is repeated on the other tapered steps and finalized by the fixing of the last riser to the main flight (riser No. 5). After each step is wedged into position, the bottom of the riser should be screwed to the tread. Sometimes, especially on wider-than-normal flights, 100 × 50 mm horizontal cross-bearers, on edge, are notched into (or cleated to) the string and newel post at each end, and fixed tight up against and under the back edge of each tapered step.

10.4 FIXING BALUSTRADES

During the second-fixing stage, the uncompleted work on the staircase can be finished. This mainly involves the balusters to the side of the staircase and the balustrade around the edge of the stairwell on the upper landing. The following steps outline the sequence of the operation.

10.4.1 Additional Newel Posts

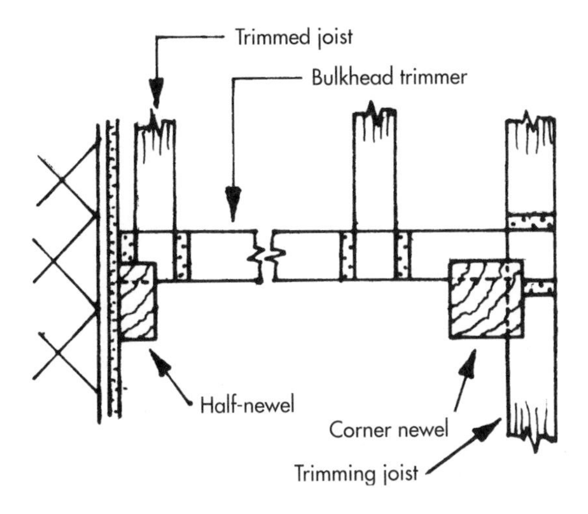

Figure 10.18(a) Plan view of additional newels

Figure 10.18(a): The number of newel posts needed to form the balustrade to the stairwell depends on individual design, but normally only one and a half are required in addition to the one already at the head of the stairs. The half-section newel is always used on good class work, because it finishes the balustrade off properly against the wall adjacent to the bulkhead trimmer and supports the return handrail on a through mortice and tenon joint. On cheap work, the handrail is housed in the wall, minus the half newel. The half newel runs down past and up against the side of the trimmer and is fixed to the finished wall surface with three counterbored screws. The full-section newel sits in the corner of the stairwell, again running down past the ceiling and up against the trimmer on one side and the trimming joist on the other. Although its rigidity will be gained from the mortice-and-tenon connections of the handrails that will joint into it at 90° to each other, it should still be counterbored and screwed to the trimmer and trimming joists. The preformed handrail mortices in the newels should be used as a datum point to ensure that handrails will be level and/or parallel to the floor.

10.4.2 Level Handrails

Figure 10.18(b): In the example used here, two handrails – additional to the first raking handrail used

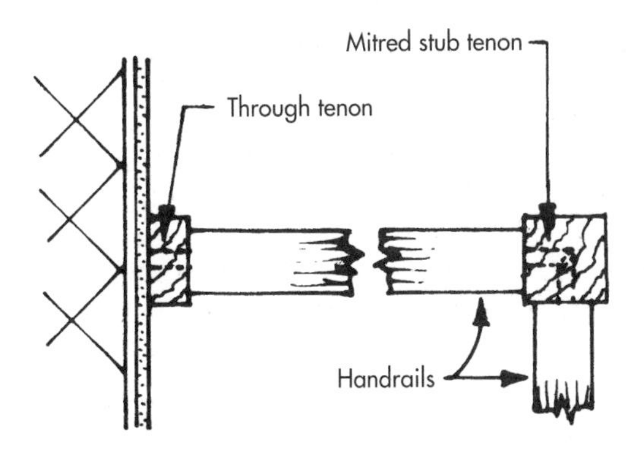

Figure 10.18(b) Plan view of level handrails

– will be required, one long, one short. If supplied from a joinery works, they may be already tenoned at one end, but not usually at both, as tolerance must be allowed for site variations. Tenons – their size taken from the newel mortices – should be cut and shouldered carefully with a fine saw. The handrail can be laid against the newels, with the shoulder butted against one, while the second shoulder is marked against the other newel. This mark can be squared around the shaped handrail by wrapping a piece of straight-edged cardboard or glasspaper tightly round the handrail until its edge overlaps precisely, then by sliding along to the mark and marking around the edge. Now mark and rip the tenon's length, then shoulder it. Where the tenons intersect in the corner newel, they will require mitring.

10.4.3 Take Care on the Return

Take care when marking the short return-handrail length that the measurement between newels may differ between the top and at floor level, depending on the plumbness of the half newel against the wall surface. All of the tenons should be draw-bored, glued and pinned with 9 mm diameter dowelling. The handrails can now be installed and, to facilitate this, the fixings of the newels will require to be slackened or – more likely – removed.

10.4.4 Apron Linings and Nosing

Figure 10.18(c): Next in sequence are the apron linings which add a finish to the rough face of the trimmer/trimming joists within the open stairwell. Traditionally, they were made from solid timber of about 21 mm thickness, but nowadays they are more common in plywood or MDF board. Their lateral position in relation to the newel's thickness is critical, because they support the landing nosing above, which in turn supports the balusters that must be dead centre of the newel in order to relate to the underside groove of the

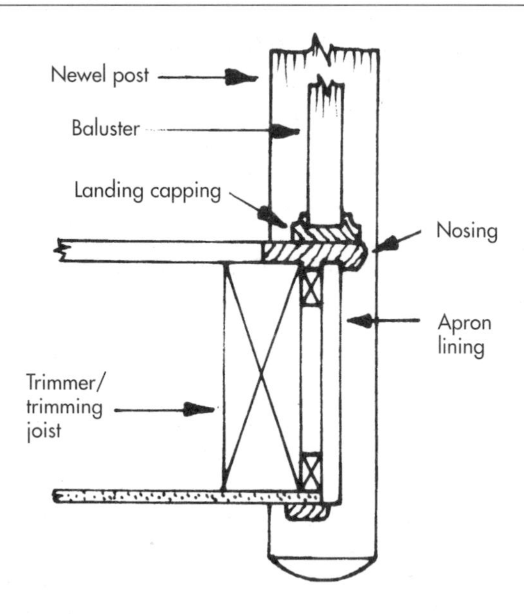

Figure 10.18(c) Vertical section through apron lining and nosing

handrail. Therefore, they need to be very near the centre of the newel, as seen in the illustration, which usually means packing them out from the joists. This is done with prepared or sawn timber grounds of whatever thickness, but say 18 mm (× 50 mm), fixed horizontally on the joists to take MDF or plywood, but vertically as so-called *soldier-pieces* at 400–600 mm centres if the apron lining is of solid timber. This is to combat the undesirable effect of *cupping* which may occur to tangentially sawn boards.

10.4.5 Nosing Pieces

Having cut the apron linings carefully in length to fit neatly and tight between newels, they are fixed at top and bottom edges with 56 mm oval or lost-head nails (punched under the surface) at maximum 600 mm centres. Next, the nosing pieces, which are to be attached to the top edge of the apron lining and the joist, being like stair nosings, are usually equal in thickness to the stair treads. Therefore, they will probably be rebated to meet the lesser thickness of the flooring and this should be checked accordingly before fixing. This done, cut carefully to length and nail into position to the joist and apron-lining edges. If the apron lining is of MDF board, it will be better to glue this connection, not nail it, as MDF does not take or hold nails very well in its edges.

10.4.6 String-capping and Landing-capping

Figure 10.18(d): Capping is the next item to be fixed and it can be either plain-edged or moulded and grooved on one or both faces. One groove fits the top edge of the string, the other houses the balusters. If

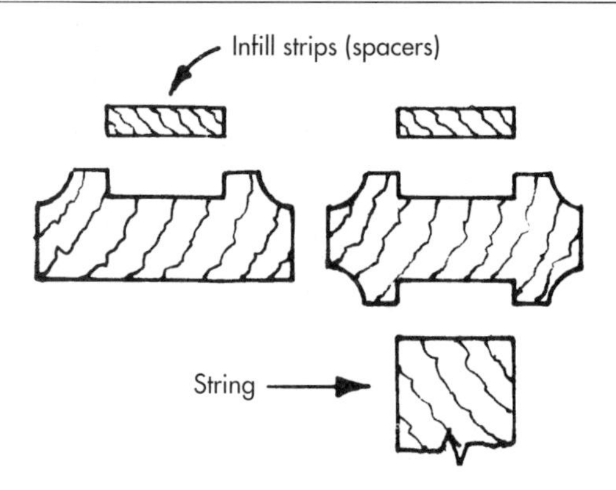

Figure 10.18(d) Sections through string and landing capping

there is only one groove, then this houses the balusters and the plain (ungrooved) face is fixed to the string. Similarly, it can also be used on the landing, as illustrated in Figure 10.18(c), fixed to the nosing with 50 mm oval nails. This has the advantage of providing a raised edge for carpet to butt up against. On the string, the capping needs to be cut to an angle to fit against the newels. This can be found with a sliding bevel, set to the angle formed by the junction of the bottom (or top) newel and the string. Set the mitre saw (or a mitre box) up with this bevel, cut carefully to length and fix with 38 mm oval nails at about 500 mm centres.

10.4.7 Balusters

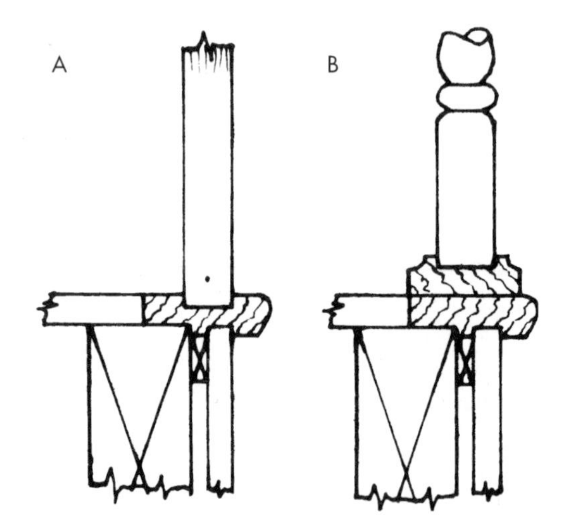

Figure 10.18(e) Baluster stick A and spindle B

Figure 10.18(e): *Balusters* are also referred to nowadays as *spindles*. Both names refer to lathe-turned ornamental posts that infill the balustrade at the side of the stairs and stairwell. *Baluster sticks* only differ by being square posts with no ornamental lathe work. Whichever is used, The Building Regulations require that they be set up with a controlled gap between them. This gap should not allow a 100 mm diameter sphere to pass through. In effect, this means that the spacing between the balusters should not be more than 99 mm. The same bevel set up to cut the capping can now be used to cut the balusters to fit the raking balustrade. First, cut one only, bevelled at each end to the precise height taken from the inner face of the newel. Try it in position in a few places and check for plumb. If acceptable, cut the estimated number to this pattern, ready for fixing.

10.4.8 Support from Above

If possible, support the centre of the handrail from above with a timber strut at right-angles to the pitch, to restrict the cumulative effect of the individually fixed balusters pushing up the handrail. Alternatively, fix a single baluster midway between newels and use a sash cramp at this point, square across the balustrade, tightened from the top of the handrail to the underside of the string.

10.4.9 Infill Strips

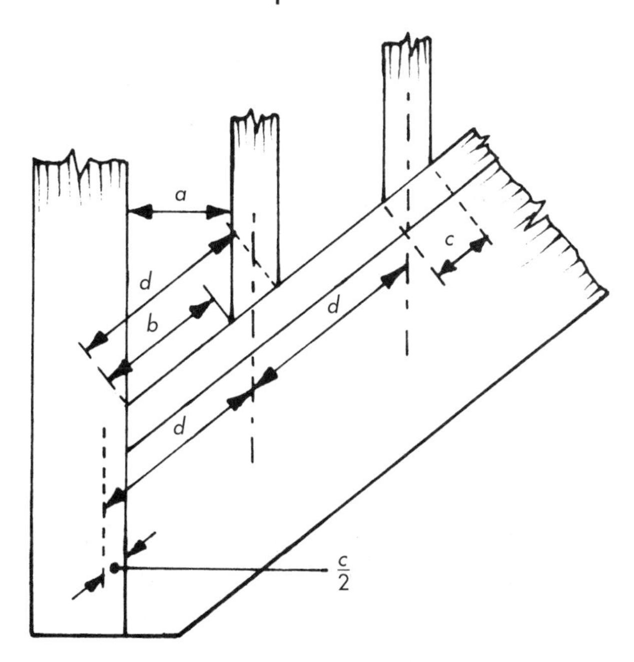

Figure 10.18(f) Working out spindle spacings

Figure 10.18(f): The correct spacing – and easy fixing – of the balusters is achieved by using short lengths of timber infill-strip between them at top and bottom. This thin strip, machined to fit the shallow groove (of say 6 × 40 mm) in the handrail and the capping (as in Figure 10.18(d)) is cut up into equal angled-lengths, worked out to achieve the required spacing gap. One way of doing this, is to set up a pre-cut baluster in the grooves of the capping and handrail near, say, the bottom newel post and slide/adjust its position until a reading of 99 mm is obtained on a right-angled measurement a. The geometrically longer measurement of the angled infill strip will now be obtainable as b, between the newel and baluster faces.

10.4.10 Creating Equal Gaps Throughout

Figure 10.18(f): To test the overall spacing of the balusters, the *raking measurement* b just obtained should have the *raking thickness* of the baluster c added to it, giving d. This value – equalling the baluster centres – should then be divided into the total measurement of the capping between the newels plus the raking thickness of half a baluster at each end, i.e. $c \div 2 \times 2 = c +$ capping length. The odds are that it will not divide equally and therefore adjustments will need to be made by lessening the divisor d and dividing again – and again, if necessary – until it works out without a remainder. The difference between the first and final divisor must then be deducted from the first-obtained measurement of the angled infill-strip b.

10.4.11 Example

Assume that the total length of capping between newels = 3.188 m, actual baluster thickness = 40 mm, desired gap a = 99 mm, initially obtained raking-measurement of infill strip b = 129 mm, raking thickness of baluster c = 52 mm. Therefore,

$$b + c = 129 + 52 = 181 \text{ mm baluster centres } d$$

The divisible length of capping between newels plus two times the half-baluster raking-thickness is

$$3.188 \text{ m} + 2 \times 26 \text{ mm} = 3.240 \text{ m}$$

Therefore

$$3240 \div 181 = 17.90$$
$$= 17 \text{ spacings with 90 mm remainder}$$

Trying again,

$$3240 \div 180 = 18 \text{ spacings exactly}$$

The difference between divisors equals 1 mm, therefore the final size for the angled infill-strips b is

$$129 - 1 = 128 \text{ mm}$$

In this example, the resultant gap between balusters will be slightly more than 98 mm. The working out also tells us that, as there are 18 spacings, there must be 17 balusters and 36 angled infill-strips.

10.4.12 Fixing

The infill strips should now be cut and the first two, one in the handrail, one in the capping groove, fixed up against the bottom newel post with a few 18 or 25 mm panel pins punched-in. Then the first baluster is fixed, skew-nailed at top and bottom with two 38 mm lost-head oval nails (punched-in) at each end. Then two more angled infill-strips are fixed, and another baluster – and so the process is repeated until the top newel is reached. On the landing, a similar technique of working out and fixing can be used for the stairwell balusters. The only variation is when a grooved capping is not used. In these situations, as was traditional, the balusters are housed into the nosing to a depth of about 6 mm (Figure 10.18(e)) and skew-nailed once from each side.

10.4.13 Newel Caps

Figure 10.18(g): The tops of newels are usually finished in three basic ways:

1. finished in themselves, shaped in a variety of simple designs such as (1) chamferred-edge, (2) round-(quadrant) edge, (3) cross-segmental top and (4) cross semi-circular top.
2. separate recessed, square caps, with projecting moulded edges (5). The caps are pinned or nailed to the tops of the newels.

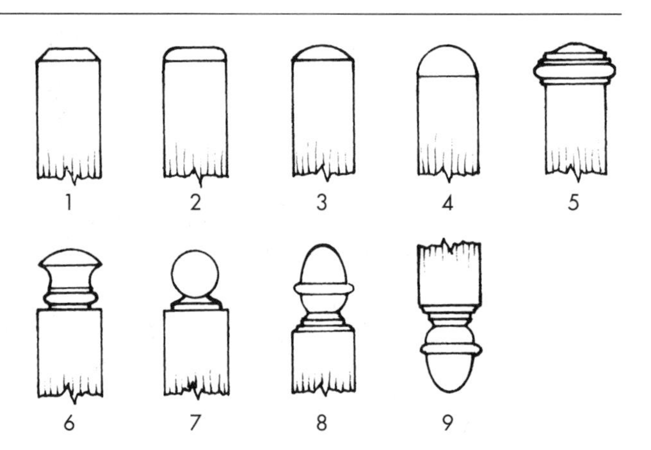

Figure 10.18(g) Newel caps

3. separate spherical ornate caps turned on the lathe with projecting spigots, ready to be glued and inserted into predrilled holes in the ends of the newel posts. Three standard shapes predominate in this range, known as (6) *mushroom cap*, (7) *ball cap* and (8) *acorn cap*.

Usually, on newel posts above ground floor level, the newel is allowed to project down below the ceiling (by minimal amounts nowadays) and should receive an identical cap (9) to that used at the top. Traditionally, these below-ceiling projections were known as *newel-drops* or *pendants*.

11

Stair Regulations Guide to Design and Construction

11.1 THE BUILDING REGULATIONS 1991

11.1.1 *Approved Document K*, 1998 Edition

This approved document (which comprises sections K1–K5) carries the seal of the Secretary of State for the purposes of the Building Regulations 1991 as amended by the Building Regulations (Amendment) Regulations 1997 and came into effect on 1 January 1998. Three categories of stairs are considered in the document:

1. *private stair*, intended to be used for only one dwelling;
2. *institutional and assembly stair*, serving a place where a substantial number of people will gather;
3. *other stair*, in all other buildings.

In producing a modified version of the approved document's K1 section here (hereinafter referred to as AD K1), covering most of the points concerning stairs and balustrades only, an attempt has been made to present a clearer picture of stair regulations as a guide, but not as a substitute.

11.1.2 Definitions

The following meanings are given to terms used in the document and a few other definitions have been added for clarity.

Alternating tread stair

Figure 11.1: A stair constructed of paddle-shaped treads with the wide portion alternating from one side to the other on consecutive treads.

Balustrade

Figure 11.2: A protective barrier comprising newel posts, handrail and balusters (spindles), which may also be a wall, parapet, screen or railing.

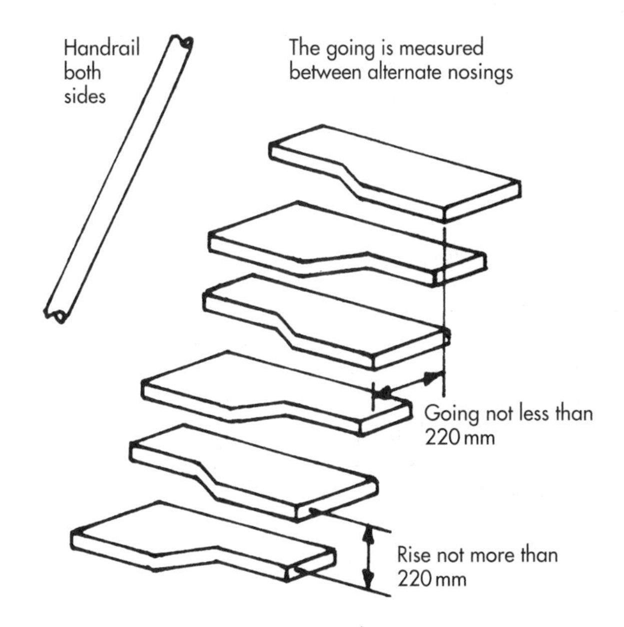

Figure 11.1 Alternating tread stair

Deemed Length

Figure 11.3: If consecutive tapered treads are of different lengths, as illustrated, each tread can be deemed to have a length equal to the shortest length of such treads. Although not now referred to in AD K1, the deemed length (DL) needs to be established on certain stairs (see Figure 11.20(a)) for the purpose of defining the extremities to which the pitch line(s) will apply.

Flight

The part of a stair or ramp between landings that has a continuous series of steps or a continuous slope.

Going

Figure 11.4: The horizontal dimension from the nosing edge of one tread to the nosing edge of the next consecutive tread above it, as illustrated.

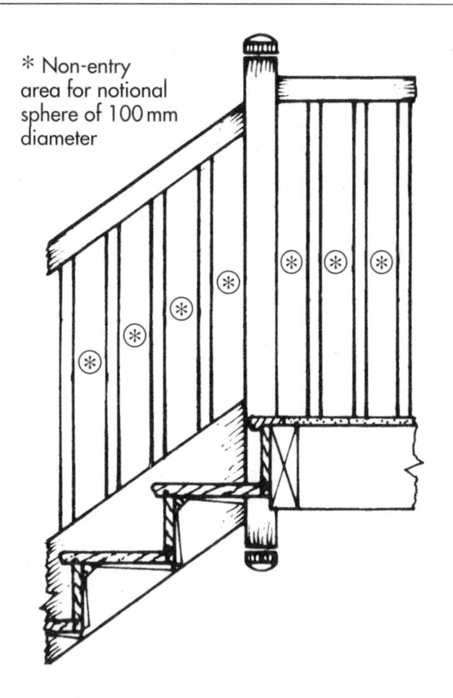

Figure 11.2 Balustrade

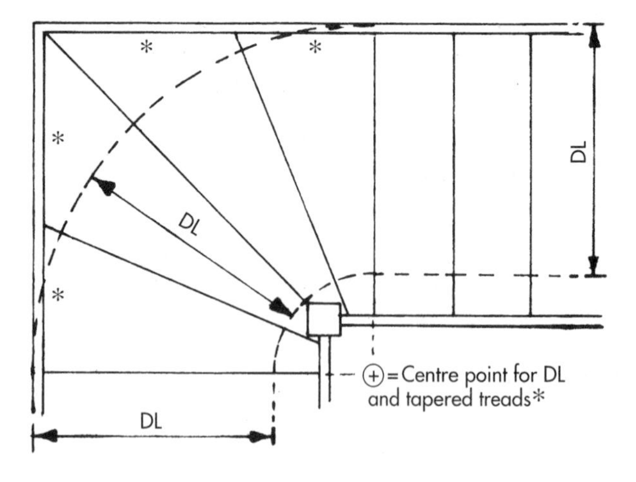

Figure 11.3 Deemed length (DL)

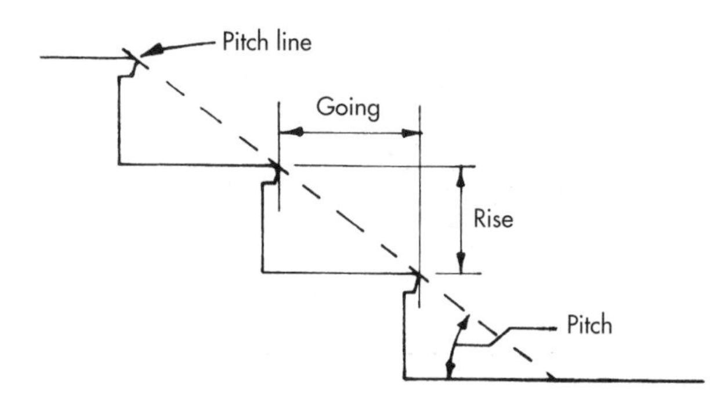

Figure 11.4 Going, rise and pitch

Going of a Landing

Figures 11.5 and 11.6: The horizontal dimension determining the width of a landing, measured at right-angles

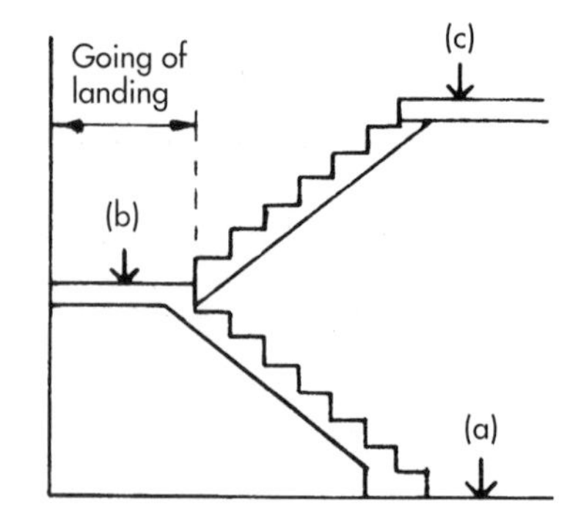

Figure 11.5 Landings

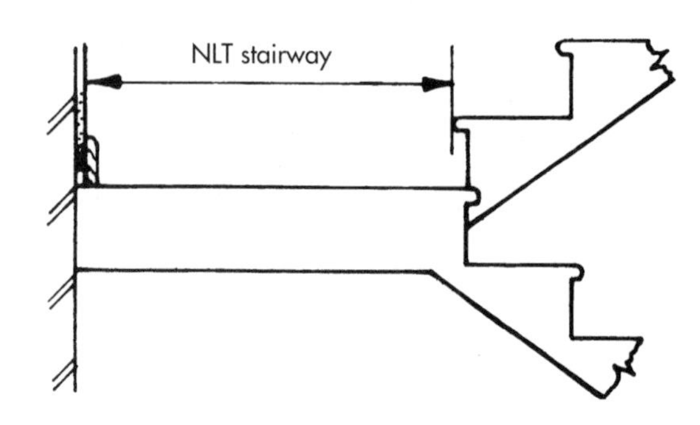

Figure 11.6 Going of landings not less than stairway

to the top or bottom step, from the nosing's edge to the wall surface or balustrade.

Helical Stair

A stair that describes a helix round a central void, traditionally known as a geometrical stair.

Nosing

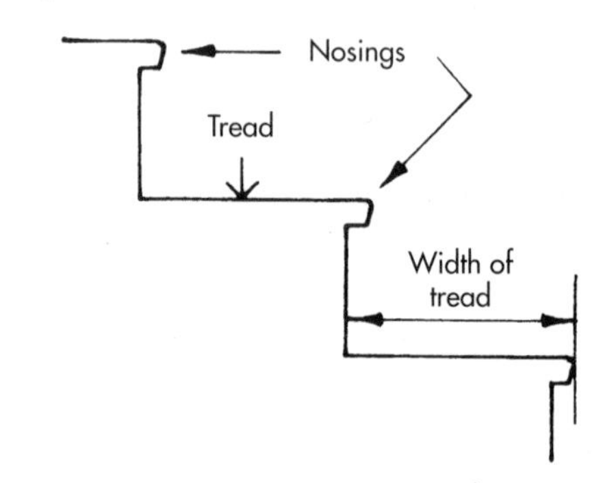

Figure 11.7 Nosings

Figure 11.7: The projecting front edge of a tread board past the face of the riser, not to be more than the tread's thickness, as an established joinery rule, and not less than 16 mm overlap with the back edge of the tread below on open-riser steps, as shown in diagram 1 in AD K1.

Pitch

Figure 11.4: This refers to the degree of incline from the horizontal to the inclined pitch angle of the stair.

Pitch Line

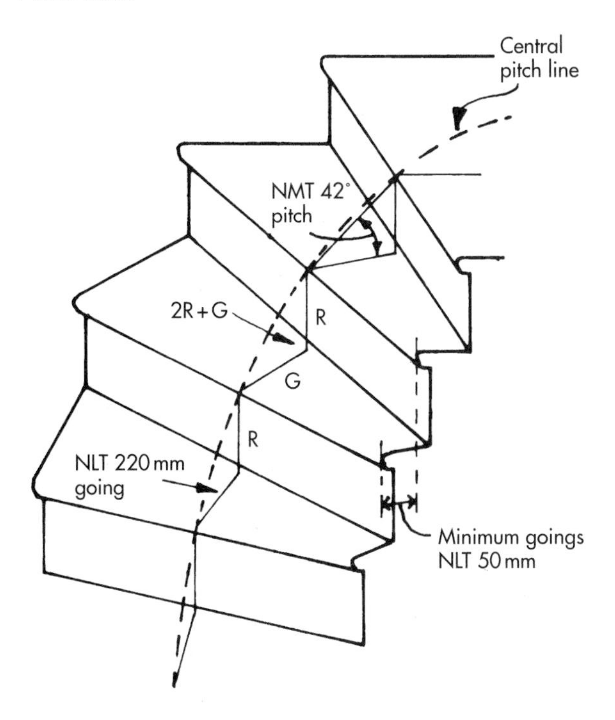

Figure 11.8 Pitch line

Figures 11.4 and 11.8: This is a notional line used for reference to the various rules, which connects the nosings of all the treads in a flight and also serves as a line of reference for measuring 2R+G on tapered-tread steps.

Rise

Figure 11.4: The vertical dimension of one unit of the total vertical division of a flight (total rise), as illustrated.

Spiral Stair

A stair that describes a helix round a central column.

Stair

A succession of steps and landings that makes it possible to pass on foot to other levels.

Tapered Tread

Figure 11.3: A step in which the nosing is not parallel to the nosing of the step or landing above it.

11.1.3 General Requirements for Stairs

Steepness of Stairs

In a flight, the steps should all have the same rise and the same going to the dimensions given later for each category of stair in relation to the 2R+G formula.

Alternative Approach

AD K1 states that the requirement for steepness of stairs can also be met by following the relevant recommendations in BS 5395: *Stairs, ladders and walkways* Part 1: 1977 *Code of practice for the design of straight stairs.*

Construction of Steps

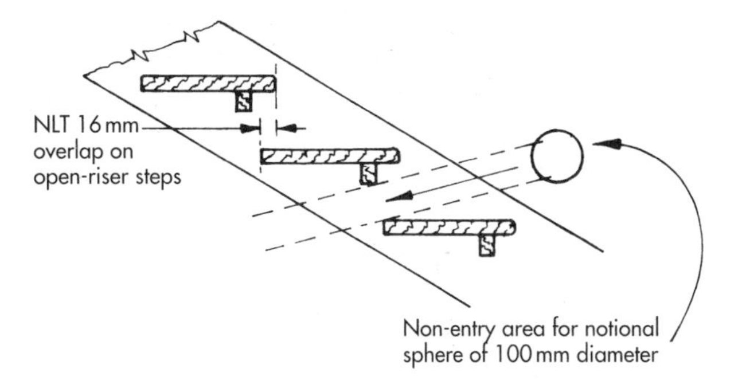

Figure 11.9 Step construction

Figure 11.9: Steps should have level treads and may have open risers, but treads should then overlap each other by at least 16 mm. For steps in buildings providing the means of access for disabled people, reference should be made to *Approved Document M: Access and facilities for disabled people.*

Open-riser Stair

Figure 11.9: All stairs that have open risers and are likely to be used by children under 5 years should be constructed so that a 100 mm diameter sphere cannot pass through the open risers.

Headroom

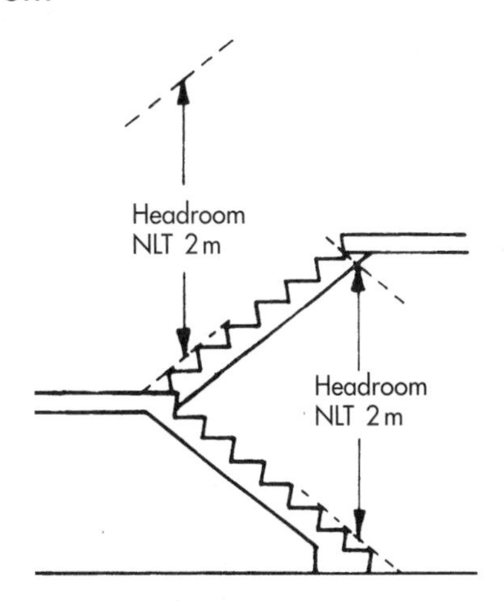

Figure 11.10 Headroom

Figures 11.10 and 11.11: Clear headroom of not less than 2 m, measured vertically from the pitch line or landing, is adequate on the access between levels, as illustrated. For loft conversions where there is not enough space to achieve this height, the headroom will

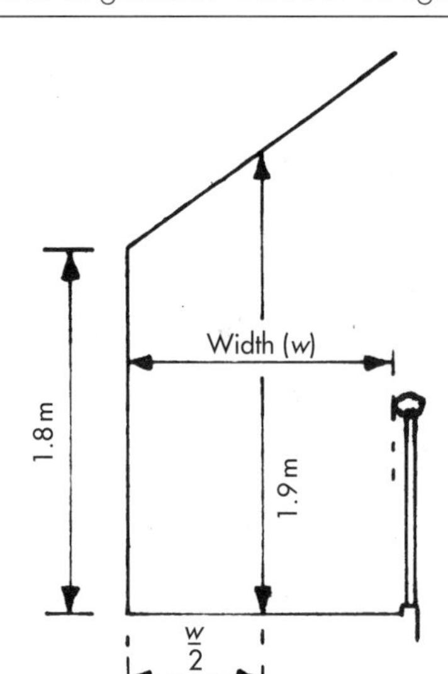

Figure 11.11 Reduced headroom for loft conversions

be satisfactory if the height measured at the centre of the stair width is 1.9 m, reducing to 1.8 m at the side of the stair.

Width of Flights

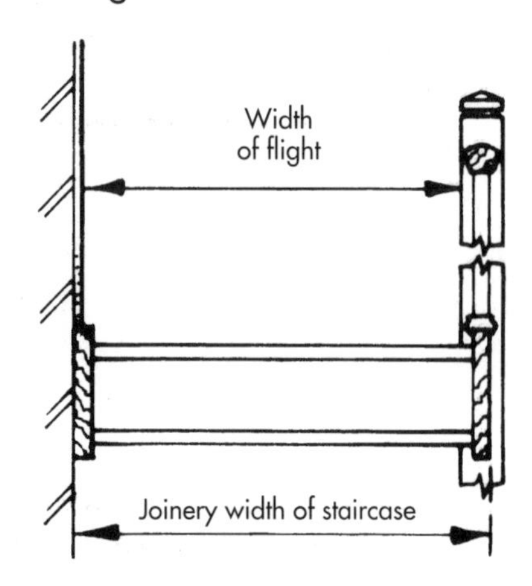

Figure 11.12 Width

Figure 11.12: Contrary to previous regulations (which gave 800 mm as the minimum unobstructed width for the main stair in a private dwelling), no recommendations for minimum stair widths are now given. However, designers should bear in mind the requirements for stairs which:

- form part of means of escape (reference should be made to *Approved Document B: Fire safety*);
- provide access for disabled people (reference should be made to *Approved Document M: Access and facilities for disabled people*).

Dividing Flights

Figure 11.13: A stair in a public building which is wider than 1800 mm should be divided into flights which are not wider than 1800 mm, as illustrated.

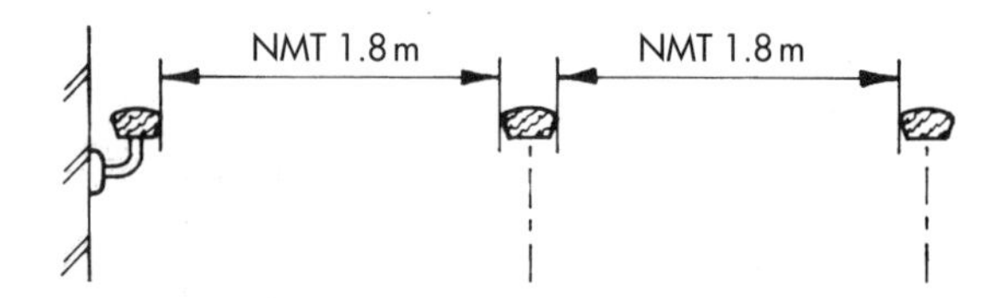

Figure 11.13 Division of flights if over 1.8 m wide

Length of Flights

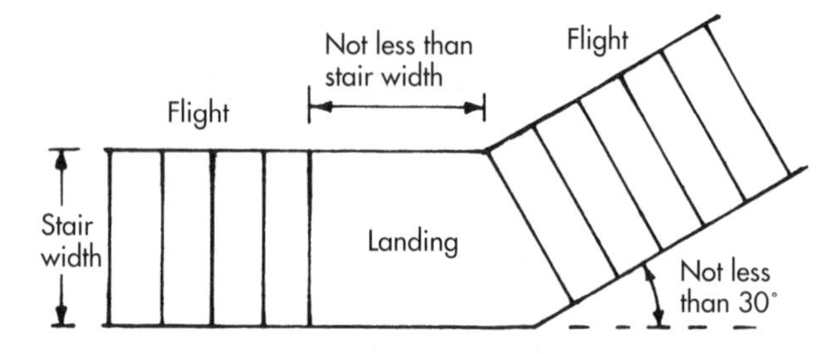

Figure 11.14 Change of direction

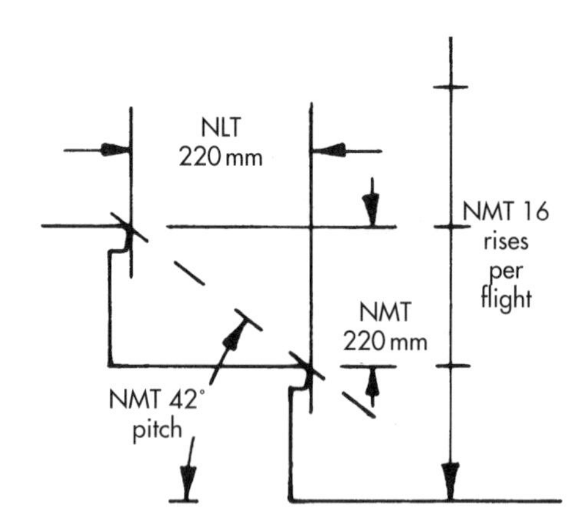

Figure 11.15 Category 1 stair

Figures 11.14 and 11.15: The number of risers in a flight should be limited to 16 if a stair serves an area used as a shop or for assembly purposes. Stairs having more than 36 risers in consecutive flights, should have at least one change of direction between flights of at least 30° in plan, as illustrated.

Landings

Figures 11.5, 11.6 and 11.16: A landing should be provided at the top and bottom of every flight. The width and length of every landing should be at least as long as the smallest width of the flight and may include part of the floor of the building. To afford safe passage, landings should be clear of permanent obstruction. A door may swing across a landing at the bottom of a flight, but only if it will leave a clear space of at least 400 mm across the full width of the flight. Doors to cupboards and ducts may open in a similar way over a landing at the top of a flight (Figure 11.16(b)). For means-of-escape requirements, reference should be made to *Approved Document B: Fire safety*. Landings should be

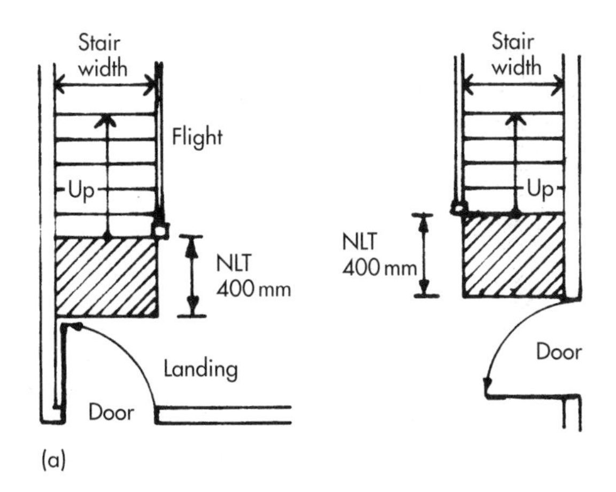

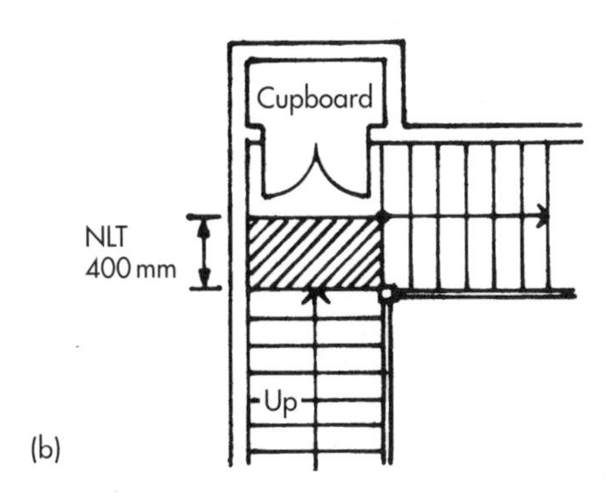

Figure 11.16(a) Landings next to doors. **(b)** Cupboard on landing

level unless they are formed by the ground at the top or bottom of a flight. The maximum slope of this type of landing may be 1 in 20, provided that the ground is paved or otherwise made firm.

11.1.4 Rise-and-going Limits for Each Category of Stair

Private Stair

Figure 11.15: Any rise between 155 mm and 220 mm can be used with any going between 245 mm and 260 mm, or any rise between 165 mm and 200 mm can be used with any going between 220 mm and 300 mm.

Institutional and Assembly Stair

Figure 11.17: Any rise between 135 mm and 180 mm can be used with any going between 280 mm and 340 mm. Note that for maximum rise for stairs providing the means of access for disabled people, reference should be made to *Approved Document M: Access and facilities for disabled people.*

Other Stair

Figure 11.18: Any rise between 150 mm and 190 mm can be used with any going between 250 mm and

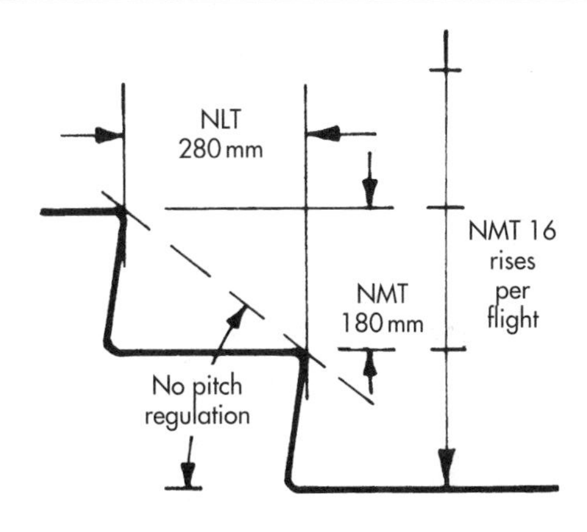

Figure 11.17 Category 2 stair

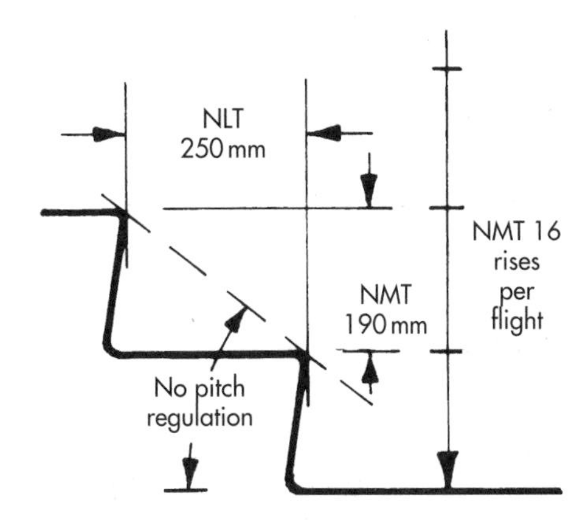

Figure 11.18 Category 3 stair

320 mm. Note that reference to *Approved Document M: Access and facilities for disabled people* also applies here.

Pitch

Figure 11.15: The maximum pitch for a private stair is 42°. Note that recommended pitch angles for the other two categories of stair are not given. However, using the criteria that are given, if the maximum rise and the minimum going were used in these categories, the maximum possible pitch for category 2 would be 33° and for category 3 would be 38°.

Note that if the area of a floor of a building in category 2 (*Institutional and assembly stair*) is less than 100 m^2, the going of 280 mm may be reduced to 250 mm. Therefore, with maximum rise and minimum going, the maximum possible pitch would be increased from 33° to 36°.

The 2R+G Design Formula

Figure 11.19: In all three categories, the sum of the going plus twice the rise of a step (traditionally established as 2R+G) should be not less than 550 mm nor more than 700 mm (subject to the criteria laid down for tapered treads).

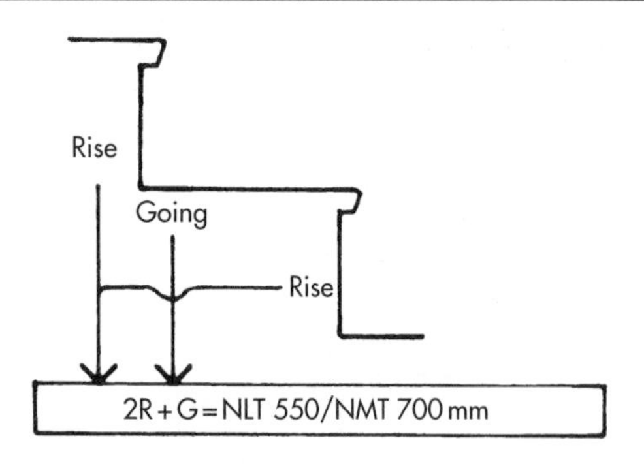

Figure 11.19 Design formula

11.1.5 Special Stairs

Tapered Treads

Figures 11.8 and 11.20: For a stair with tapered treads, the going and 2R+G should be measured as follows.

1. If the width of the stair is less than 1 m, it should be measured in the middle, tangentially where the curved pitch line touches the nosings (Figure 11.8).
2. If the width of the stair is 1 m or more, it should be measured at 270 mm from each side, tangentially where each curved pitch line touches the nosings (Figure 11.20(c)). The minimum going of tapered treads should not be less than 50 mm, measured at

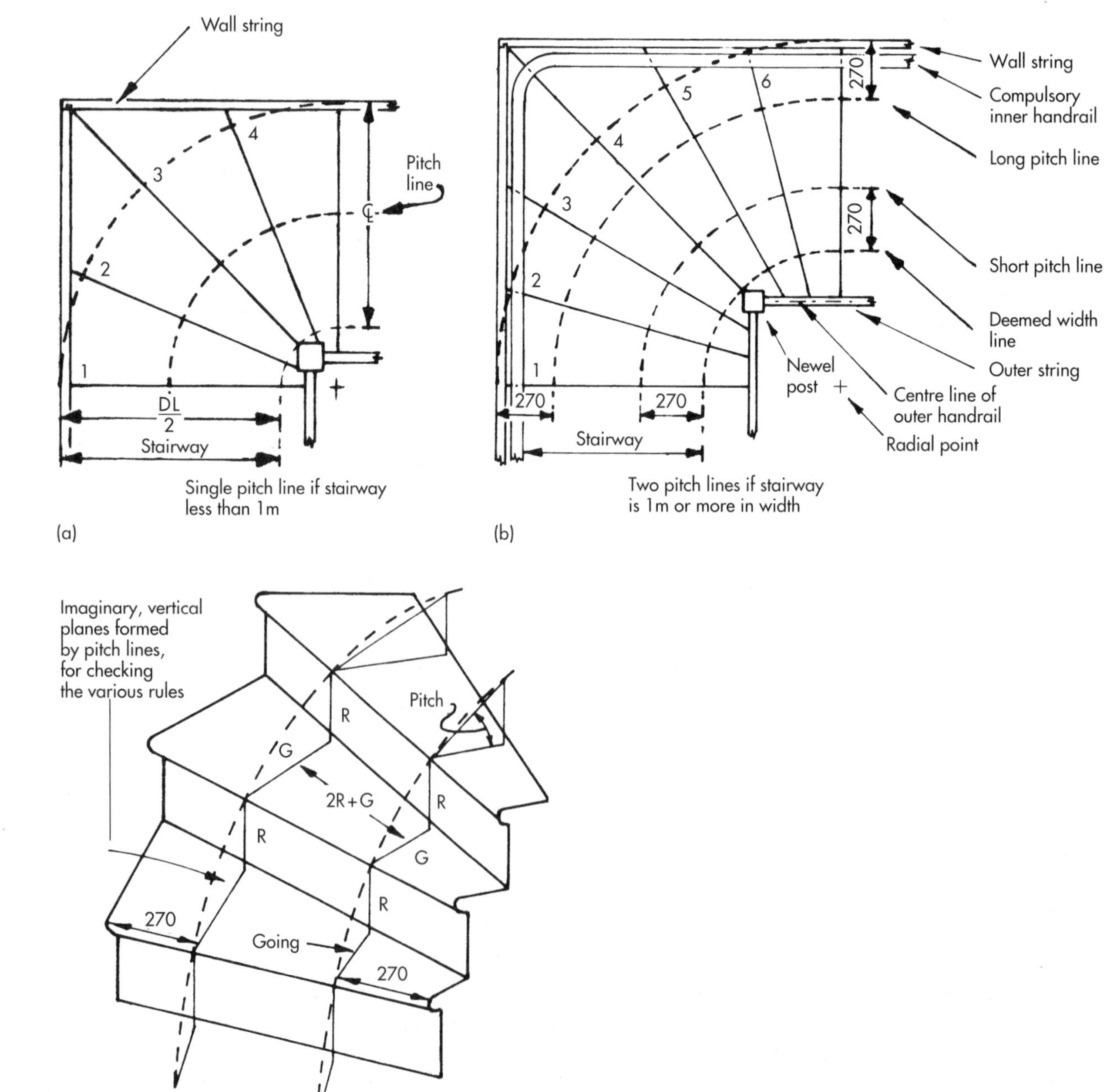

Figure 11.20 (a) Single pitch line. (b) Two pitch lines if stairway is 1 m or more in width. (c) Imaginary vertical planes

right angles to a nosing in relation to the nosing above. Where consecutive tapered treads are used, a uniform going should be maintained. Where a stair consists of straight *and* tapered treads, the going of the tapered treads should not be less than the going of the straight flight.

Note that BS 585: *Wood stairs* Part 1: 1989 *Specification for stairs with closed risers for domestic use, including straight and winder flights and quarter or half landings* is given in AD K1 as a British Standard which will offer reasonable safety in the design of stairs.

Spiral and Helical Stairs

It is further recommended in AD K1 that stairs designed in accordance with BS 5395: *Stairs, ladders and walkways* Part 2: 1984 *Code of Practice for the design of helical and spiral stairs*, will be adequate. Stairs with goings less than shown in this standard may be considered in conversion work when space is limited and the stair does not serve more than one habitable room.

Alternating Tread Stairs

Figure 11.1: This type of stair is designed to save space and has alternate handed steps with part of the tread cut away; the user relies on familiarity from regular use for reasonable safety. Alternating tread stairs should only be installed in one or more straight flights for a loft conversion and then only when there is not enough space to accommodate a stair which satisfies the criteria already covered for *Private stairs*. An alternating tread stair should only be used for access to one habitable room, together with, if desired, a bathroom and/or a WC. This WC must not be the only one in the dwelling. Steps should be uniform with parallel nosings. The stair should have handrails on both sides and the treads should have slip-resistant surfaces. The tread sizes over the wider part of the step should have a maximum rise of 220 mm and a minimum going of 220 mm and should be constructed so that a 100 mm diameter sphere cannot pass through the open risers.

Fixed Ladders

A fixed ladder should have fixed handrails on both sides and should only be installd for access in a loft conversion – and then only when there is not enough space without alteration to the existing space to accommodate a stair which satisfies the criteria already covered for *Private stairs*. It should be used for access to only one habitable room. Retractable ladders are not acceptable for means of escape. For reference to this, see *Approved Document B: Fire safety*.

Handrails for Stairs

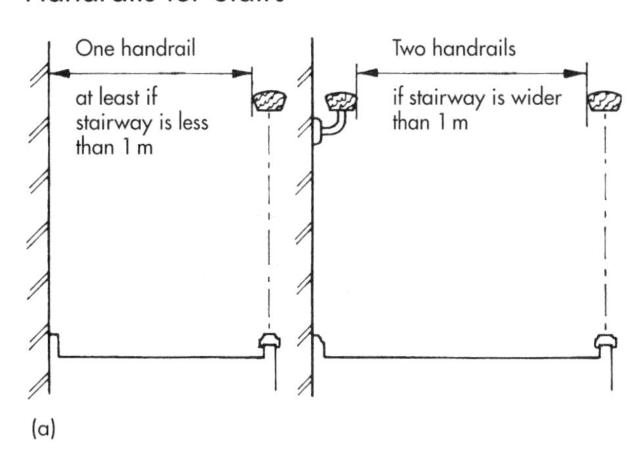

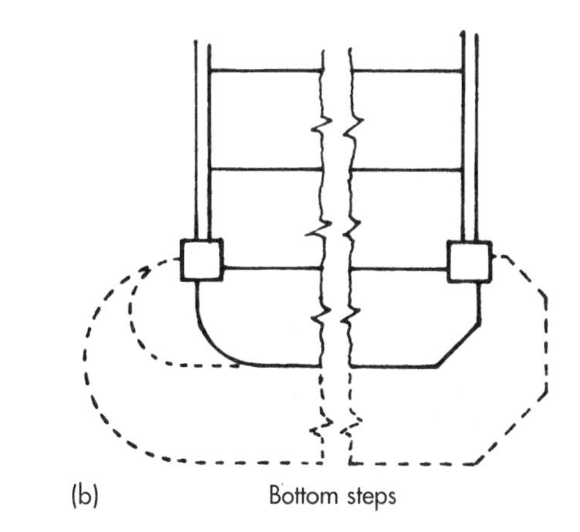

Figure 11.21(a) Handrails for stairs. (b) Bottom steps

Figure 11.21: Stairs should have a handrail on at least one side if they are less than 1 m wide and should have a handrail on both sides if they are wider. Handrails should be provided beside the two bottom steps in public buildings and where stairs are intended to be used by people with disabilities. In other places, handrails need not be provided beside the two bottom steps (Figure 11.21(b)).

Handrail Heights

Figure 11.22: In all buildings, handrail heights should be between 900 mm and 1000 mm, measured to the top of the handrail from the pitch line or floor. Handrails can form the top of a guarding, if the heights can be matched.

Guarding of Stairs

Figure 11.2: As illustrated, flights and landings should be guarded at the sides:

- in dwellings when there is a drop of more than 600 mm;
- in other buildings when there are two or more risers.

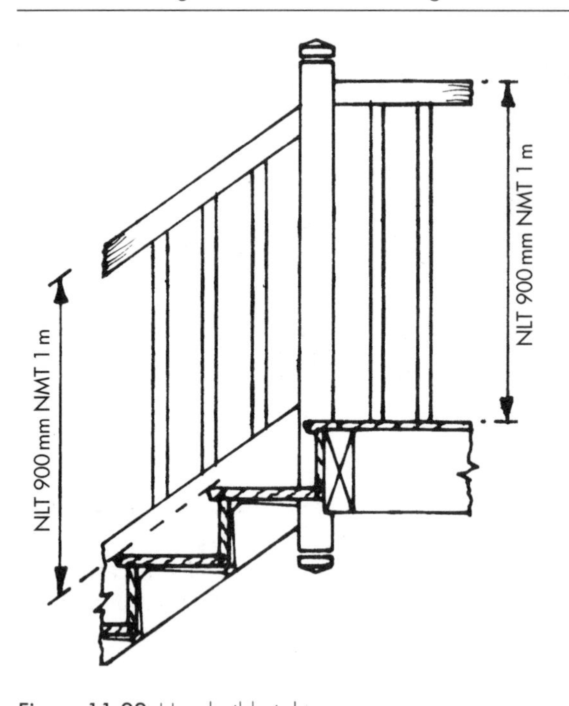

Figure 11.22 Handrail heights

The guarding to a flight should prevent children being held fast by the guarding, except on stairs in a building which is not likely to be used by children under 5 years. In the first case, the construction should be such that

- a 100 mm diameter sphere cannot pass through any openings in the guarding;
- children will not readily be able to climb the guarding (this in effect means that horizontal ranch-style balustrades used in recent years should not now be used).

The height of the guarding is given in Figure 11.22 for single-family dwellings in category 1. External balconies should have a guarding of 1100 mm.

12

Constructing Traditional and Modern Roofs

12.1 INTRODUCTION

Roofing in its entirety is an enormous subject, but the practical issues that involve the first-fixing carpenter on the most common types of dwelling-house roof – dealt with here – are less formidable.

12.1.1 Types of Roof

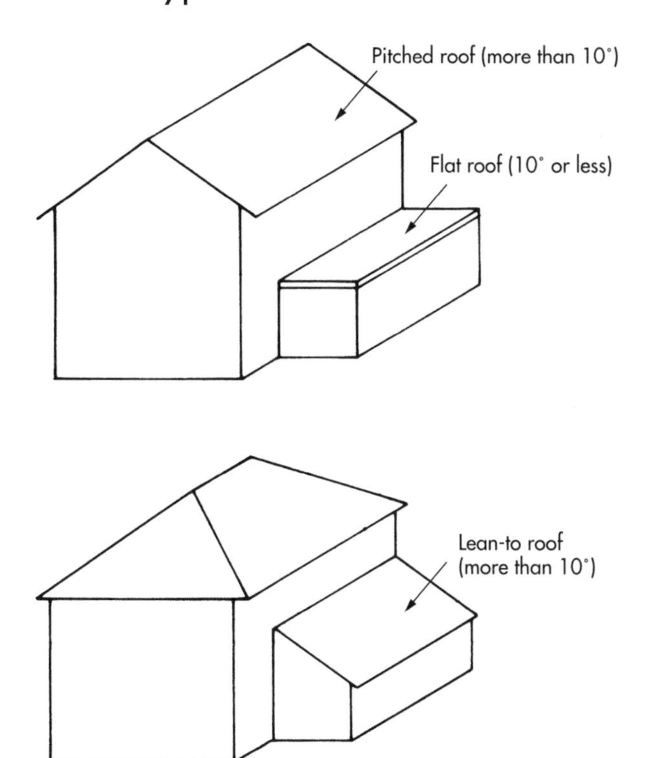

Figure 12.1 Types of roof

Figure 12.1: The three types of roof to contend with are *flat*, *lean-to* and *pitched*. Timber is used for the skeleton structure to form a carcase and this is covered with various impervious materials. Flat roofs are usually covered with sheet-boarding and (for economy) bituminous felt in built-up layers. Pitched and lean-to roofs are covered with quarried slates (which are very expensive) or non-asbestos-cement slates, blue/black in colour, a cheaper lookalike, clay tiles (also expensive) or, more commonly, concrete tiles on battens and roofing (sarking) felt.

12.1.2 Knowledge Required

Roofing carpenters nowadays, ideally, require a knowledge of both modern *and* traditional roof construction – and a sound knowledge of at least one of the various methods used for finding lengths and bevels of roofing members. The various methods dealt with later include geometry (to develop an understanding of finding bevels and lengths, but not meant to be used in practice), steel roofing square or metric rafter square, the Roofing Ready Reckoner and last, but not least, the Roofmaster, a revolutionary device invented recently by Kevin Hodger, a carpentry lecturer at Hastings College.

12.1.3 Traditional Roofing

This consists basically of rafters pitched up from wall plates to a ridge board, such rafters being supported by purlins and struts which transfer the load to either straining pieces or binders and ceiling joists to an internal load-bearing wall.

12.1.4 Modern Roofing

Modern roofing, using trussed-rafter assemblies, often uses smaller sectional-sized timbers and normally only requires to be supported at the ends. This frees the designer from the need to provide intermediate load-bearing walls and dispenses with purlins and ridge boards. Other differences are that traditional roofing is usually an entire site operation, whereas trussed-rafter roofing involves prefabrication under factory conditions and delivery of assembled units to the site for a much-reduced site-fixing operation.

12.2 BASIC ROOF DESIGNS

12.2.1 Introduction

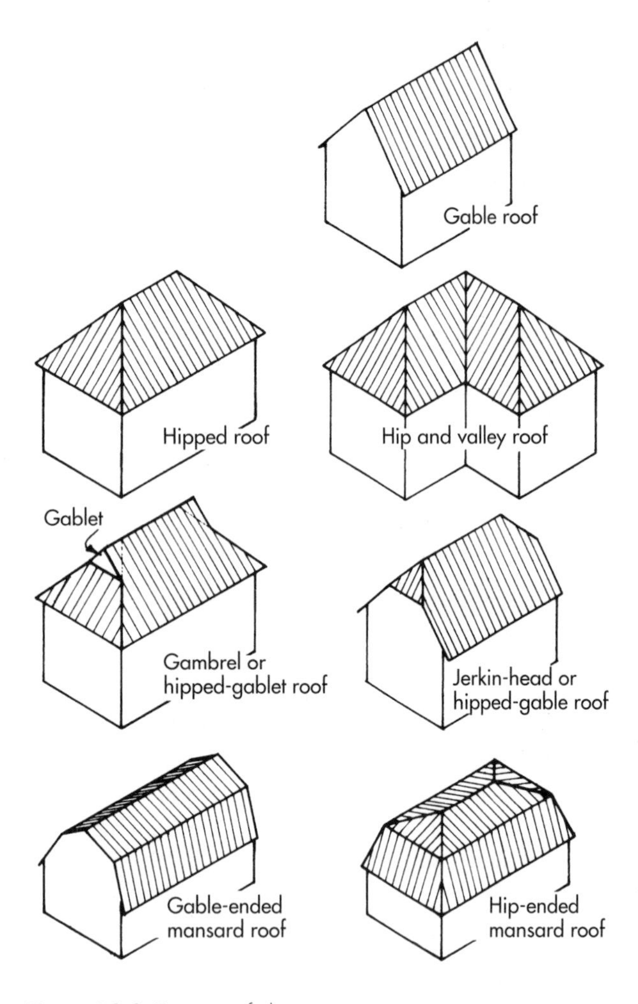

Figure 12.2 Basic roof designs

Figure 12.2: Traditional pitched roofs, by virtue of having been in existence for centuries, predominantly outnumber modern roofs and therefore can be easily spotted for reference and comparison. Their design was often used to advantage in complementing and enhancing the beauty of a dwelling and variations on basic designs can be seen to be infinitely variable. Still occasionally required on individual, one-off dwellings being built, the main features of traditional roofs are described below.

12.2.2 Gable Roof

This design is now widely used in modern roofing because of its simplicity and therefore relatively lower cost. As illustrated, triangular ends of the roof are formed by the outer walls, known as gable ends. Traditionally, purlins took their end-bearings from these walls.

12.2.3 Hipped (or Hip-ended) Roof

This design is also used in modern roofing, but to a lesser extent. The hipped ends are normally the same pitch as the main roof and therefore each hip is 45° in plan. In traditional roofing, the purlins are continuous around the roof and all the old roofing skills are needed. In modern roofing, the hip-ends are either constructed traditionally (by cut and pitch methods) or by using hip trusses supplied with the main trusses; either way, a method for finding lengths and bevels will be required at this stage.

12.2.4 Hip and Valley Roof

As illustrated in Figure 12.2, valleys occur when roofs change direction to cover offshoot buildings.

12.2.5 Gambrel Roof and Jerkin-head Roof

As illustrated, these roofs include small design innovations to the basic hipped and gable roofs. Gambrel roofs can be built traditionally as normal hipped roofs with the full-length hips running through, under the gablets, and the ridge board protruding each end to accommodate the short cripple rafters – and jerkin-head roofs simply have shorter hips. Both types can be built with modern trussed-rafter assemblies.

12.2.6 Mansard Roofs

Apart from their individual appearance being a reason for using such a roof, the lower, steeper roof-slopes, which were vertically studded on the inside, acted as walls and accommodated habitable rooms in the roof space. The upper, shallower roof slopes had horizontal ceiling joists acting as ties, giving triangular support to the otherwise weak structure. Technically, these roofs usually incorporated king-post trusses superimposed on queen-post trusses. This design of roof has been built using modern techniques, especially involving steel beams, taking their bearings from gable ends.

12.3 ROOF COMPONENTS AND TERMINOLOGY

12.3.1 Wall Plates

Figure 12.3(a): 100 × 75 mm or 100 × 50 mm sawn timber bearing plates, are laid flat and bedded on mortar to a level position, flush to the inside of the inner wall and running along the wall to carry the feet of all the

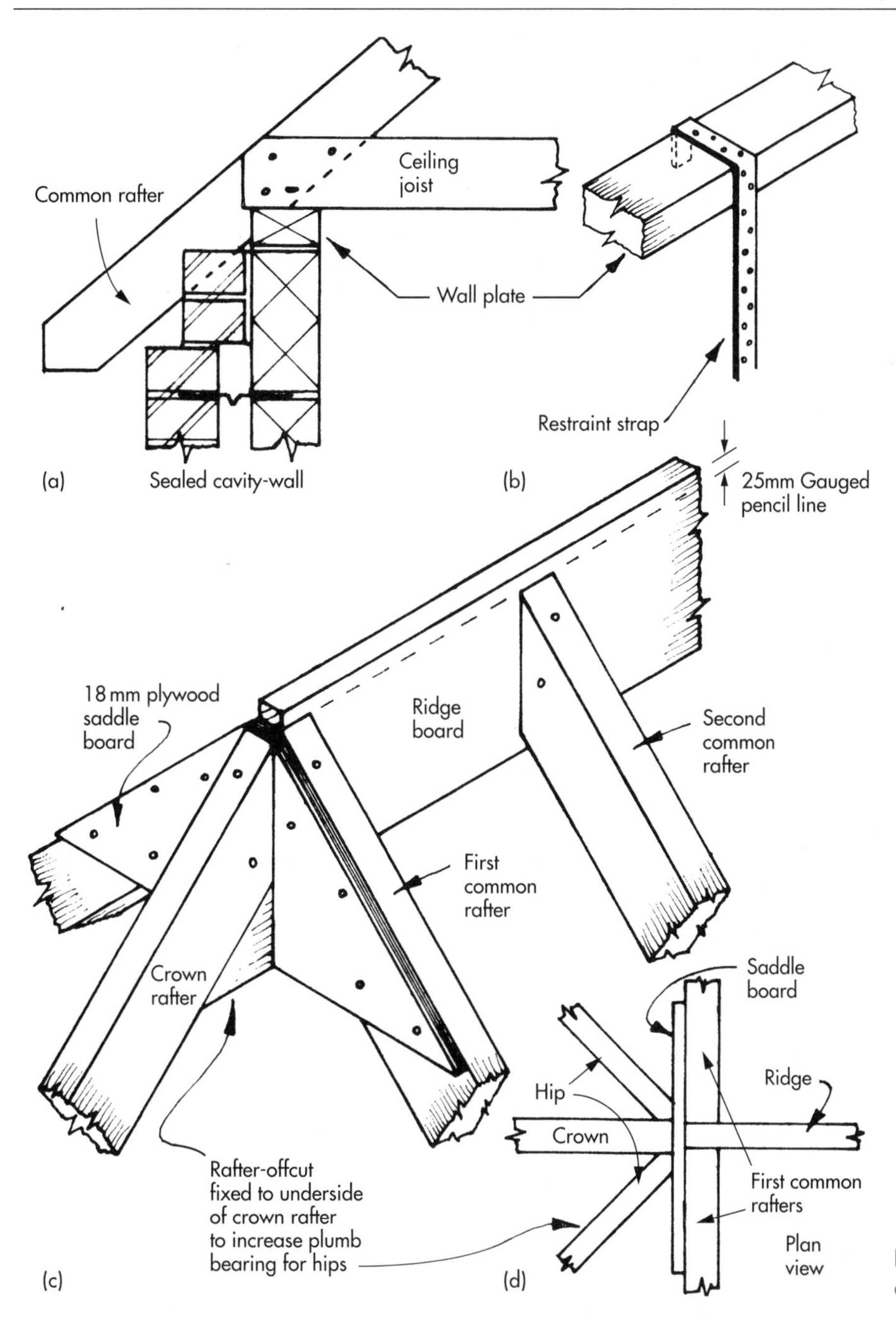

Figure 12.3(a)–(d) Principal roof components

rafters and the ends of the ceiling joists. Nowadays, the wall plates must be anchored down with restraint straps.

12.3.2 Restraint Straps

Figure 12.3(b): Vertical, galvanized steel straps, 2.5 mm thick, 30 mm wide, with 6 mm holes at 15 mm offset centres and lengths up to 1.5 m, are fixed over the wall plates and down the inside face of the inner skin of blockwork at maximum 2 m intervals. Additional straps should be used to reinforce any half-lap wall plate joint. Horizontal straps of 5 mm thickness are used across the ceiling joists and rafters, to anchor the gable-end walls. They should bridge across at least three of these structural timbers and have noggings and end packing fixed between the bridged spaces. These noggings should be at least 38 mm on the face-side by half the depth of joist or rafter.

12.3.3 Ceiling Joists

Figure 12.3(a): Like floor joists, these should span the shortest distance, rest on and be fixed to the wall plates, as well as to the foot of the rafter on each side – thereby acting as an important tie and also providing a skeleton structure for the underside-boarding of the ceiling. Usually, 100 × 50 mm sawn timbers are used.

12.3.4 Common Rafters

Figures 12.3(a) and (c): Again, 100 × 50 mm sawn timber is normally used for these load-bearing ribs that pitch up from the wall plate on each side of the roof span, to rest opposite each other and be fixed to the wall plate at the bottom and against the ridge board at the top.

12.3.5 Ridge Board

Figure 12.3(c): This is the spine of the structure at the apex, running horizontally on edge in the form of a sawn board of about 175 × 32 mm section (deeper on steep roofs, depending upon the depth of the common-rafter splay cut + 25 mm allowance) against which the rafters are fixed.

12.3.6 Saddle Board

Figures 12.3(c) and (d): The saddle board is a purpose-made triangular board (usually of 18 mm WBP exterior plywood), like a gusset plate, fixed at the end of the ridge board and to the face of the first pair of common rafters. It supports the hips and crown rafter of a hipped end.

12.3.7 Crown or Pin Rafter

Figures 12.3(c) and (d): This is the central rafter of a hipped end.

12.3.8 Alternative Hip-arrangement

This is shown in plan and isometric views in Figure 12.4.

12.3.9 Terminology

Figure 12.4(b): This isometric view of a single-line roof carcase shows the majority of components in relation to each other and defines such terminology as *eaves*, *verge*, *gable end*, etc.

12.3.10 Angle Tie

Figure 12.5(a): Sometimes a piece of 75 × 50 mm or 50 × 50 mm timber, acting as a corner tie across the wall plates, replaces the traditional and elaborate dragon-tie beam used to counteract the thrust of the hip rafter. Alternatively, a simplified, modern, metal dragon-tie (Figure 12.5(b)) may be used. Either one of these restraints to the hip is recommended, especially on larger-than-normal roofs.

12.3.11 Hip Rafters

Figure 12.5(a): Similar to ridge boards, hip rafters pitch up from the wall-plate corners of a hipped end to the saddle board at the end of the ridge. They act as a spine for the location and fixing of the jack rafter heads.

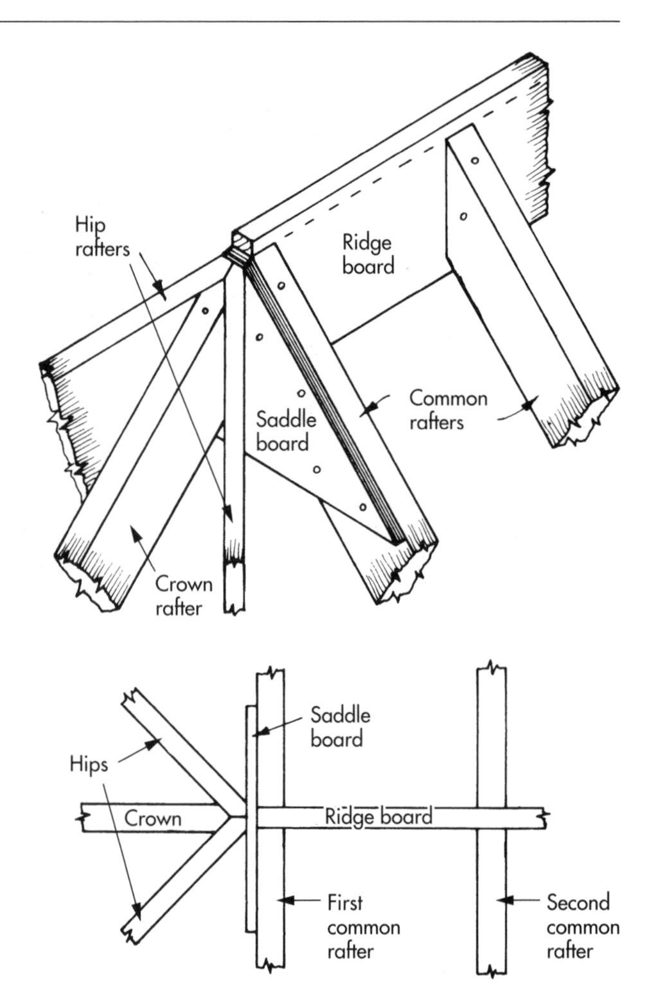

Figure 12.4(a) Alternative hip-arrangement

12.3.12 Jack Rafters

Figure 12.5(a): These are rafters with a double (compound) splay-cut at the head, fixed in diminishing pairs on each side of the hip rafters.

12.3.13 Valley Rafters

Figure 12.5(c): Valley rafters are like hip rafters, but form an internal angle in the roof-formation and act as a spine for the location and fixing of the cripple rafters. Sectional sizes are similar to ridge board and hip rafters.

12.3.14 Cripple Rafters

Figure 12.5(c): These are pairs of rafters, diminishing like jack rafters, spanning from ridge boards to valley rafter and gaining their name traditionally by being cut off at the foot.

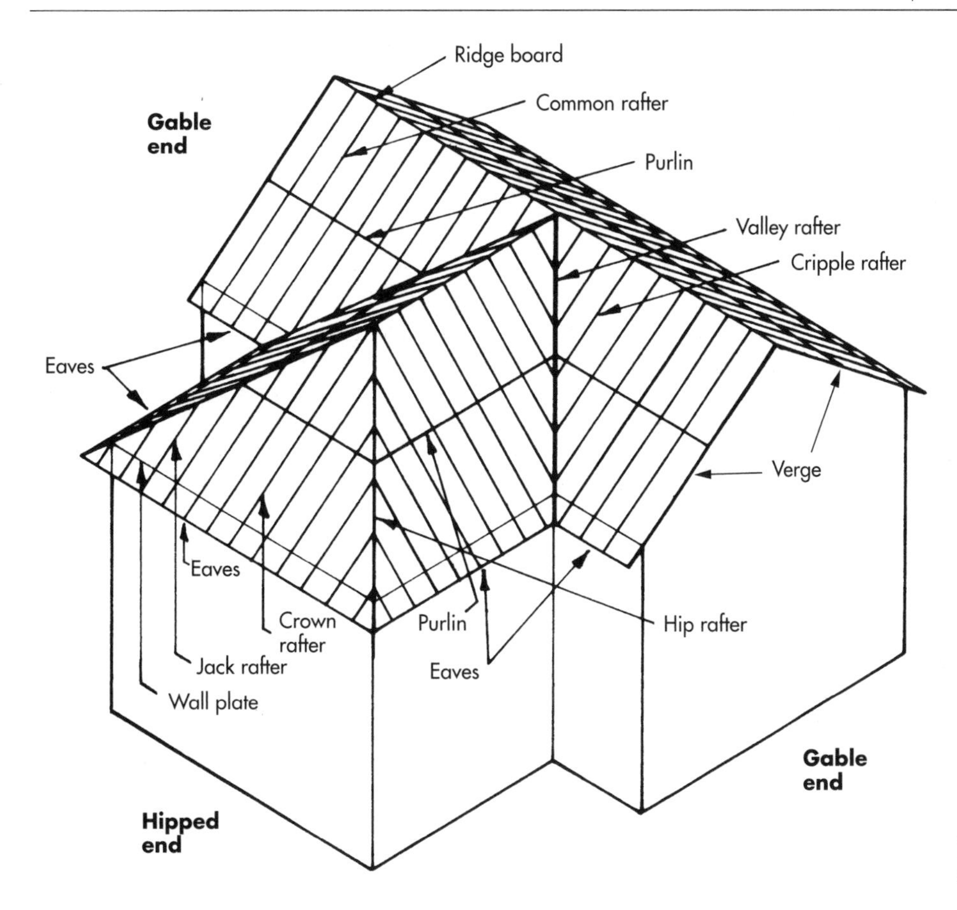

Figure 12.4(b) Isometric view of single-line roof carcase

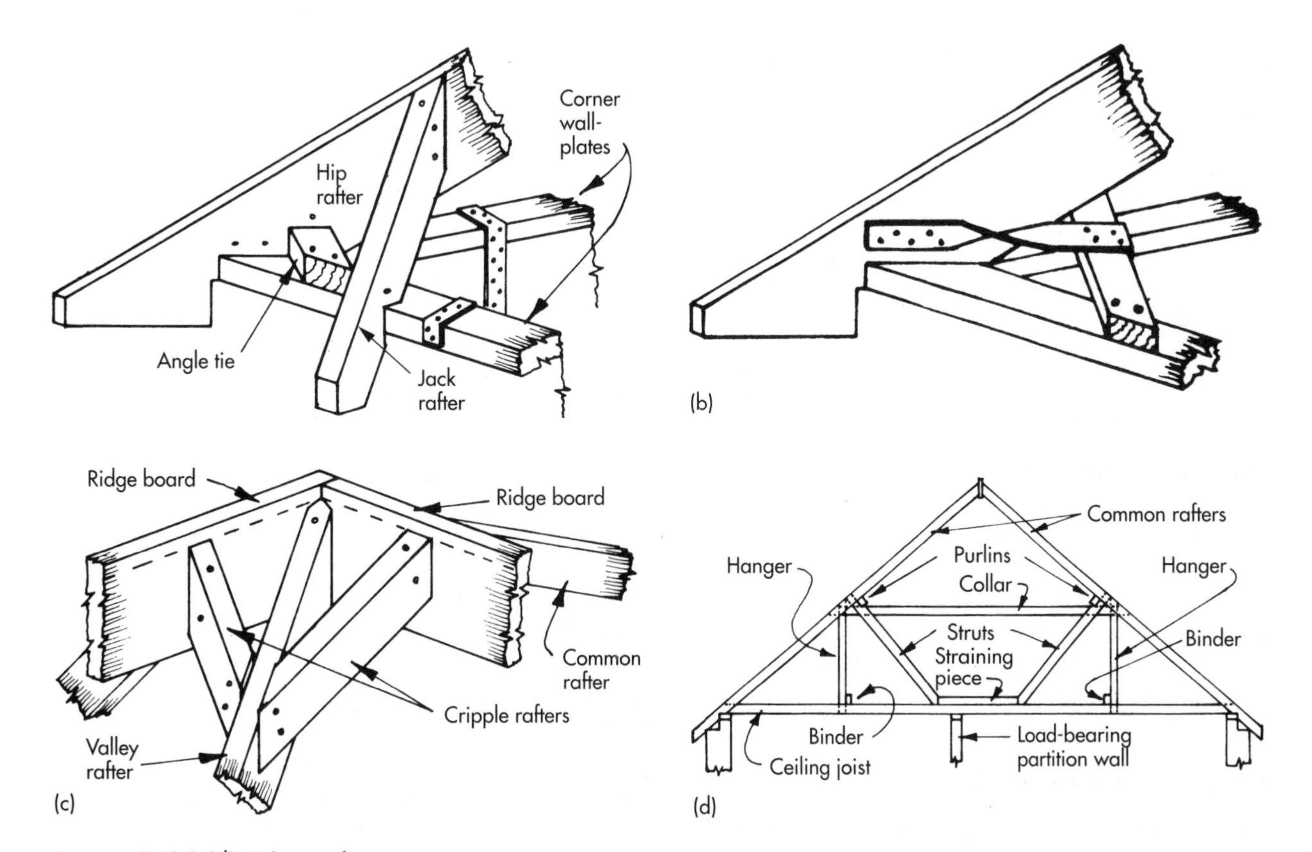

Figure 12.5(a)–(d) Other roof components

12.3.15 Purlins

Figure 12.5(d): Purlins are horizontal beams, about 100–150 × 75 mm sawn, that support the rafters midway between the ridge and the wall plate, when the rafters exceed 2.5 m in length.

12.3.16 Struts

Figure 12.5(d): These are 100 × 50 mm sawn timbers that support the purlins at about every fourth or fifth pair of rafters. This arrangement transfers the roof load to the ceiling joists and, therefore, requires a load-bearing wall or partition at right-angles to the joists and somewhere near the mid-span below.

12.3.17 Straining Pieces

Figure 12.5(d): These are basically sole plates of 100 × 50 mm or 100 × 75 mm section, fixed to the ceiling joists between the base of the struts. They support and balance the roof thrust and allow the struts to be set at 90° to the roof slope.

12.3.18 Collars

Figure 12.5(d): Collars are 100 × 50 mm sawn ties, fixed to each rafter, sometimes used at purlin level to give extra resistance to the roof-spread.

12.3.19 Binders

Figure 12.5(d): These are 100 × 50 mm timbers fixed on edge in the roof space with skew-nails or modern framing anchors. They are set at right-angles to the ceiling joists, to give support and counteract deflection of the joists if the span exceeds 2.5 m.

12.3.20 Hangers

Figures 12.5(d) and 12.6(a): These are 100 × 50 mm ties that hang vertically from a rafter side-fixing position near the purlin, to a side-fixing position on the ceiling joist and the binder, usually close to the struts.

12.3.21 Roof Trap or Hatch

Figure 12.6(a): The roof trap is a trimmed and lined opening in the ceiling joists, with a hinged or loose trap door. It provides access to the roof void for maintenance of storage tank and pipes, etc.

12.3.22 Cat Walk

Figure 12.6(a): This consists of one or two 100–150 × 25 mm sawn boards fixed across the ceiling joists from the edges of the trap, to provide an access path through the roof (usually to the storage tank). A line of chipboard floor-panels could be used as a modern alternative.

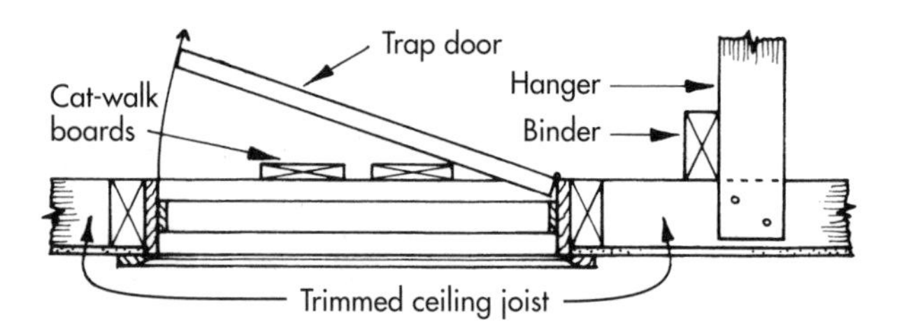

Figure 12.6(a) Section through roof trap

12.3.23 Lay Boards

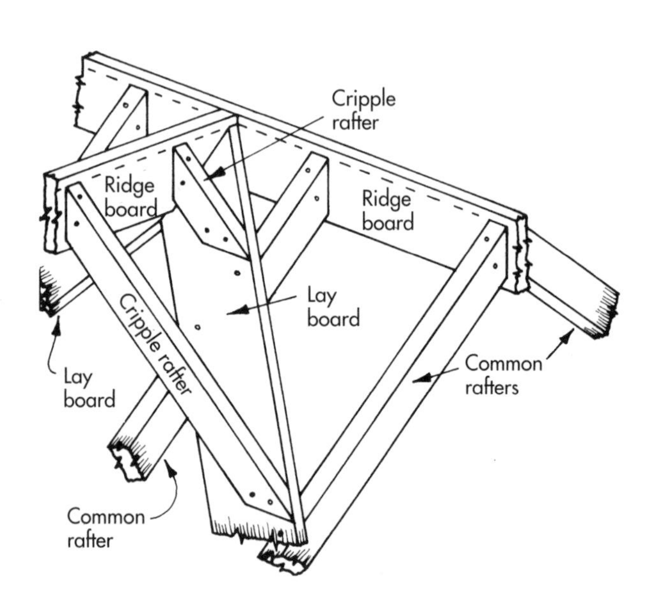

Figure 12.6(b) Lay boards

Figure 12.6(b): These are location sole-plates of about 175 × 25 mm sawn section, laid flat and diagonally on the ribbed roof structure, to receive the splay-cut feet of cripple rafters forming a valley. This is a popular alternative to using valley rafters, as it is stronger and involves less work.

12.3.24 Eaves

Figure 12.6(c): This is the lowest edge of the sloping roof, which usually overhangs the structure from as little as the fascia-board thickness up to about 450 mm. This is measured horizontally and is known as the *eaves' projection*, which may be open (showing the ends of the rafters on the underside, as a feature) or closed (by the addition of a soffit board).

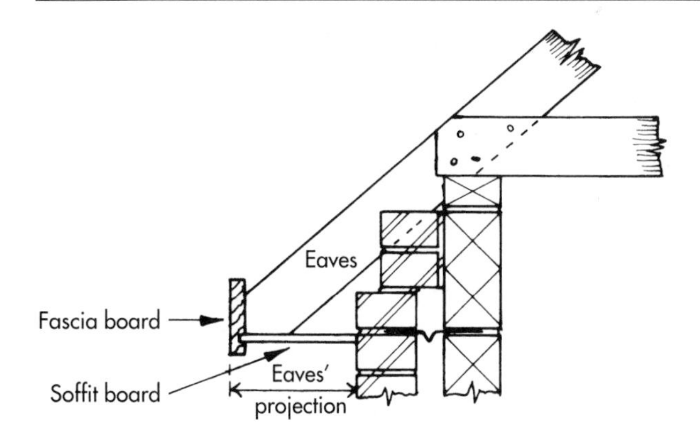

Figure 12.6(c) Fascia, soffit and eaves

12.3.25 Fascia Board

Figure 12.6(c): The fascia is a prepared board of about ex. 175 × 25 mm, fixed with 75 or 100 mm cut, clasp or oval nails to the plumb cuts of the rafters at the eaves. It provides a visual finish, part of the closing-in at the eaves, and a fixing board for the guttering.

12.3.26 Soffit Board

Figure 12.6(c): This can be fixed to cradling brackets on the underside of closed eaves, as illustrated separately in Figure 12.7(a), between a groove in the fascia board and the wall. This board can be of 9–12 mm WBP exterior plywood or non-asbestos fibreboard. The soffit of the eaves must provide cross-ventilation, equivalent in area to a continuous gap of 10 mm width along each side of the roof.

12.3.27 Cradling

Figure 12.7(a): Cradling is a traditional way of providing wall-side fixing points for the soffit board. Purpose-made, L-shaped brackets made up from 50 × 25 mm sawn battens, with simple half-lap corner joints clench-nailed together, are fixed to the sides of the rafters with 63 mm round-head wire nails.

12.3.28 Sprocket Pieces

Figures 12.7(a) and (b): These are long wedge-shaped pieces of ex. rafter material, fixed on top of each rafter at the eaves to create a bell-shape appearance or upward tilt to the roof slope. As shown in Figure 12.7(b), this is also achieved by fixing offcuts of rafter to the rafter sides. Apart from aesthetic reasons, this is done to reduce a steep roof slope, in order to ease the flow of rainwater into the guttering.

12.3.29 Tilting Fillets

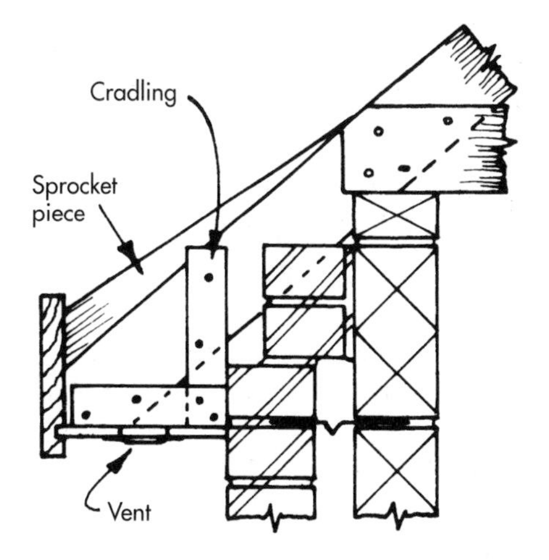

Figure 12.7(a) Cradling

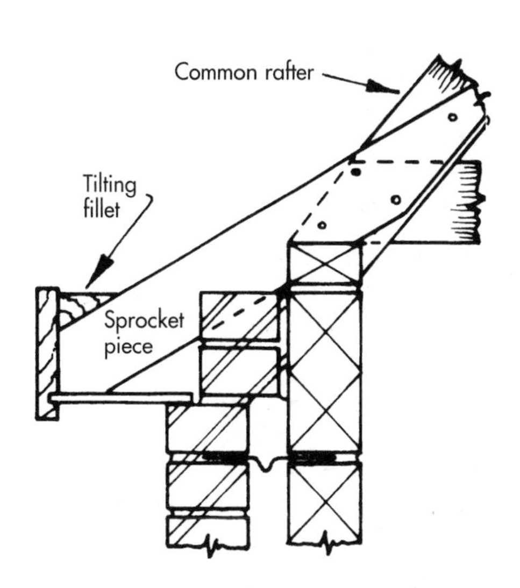

Figure 12.7(b) Sprocket piece and tilting fillet

Figures 12.7(b) and (c): Tilting fillets are timber battens of triangular-shaped cross-section fixed behind the top of a raised fascia board to give it support. The fascia board has been raised primarily to tilt and close the double-layered edge of the tiles or slates, but the increased depth of board also helps the plumber create the necessary *falls* in the guttering. Tilting fillets are also used in other places, such as on the top edges of valley boards (Figure 12.7(c)) and back gutters, etc.

12.3.30 Valley Boards

Figure 12.7(c): Traditionally, 25 mm sawn boards were used to form a gutter in the valley recess, each board being about 225 mm wide. A tilting fillet was fixed at the top edge of each board, ready to be included in the lining of the valley with sheet lead.

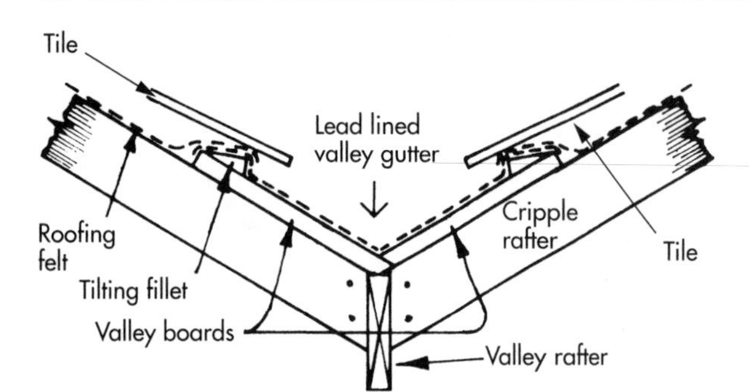

Figure 12.7(c) Valley gutter

12.3.31 Glass-reinforced Plastic Valleys

Modern valley linings are now manufactured to a pre-formed shape, using glass-reinforced plastic (GRP). The moulded shape includes a double-roll on each edge to simulate the weathering-upstand of a traditional tilting fillet. These edges should rest on continuous tiling battens, fixed to the cripple rafters; valley boards are not needed.

12.3.32 Verge

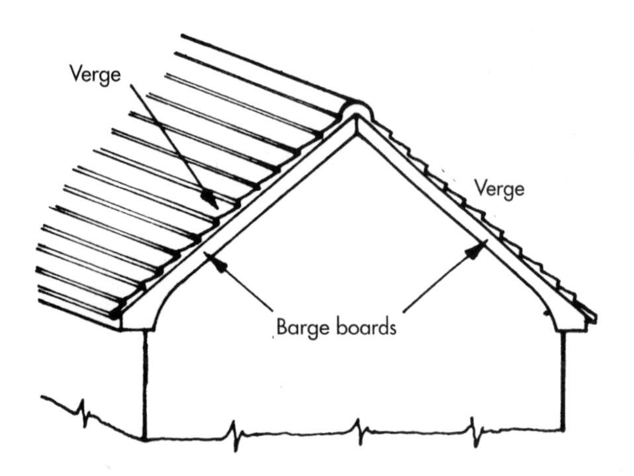

Figure 12.7(d) Verge and barge boards

Figure 12.7(d): This is the edge of a roof on a gable end. It may have a minimum tile-projection and no barge boards, or a greater projection (about 150–200 mm) with barge boards and soffit boards to the sloping underside.

12.3.33 Barge Boards

Figure 12.7(d): These are really fascia boards inclined like a pair of rafters and fixed to a small projection of roof at the gable-end verges. When projecting from the wall, a soffit board will be required, involving a boxed-shape at the eaves on each side.

12.3.34 Tile or Slate Battens

These are usually 38 × 18 mm, but can be 38 × 25 or 50 × 25 mm sawn, tantalized (pressurized preservative treatment) battens fixed at gauged spacings on top of the lapped roofing felt. The fixing of these battens is normally done by the slater and tiler, not the carpenter.

12.3.35 Framing Anchors

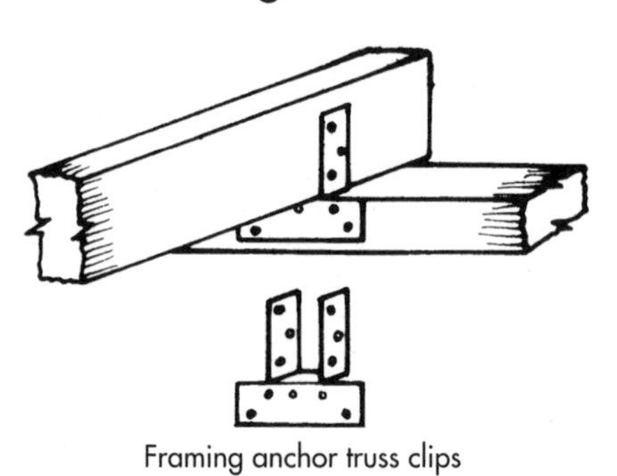

Figure 12.7(e) Framing anchors

Figure 12.7(e): Galvanized steel framing anchors of various designs for various situations may be used to replace traditionally nailed fixings in certain places. The one shown here replaces skew-nailing of the ceiling joist to the wall plate. The recommended fixings to be used (in *every* hole of the framing anchor) are 3 mm in diameter by 30 mm long sherardized clout nails.

12.4 BASIC SETTING-OUT TERMS

12.4.1 Introduction

All the bevels and lengths in roofing can be worked out by various methods – all of which are based on the principles of geometry. Although the actual method of using drawing-board geometry is not practical in a site situation, it is given in Figure 12.9 as an introduction to understanding how the various bevels and lengths are found.

First, as illustrated in Figure 12.8, the following basic setting-out terms must be appreciated in relation to the sectional view through the pitched roof.

12.4.2 Span

This is an important distance measured in the direction of the ceiling joists at wall-plate level, from the outside

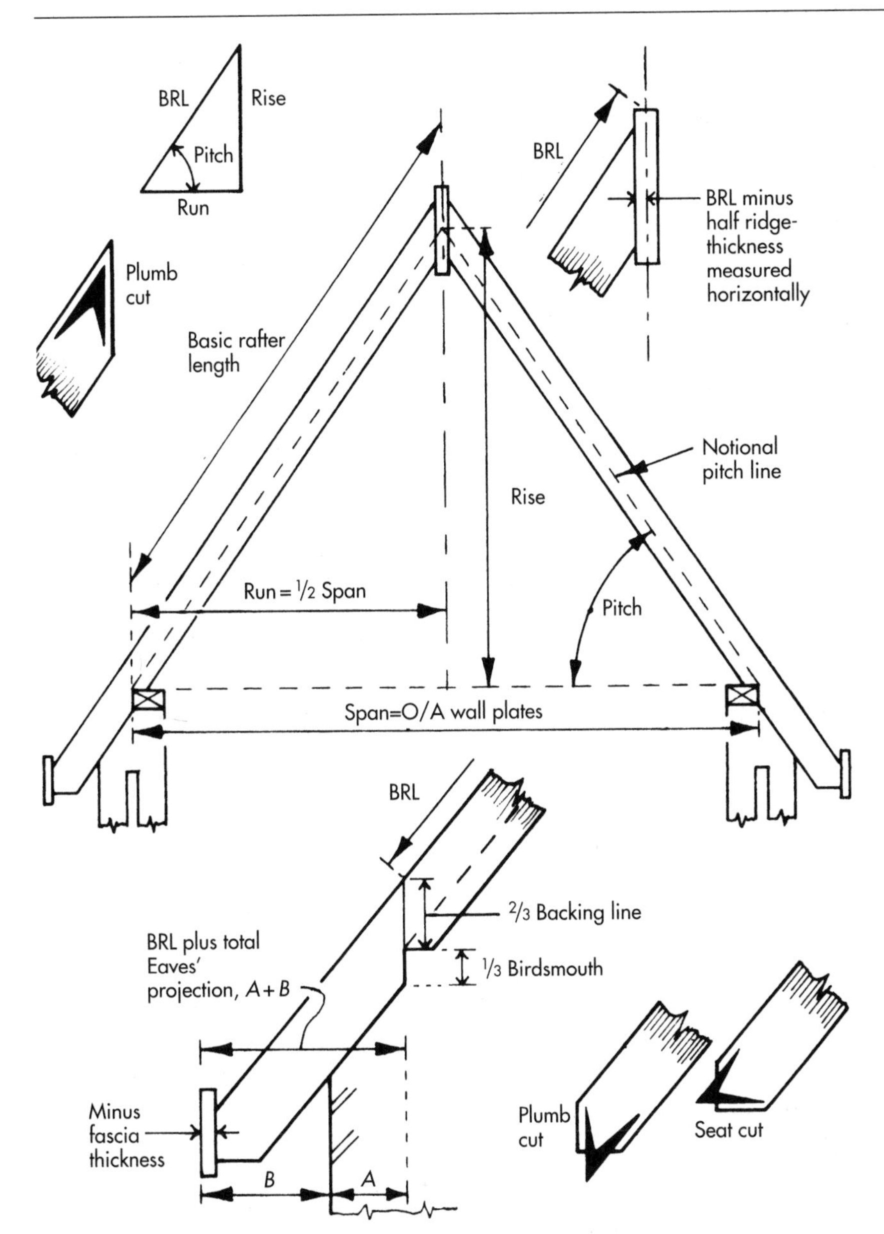

Figure 12.8 Basic setting-out terms (BRL = basic rafter length)

of one wall plate to the outside of the other, i.e. overall (O/A) wall plates.

12.4.3 Run

For the purpose of reducing the isosceles roof-shape to a right-angled triangle with a measureable base-line, the span measurement is divided by two to produce what is known as the *run*.

12.4.4 Rise

This represents the perpendicular of the triangle, measured from wall-plate level up to the apex of the imaginary hypotenuse or notional pitch lines running at two-thirds rafter-depth, through the sides of the rafters, from the outside arris of the wall plates.

12.4.5 Pitch

This is the degree angle of the roof slope. The known rise and the run gives the pitch angle and the basic rafter length – or the known pitch angle and the run gives the rise and the basic rafter length.

12.4.6 Backing Line

This is an important plumb line marked at the base of the setting-out rafter (pattern rafter), marked down two-thirds of its depth to the top of the birdsmouth cut, acting as a datum or reference point for the rafter's length on one side and the total eaves' projection on the other.

12.4.7 Birdsmouth

This is a notch cut out of the rafter to form a seating on the outside edge of the wall plate.

12.5 GEOMETRICAL SETTING-OUT OF A HIPPED ROOF

12.5.1 Bevels and Length of Common Rafter

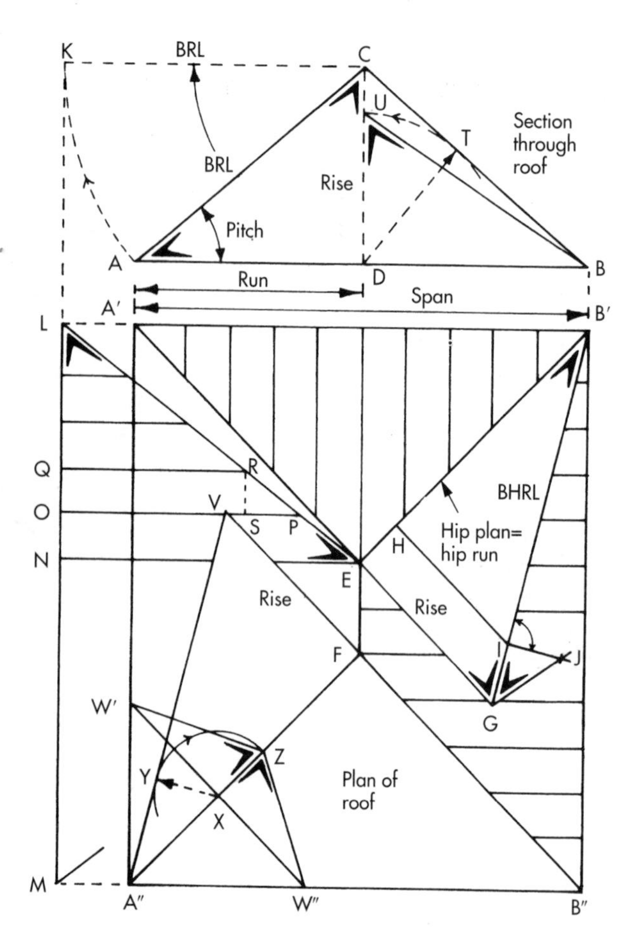

Figure 12.9 Geometry for hipped roof

Figue 12.9: Draw triangle ABC (section through roof) where: AB = span, AD = run (half span), CD = rise, angle DAC = seat cut, angle ACD = plumb cut, line AC or CB = basic rafter length (BRL).

12.5.2 Bevels and Length of Hip Rafter

Figure 12.9: Draw plan of roof, showing two hipped ends (these being always drawn at angles of 45° in equal pitched roofs), denoted as A′E, B′E and A″F, B″F. Now, at right angles to B′E, draw line EG, equal in length to the rise at CD. Join G to B′: angle EB′G = seat cut, angle B′GE = plumb cut, line B′G = basic hip rafter length (BHRL).

12.5.3 Hip Edge Cut

Figure 12.9: Draw line HI, parallel to EG at any distance from E; then draw line IJ, equal to distance EH, at right-angles to B′I. Draw line JG: angle JGI is the edge cut.

12.5.4 Jack (and Cripple) Rafter Bevels and Lengths

Figure 12.9: With radius CA, equal to basic rafter length, describe an arc from A to K. Project K down vertically to form line LM. Join L to E to give an elevated, true shape of roof side. Draw single lines to represent rafters at 400 mm (scaled) centres: angle LEN = jack/cripple edge cut. (The jack/cripple side cut is the same as the common rafter plumb cut.) Line OP = the basic length of the first jack rafter; line QR = the second diminishing jack rafter.

12.5.5 Diminish of Jack (and Cripple) Rafter Lengths

Figure 12.9: The perpendicular line SP, of the right-angled triangle RSP, is equal to the constant diminish of the jack (or cripple) rafters.

12.5.6 Purlin Bevels

Figure 12.9: Angle QLR, on the elevated plan view = purlin edge cut. The purlin side cut is developed within the sectional view through the roof. With compass at D and CB as a tangent at T, describe an arc to cut CD at U, and join U to B: angle DUB = purlin side cut.

12.5.7 Dihedral Angle or Backing Bevel

Figure 12.9: This is for the top edge of hip boards and might only be used nowadays if the roof – on a less cost-conscious, quality job – were to be boarded or sheathed with 12 mm plywood, prior to felting, battening and slating or tiling. On the plan view, establish triangle A″VF, as at B′GE. Draw a 45° line at any point, marked W′W″. With compass at X and A″V as a tangent at Y, describe an arc to cut A″F at Z. Join Z to W′ and W″: angle W′ZX or XZW″ is the required backing bevel.

12.6 ROOFING READY RECKONER

12.6.1 Introduction

The first practical method to be considered for finding the bevels and lengths in roofing, is by reference to a small limp-covered booklet entitled *Roofing Ready Reckoner* by Ralph Goss, published by Blackwell Scientific Ltd, ISBN 0632021969. The tables are given separately in metric and Imperial dimensions and are quite easy to follow, once a few basic principles have been grasped.

12.6.2 Choice of Tables

Figure 12.10: The tables cover a variety or roof pitches up to 75°, giving the various bevels required and the diminish for jack rafters. Basic rafter lengths (BRL) and basic hip rafter lengths (BHRL) must be worked out from the tables – which show a given measurement for the hypotenuse of the inclined rafter, in relation to base measurements of metres, decimetres and millimetres (or, in separate tables given, feet, inches and eighths of an inch) contained in the run of the common rafter.

12.6.3 Determining Run and Pitch

To use this method, first the span of the roof must be measured from the bedded wall-plates and halved to give the run. Also, the pitch must be known – and if not specified, should be taken, with the aid of a protractor, from the elevational drawings.

12.6.4 Example Workings Out

As an example, take a hipped roof of 36° pitch, with a span of 7.460 m. Halve this to give a run of 3.730 m. Then, referring to the tables taken from the booklet and illustrated here in Figure 12.10, work out the lengths of the common rafters and the hip rafters.

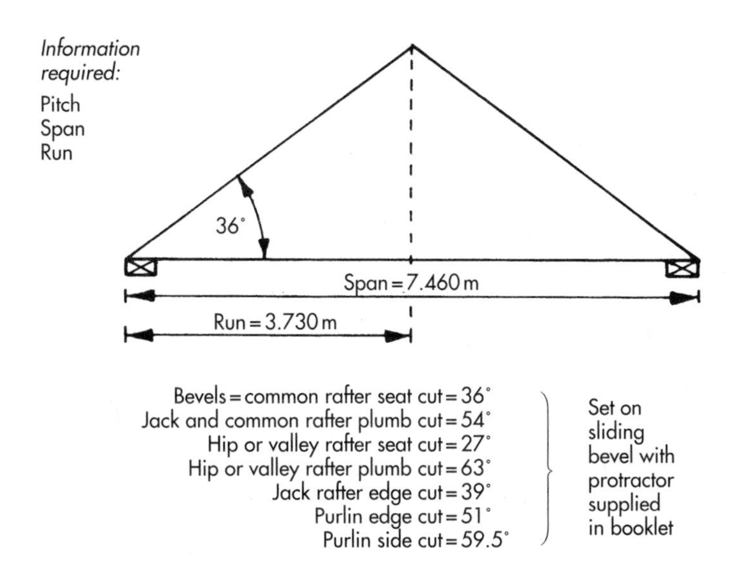

Table for 36° pitch

Run of rafter	0.1	0.2	0.3	0.4	0.5	0.6	0.7	0.8	0.9	1.0
Rafter length	0.124	0.247	0.371	0.494	0.618	0.742	0.865	0.989	1.112	1.236
Hip length	0.159	0.318	0.477	0.636	0.795	0.954	1.113	1.272	1.431	1.590

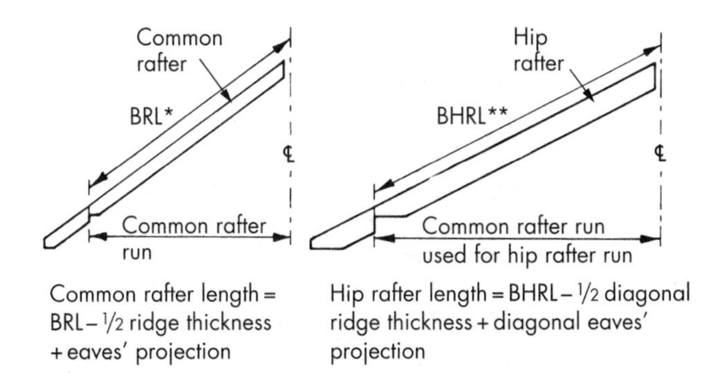

Figure 12.10 *Roofing Ready Reckoner*

Common Rafter

Length of common rafter for 1 m of run
= 1.236 m

Length of common rafter for 3 m of run
= 3.708 m

Length of common rafter for 0.7 m of run
= 0.865 m

Length of common rafter for 0.03 m of run
= 0.0371 m

Length of common rafter for 3.730 m of run
= 4.6101 m

Therefore the basic rafter length (BRL*) = 4.610 m

Speedier Working Out

It must be mentioned that a speedier mathematical method would be to multiply the total common rafter run by the 1 m run-of-rafter figure given in the booklet's tables, i.e.:

$3.730 \times 1.236 = 4.610\,\text{m} = \text{BRL}$

Hip Rafter

Length of hip rafter for 1 m of run = 1.590 m

Length of hip rafter for 3 m of run = 4.770 m
Length of hip rafter for 0.7 m of run = 1.113 m
Length of hip rafter for 0.03 m of run = 0.0477 m

Length of hip rafter for 3.730 m of run = 5.9307 m

Therefore, the basic hip rafter length (BHRL**) = 5.931m

Speedier Working Out

This involves multiplying the total common rafter run by the 1 m run-of-rafter figure given directly under that column for the length of hip, in the booklet's tables, i.e.

$3.730 \times 1.590 = 5.9307\,\text{m}$ (call this 5.931 m BHRL)

12.6.5 Jack Rafter Diminish

Figure 12.11: Jack and cripple rafters should diminish in length by a constant amount in relation to the length of the main rafter (crown or common). The *Roofing Ready Reckoner* gives a set of figures for each different pitch, shown on the same page as the bevels and tables, to deal with the decrease according to the spacing of the rafters. Assuming rafters to be spaced at 400 mm centres on the 36° pitched roof described in the text for Figure 12.10, then the diminish, as illustrated in Figure 12.11, would be 494 mm.

12.6.6 Imperial-dimensioned Tables

Figure 12.12: As a comparison – and perhaps as an alternative for any die-hards in the industry, still using feet and inches – the following example is given, using the imperial tables in the *Ready Reckoner*, expressed in

Information required: Rafter centres (spacings)

Jack rafters 400 mm centres decrease 494 mm
Jack rafters 500 mm centres decrease 618 mm
Jack rafters 600 mm centres decrease 742mm

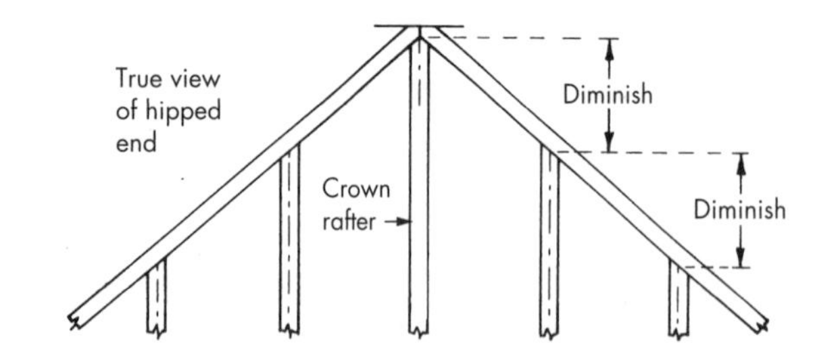

Figure 12.11 Jack rafter diminish for 36° pitch

Table for 36° pitch

Run of rafter (in)	$\frac{1}{2}$	1	2	3	4	5	6	7	8	9	10	11
Rafter length	$\frac{5}{8}$	$1\frac{1}{4}$	$2\frac{1}{2}$	$3\frac{3}{4}$	5	$6\frac{1}{8}$	$7\frac{3}{8}$	$8\frac{5}{8}$	$9\frac{7}{8}$	$11\frac{1}{8}$	$12\frac{3}{8}$	$13\frac{5}{8}$
Hip length	$\frac{3}{4}$	$1\frac{5}{8}$	$3\frac{1}{8}$	$4\frac{3}{4}$	$6\frac{3}{8}$	8	$9\frac{1}{2}$	$11\frac{1}{8}$	$12\frac{3}{4}$	$14\frac{1}{4}$	$15\frac{7}{8}$	$17\frac{1}{2}$

Run of rafter (ft)	1	2	3	4	5	6	7	8	9	10
Rafter length	$1\text{-}2\frac{7}{8}$	$2\text{-}5\frac{5}{8}$	$3\text{-}8\frac{1}{2}$	$4\text{-}11\frac{3}{8}$	$6\text{-}2\frac{1}{8}$	7-5	$8\text{-}7\frac{7}{8}$	$9\text{-}10\frac{5}{8}$	$11\text{-}1\frac{1}{2}$	$12\text{-}4\frac{3}{8}$
Hip length	$1\text{-}7\frac{1}{8}$	$3\text{-}2\frac{1}{8}$	$4\text{-}9\frac{1}{4}$	$6\text{-}4\frac{1}{4}$	$7\text{-}11\frac{3}{8}$	$9\text{-}6\frac{1}{2}$	$11\text{-}1\frac{1}{2}$	$12\text{-}8\frac{5}{8}$	$14\text{-}3\frac{5}{8}$	$15\text{-}10\frac{3}{4}$

Jack rafters 16 in centres decrease $19\frac{3}{4}$ in
Jack rafters 18in centres decrease $22\frac{1}{4}$ in
Jack rafters 24in centres decrease $29\frac{5}{8}$ in

Figure 12.12 Imperial tables

feet, inches and eighths of an inch. Again, take a hipped roof of 36° pitch, but with a span of 24 ft 5 in (24 feet, 5 inches). Halve this to give a run of 12 ft $2\frac{1}{2}$ in. Then, referring to the tables taken from the booklet and illustrated here, work out the lengths of the common rafters and hip rafters as follows:

Length of common rafter for 10 ft of run = 12 ft $4\frac{3}{8}$ in
Length of common rafter for 2 ft of run = 2 ft $5\frac{5}{8}$ in
Length of common rafter for 2 in of run = $2\frac{1}{2}$ in
Length of common rafter for $\frac{1}{2}$ in of run = $\frac{5}{8}$ in

Length of common rafter for 12 ft $2\frac{1}{2}$ in of run = 15 ft $1\frac{1}{8}$ in

Length of hip rafter for 10 ft of run = 15 ft $10\frac{3}{4}$ in
Length of hip rafter for 2 ft of run = 3 ft $2\frac{1}{8}$ in
Length of hip rafter for 2 in of run = $3\frac{1}{8}$ in
Length of hip rafter for $\frac{1}{2}$ in of run = $\frac{3}{4}$ in

Length of hip rafter for 12 ft $2\frac{1}{2}$ in run = 19 ft $4\frac{3}{4}$ in

12.6.7 Conclusion

It must be realized that, because of the involvement with fractions of an inch, the imperial method of working out lengths of rafters and hips, unlike the metric method, does not offer a speedier working out. However, it can be done by using inches as the basic units, providing the fractions of an inch are changed to decimals, whereby $\frac{1}{2}$ in = 0.5, $\frac{1}{4}$ in = 0.25, $\frac{1}{8}$ in = 0.125, $\frac{1}{16}$ in = 0.0625 and $\frac{1}{32}$in = 0.03125.

Therefore, on the last working out, the pitch was 36°, the common rafter run was 12 ft $2^1/_2$ in. Converted to inches and decimals, this would become 146.5 in. This figure is then multiplied by 1.236 (the 1 m run-of-rafter figure for 36°), therefore:

$$146.5 \times 1.236 = 181.074\,\text{in} = 15\,\text{ft}\ 1.074\,\text{in}$$

Therefore the length of common rafter for a 12 ft $2\frac{1}{2}$ in run is 15 ft $1\frac{1}{16}$ in. The $\frac{1}{16}$ in less than the previous working out is insignificant in roofing.

12.7 METRIC RAFTER SQUARE

12.7.1 Introduction

Figure 12.13: The next method to be considered involves the use of a traditional instrument known as a *steel roofing square*, now metricated and called the *metric rafter square*. The one referred to here is manufactured by I & D Smallwood Ltd, and comes with an explanatory booklet on its use. Other, similar rafter squares on the market, adequate for the job, do not have the protractor facility incorporated in the Smallwood square, which is personally preferred by the author. The booklet, which is well illustrated, clearly explains the elements of roofing and the application of the square.

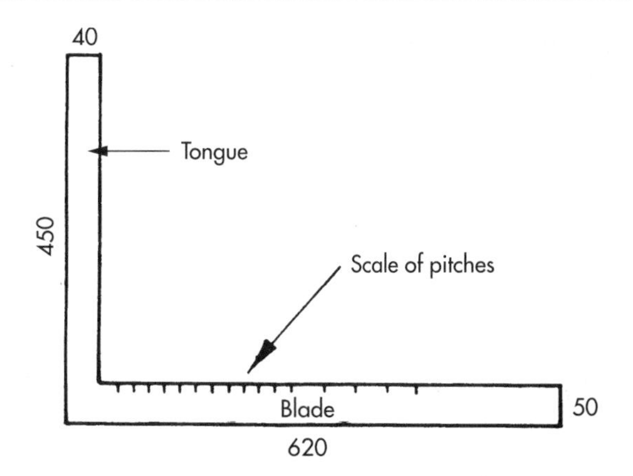

Figure 12.13 Metric rafter square model No. 390 or 390B

12.7.2 Protractor Facility

The various settings for different roof bevels – which are not easy to remember – are usefully given on the faces of the tongue. These are related to traditional settings, based on geometric principles. The main innovation, though, is that the square has a protractor facility in the form of a scale of pitches on the inner edge of the blade, enabling any pitch angle up to 85° to be set up quickly and easily.

12.7.3 Common Rafter Plumb and Seat Cuts

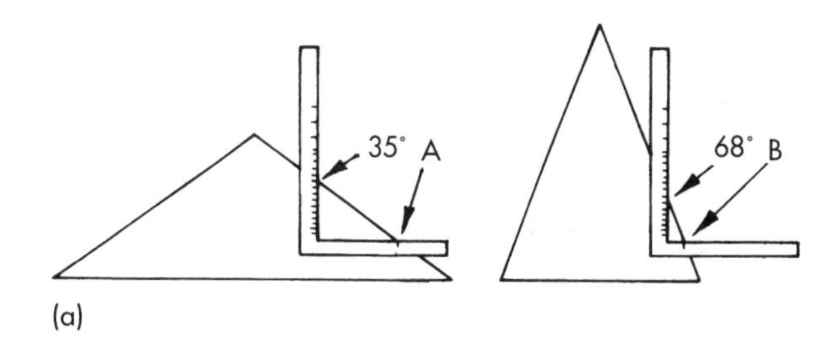

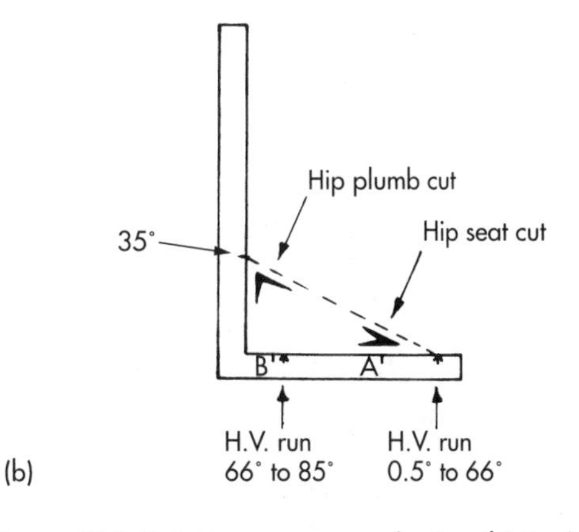

Figure 12.14(a) Using protractor facility. **(b)** Finding hip bevels

Figure 12.14(a): By using the protractor facility to set up the pitch, common-rafter plumb and seat cuts are quickly found. This is achieved by using a common-rafter-run point A, set at 250 mm on the inner edge of the tongue, as illustrated, and by rotating the square until the degree figure for the roof pitch registers on the inner edge of the blade. This works for pitch angles up to 66°. Should greater angles than this be required, then a common-rafter-run point B, set at 50 mm on the inner edge of the tongue, is used in relation to the scale of pitches on the blade. This will give angles from 66° to 85°.

12.7.4 Hip (or Valley) Rafter Plumb and Seat Cuts

Figure 12.14(b): These bevels are also found by using the protractor facility related to the roof pitch. This is achieved by using a hip or valley-rafter-run point, marked 'H.V. run' on the square, set at 354 mm on the inner edge of the tongue and, by rotating the square until the degree figure for the roof pitch registers on the inner edge of the blade. This will give hip or valley plumb and seat cuts for roofs up to 66°. Should the roof be steeper than this, then a hip or valley-rafter-run point, set at 71 mm on the inner edge of the tongue for roofs from 66° to 85°, is used in relation to the scale of pitches on the blade.

12.7.5 Common and Hip (or Valley) Rafter Lengths

As is normal practice, these rafter lengths are worked out in relation to the run. Some figures for this are given on both sides of the blade, against the pitch required, related to a 1 m run. A more definitive set of figures are given in the booklet, varying from 1° up to 89.5° pitch, in increments of 0.5°. Hence, a small and simple calculation will be necessary, whereby, as with the *Roofing Ready Reckoner*, the total common rafter run is multiplied by the 1 m run-of-rafter figure given on the blade or in the booklet, there being, as before, one figure for common-rafter lengths and another for hip (or valley) rafter lengths.

12.7.6 Stair Gauge Fittings (British Pattern)

Figure 12.15 Stair gauge fittings model No. 385

Figure 12.15: To increase the square's use from an instrument to a working tool, stair gauge fittings, as illustrated, are an available option for this – or any roofing square. The fittings are attached to the square's edges to act as stops, for setting and marking repetitive bevels, thus avoiding the use of a separate carpenter's bevel.

12.8 ALTERNATIVE METHOD FOR THE USE OF THE METRIC RAFTER SQUARE

12.8.1 The Scaled Method

A method used traditionally for its speed and simplicity in finding the main rafter lengths and bevels was known as the scaled method. This lends itself easily to metrication, from its original use in Imperial dimensions – whereby inches represented feet and twelfths-of-an-inch represented inches. Figure 12.16 should help in grasping this method, by visualizing the roofing square within the roof, either by imagining the square scaled up to roof-size or the roof scaled down to square-size.

12.8.2 Stair Gauge Fittings (American Pattern)

Figure 12.17: These American pattern stair gauge fittings (also illustrated on the sides of the square in Figure 12.18(b)) are almost an essential requirement – if you can obtain them – for this particular roofing method. This is because they relate accurately to pre-set measurements on the edges of the blade or tongue of the square; they also enable very accurate readings of hypotenuse measurements to be made.

As the name of these fittings implies, they were originally meant to be used on the square for setting out the

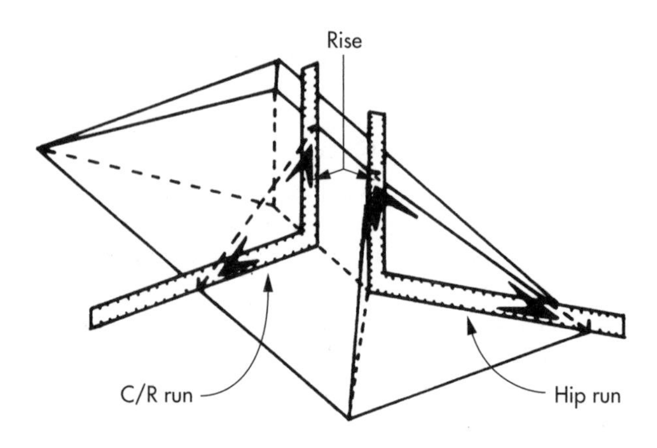

Figure 12.16 Visualizing the square within the roof (C/R = common rafter)

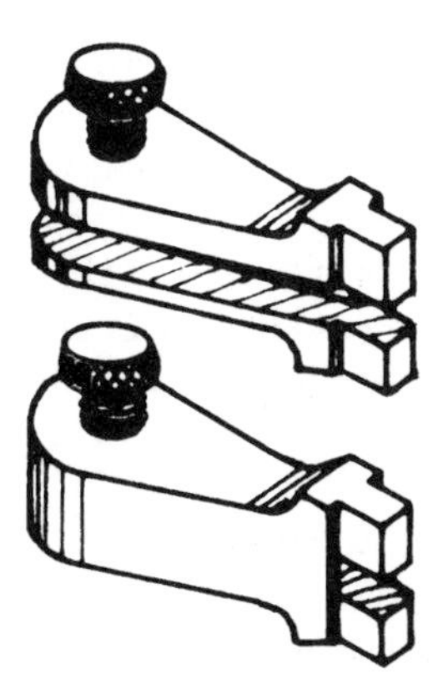

Figure 12.17 American-pattern stair-gauge fittings

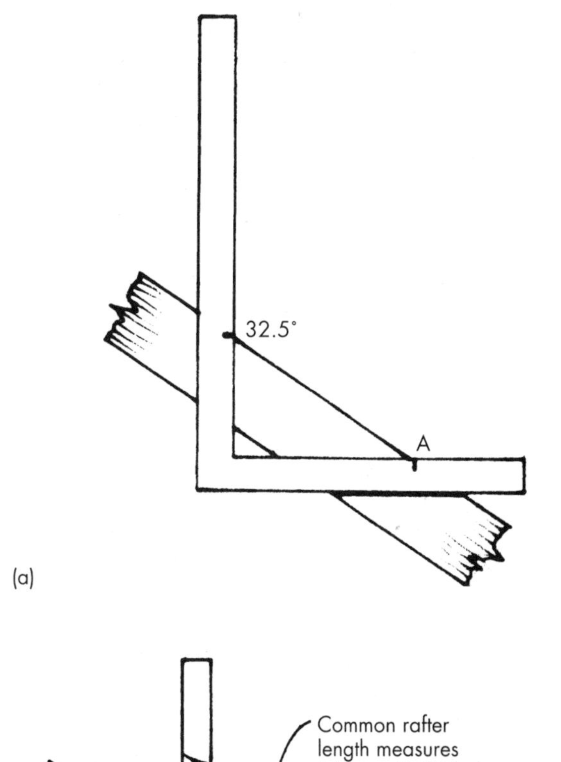

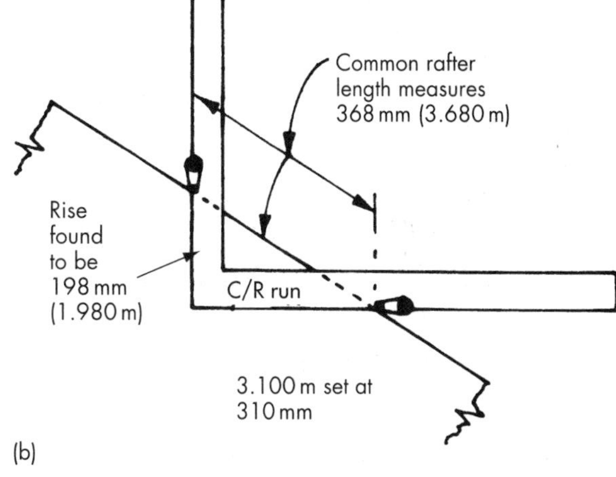

Figure 12.18(a) First step to finding the rise by scaled method. (b) Second step to find the rise, bevels and length

steps of a staircase or the shuttering for concrete stairs – a use for which, of course, they can still be put.

12.8.3 Finding the Rise

Figure 12.18(a): With the scaled method, it will be necessary to determine the rise before proceeding further. The key to this is the protractor facility, which will simplify the task. First, use the square as already described in the text for Figure 12.14, by laying it on a straight piece of rafter or straightedge material, relating to common-rafter-run point A on the inside edge of the tongue and – in this example – 32.5° pitch on the inside edge of the blade. Mark the bottom, outer edge of the tongue with a marking knife, chisel or a sharp pencil. This will give the seat cut and, being greater in length on this shallow pitch than the plumb cut would be on a given width of straightedge material, the plumb-cut mark available against the blade will not be needed.

12.8.4 Common Rafter Length, Plumb and Seat Cuts

Figure 12.18(b): The next step is to attach a stair gauge fitting to the outer edge of the blade, set carefully at, say, 310 mm, representing a scaled run of roof of 3.100 m. Line up the blade to the previously found seat cut bevel marked on the straightedge, as illustrated, then carefully attach the other stair gauge fitting to the outer edge of the tongue against the straightedge. This will produce the scaled rise at 198 mm. These settings determine the plumb and seat cuts and provide a scaled measurement of the common rafter length. The scale used is one-tenth full size (1:10) and is easily achieved by moving the decimal point by one place:

3.100 m run divided by 10 = 0.310 mm on the blade
0.198 mm on the tongue times 10 = 1.980 m rise

The scaled measurement of the rafter length on the hypotenuse – measured between the sharp arrises of the stair gauge fittings – is also brought up to size by multiplying by 10, i.e., 0.368 mm × 10 = 3.680 m.

12.8.5 Finding the Hip-rafter Run

Figure 12.19: To find the hip rafter length, first find the hip rafter run. As this represents – in an equal pitched roof – a 45° diagonal line in plan, contained within the run of the common rafter and the run of the crown rafter, forming a square, it follows that the scaled common-rafter run, set on both the blade and the tongue, as illustrated, gives the scaled measurement of the hip run at 45° on the hypotenuse.

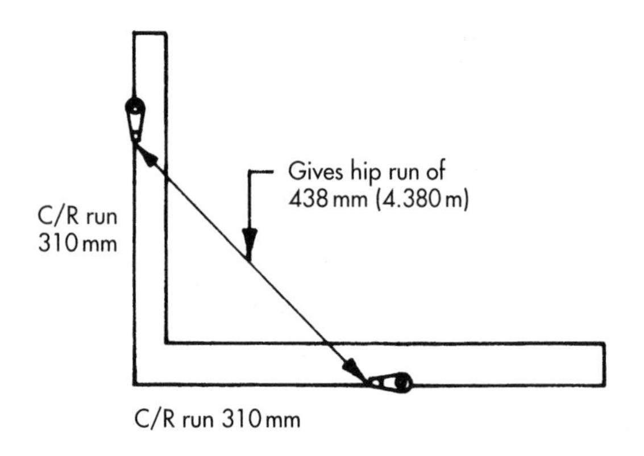

Figure 12.19 Settings to find the hip run

12.8.6 Hip Rafter Length, Plumb and Seat Cuts

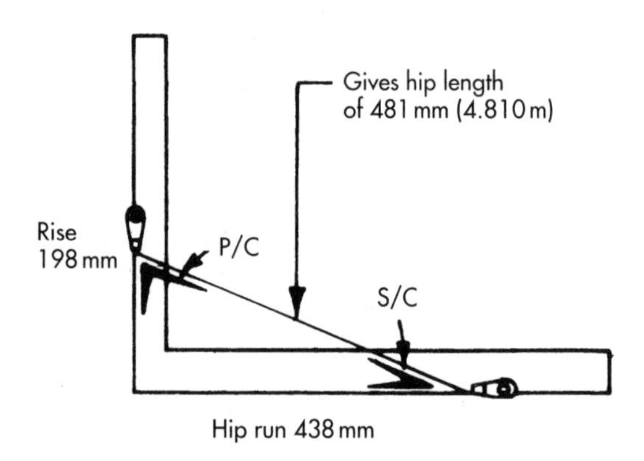

Figure 12.20 Hip rafter length, plumb and seat cuts

Figure 12.20: Now alter the position of the stair gauge fittings and set the scaled hip run on the blade, and the scaled rise on the tongue, as illustrated. This will give the scaled hip rafter length, the plumb cut (P/C) and the seat cut (S/C).

12.9 BEVEL-FORMULAS FOR ROOFING SQUARE

12.9.1 Hip Edge Cut

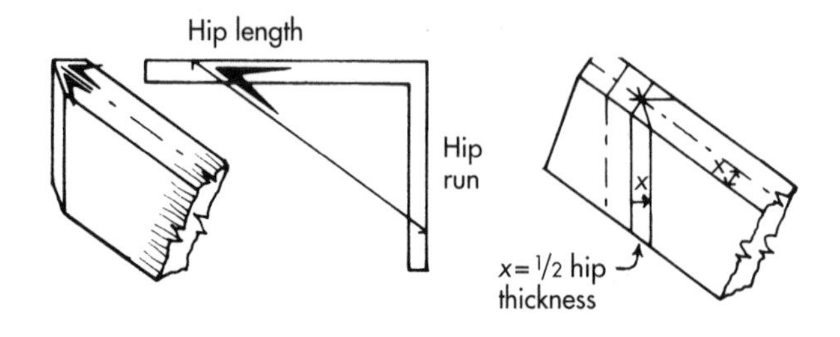

Figure 12.21 Hip edge cut

Figure 12.21: This equals the scaled hip length on the blade and the scaled hip run on the tongue. As illustrated, the bevel is found on the blade. This is applied to the edges of the hip plumb cut and enables the hips to fit into the heads of the crown rafter and the first commons, against a saddle board – or, alternatively, to fit against each other and the saddle board. A simpler way of finding the edge bevel, is to measure and mark half the hip's thickness x in from the second plumb cut each side. This gives the three points on the top edge for marking the edge bevels. To understand this more fully, see 'setting out hip rafters' after the Roofmaster method (page 132).

12.9.2 Hip Backing Bevel (Dihedral Angle)

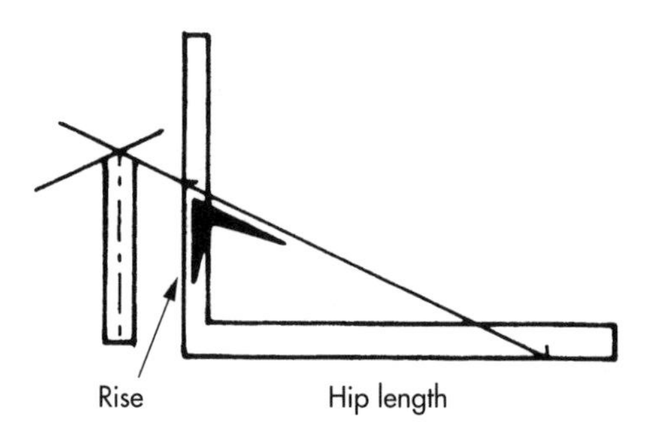

Figure 12.22 Hip backing bevel

Figure 12.22: This equals the scaled hip length on the blade and the scaled rise on the tongue. As illustrated, the bevel is found on the tongue. This was used traditionally when the roof was to be boarded and was applied by planing the top edge from both sides to a top centre line.

12.9.3 Jack or Cripple Side (Plumb) Cut

This is the same formula as for the common rafter plumb cut, with the scaled common rafter run set on the blade, the scaled rise on the tongue and the required bevel found on the tongue.

12.9.4 Jack or Cripple Edge Cut

Figure 12.23: This combines with the above cut to make a compound angle to fit against the hip or valley rafters. The formula equals the scaled common-rafter length on the blade and the common rafter run on the tongue. The bevel, A, is found on the blade.

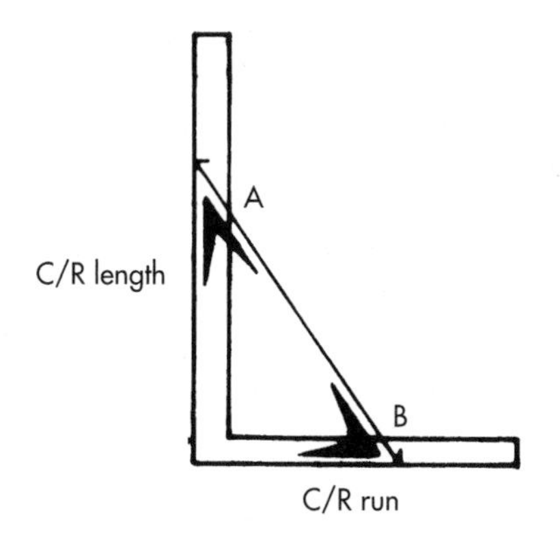

Figure 12.23 Jack edge cut (bevel A) and purlin edge cut (bevel B)

12.9.5 Purlin Edge Cut

Figure 12.23: This bevel is applied to the surface that the underside of the rafters rest on, at the junction of the hips or valleys, to form a mitred edge against the sides and under the centre of the hip or valley rafters. The formula is the same as for the jack or cripple edge cut, except that the required bevel B is found on the tongue.

12.9.6 Purlin Side Cut

Figure 12.24: This combines with the above cut, to complete the mitred faces against and under the hips or valley rafters. The formula equals the scaled common rafter length on the blade and the rise on the tongue. The bevel required is found on the tongue.

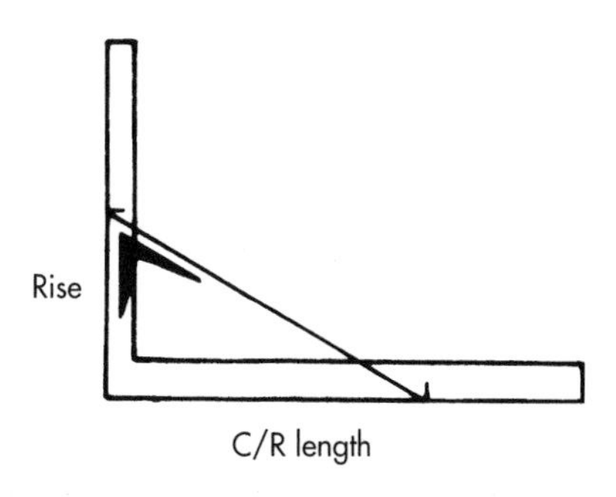

Figure 12.24 Purlin side cut

12.9.7 Purlin Lip Cut

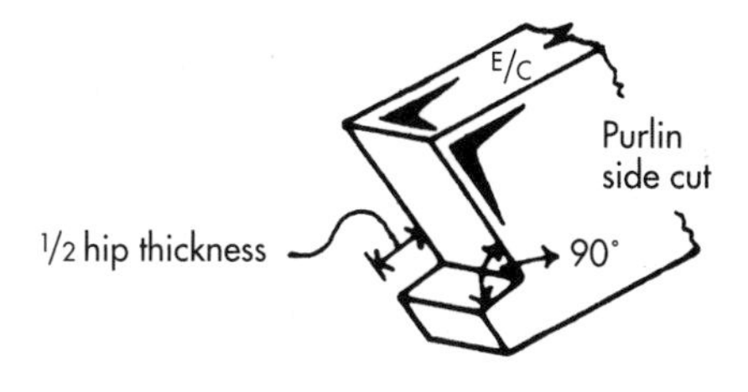

Figure 12.25 Purlin lip cut (E/C = edge cut)

Figure 12.25: As illustrated, this is simply marked at 90° to the purlin side cut in relation to the amount of hip projection below the rafters.

12.9.8 Jack or Cripple Rafter Diminish

Figure 12.26: As illustrated, this is found in two stages. First, by setting the jack edge-cut formula on the square

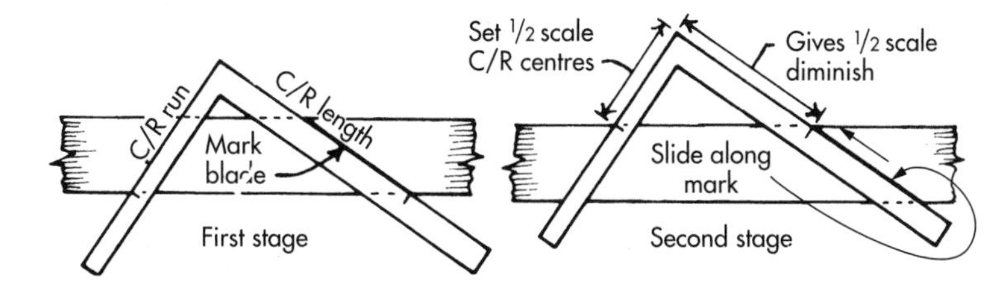

Figure 12.26 Jack rafter diminish

Table 12.1 Bevel formulas

Component	Required bevel	Blade setting	Tongue setting	Side for marking
Common rafter	plumb cut	C/R run	rise	tongue
Common rafter	seat cut	C/R run	rise	blade
Hip rafter	run of hip	C/R run	C/R run	hypotenuse measurement
Hip rafter	plumb cut	hip run	rise	tongue
Hip rafter	seat cut	hip run	rise	blade
Hip rafter	edge cut	hip length	hip run	blade
Hip rafter	backing bevel	hip length	rise	tongue
Jack rafter	side cut	C/R run	rise	tongue
Jack rafter	edge cut	C/R length	C/R run	blade
Purlin	edge cut	C/R length	C/R run	tongue
Purlin	side cut	C/R length	rise	tongue
Lay board	plumb cut	C/R length	C/R run	blade
Lay board	seat cut	C/R length	C/R run	tongue

and marking this – against the edge of the blade – on a straightedge. Then by sliding the blade along the mark until the projecting tongue registers the common-rafter centres at a half full-size scale (say 400 mm centres divided by 2 = 200 mm on the tongue). The measurement now showing on the projecting blade will be equal to a half full-size scale of the diminish.

12.9.9 Bevel formulas

Table 12.1 on page 127 gives bevel formulas in quick reference style.

12.10 ROOFMASTER SQUARE

12.10.1 Introduction

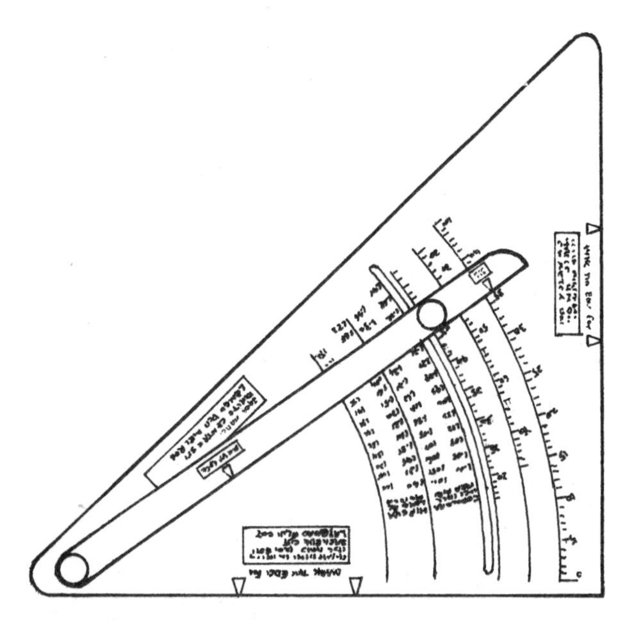

Figure 12.27 Roofmaster square

Figure 12.27: The final method to be considered as an alternative to the roofing square and the Roofing Ready Reckoner, is the innovative, recently marketed tool (or instrument) known as the Roofmaster. It consists of an anodized aluminium blade, resembling a 45° set square, engraved with easy-to-read laser-etched figures and markings on each face side. Attached to the blade is a pivoting and lockable, double-sided fence (or arm), through which the blade is slid when locking the fence to a required setting. The instrument comes with a booklet which is well illustrated and clearly explains the application and use.

12.10.2 Basic Concept

The basic concept of this revolutionary square, is that the only knowledge needed to access all the different bevels required in the cutting of a roof, is the pitch angle. If, for example, the pitch to be used is 35°, that would be the only reference required to set up the Roofmaster as many times as necessary, and *when* necessary, to determine the different bevels, such as common-rafter plumb and seat cuts, hip-rafter plumb and seat cuts, purlin side and edge cuts, jack and cripple rafter plumb and edge cuts, etc.

12.10.3 Main Features

Three main features are engraved on each side of the blade as follows.

- A set of separate, graduated segmental arcs, each one is referenced to a specific roof member and calibrated to give the required bevel, which is obtained simply by setting the adjustable fence to the required pitch angle numbered on the selected arc.
- Information panels on each face side of the blade indicate which edge to mark for the selected angle cut. This valuable facility eliminates yet another area of confusion associated with other roofing squares.
- Tables for length of common, hip and valley rafters per metre of run, radiating around the remaining segmental arcs, are calibrated to line up the multiplier figure for use in obtaining the true or basic length, obtained simply – once again – by setting the adjustable fence to the required pitch angle numbered on the pitch-angle arc.

12.10.4 Handling the Roofmaster

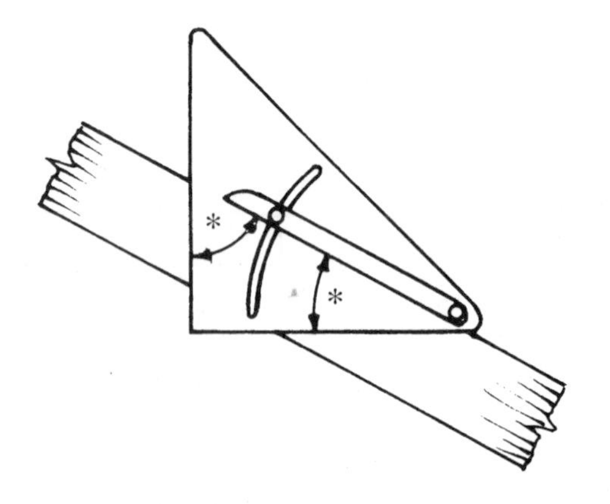

Figure 12.28 Acute angles required*

Figure 12.28: As with other roofing squares, the angles required will always be acute angles contained within the right-angled triangle formed with this tool, by the right-angle of the blade and the varying position of the adjustable fence, acting as the triangle's hypotenuse. As illustrated, the setting edge of the fence, which is clearly marked, always butts against the edge of the roof member.

12.10.5 Reverse and Opposite Marking-positions

Figure 12.29: An important operational fact to realize is that once the fence has been locked into the required position, the Roofmaster can be used on opposite side-edges of the timber to mark the same bevel (Figures 12.29(a) and (b)) or, if necessary, can be reversed (turned over) to give either left- or right-hand cuts (Figure 12.29(c)).

12.10.6 Finding Cutting Angles

To apply the plumb and seat cuts to a common rafter for a roof to be pitched, say, at 35°, first, select the arc designated for common-rafter cuts of 16–45°; second, lock the adjustable arm on number 35; third, hold the tool with the setting edge of the fence firmly against the rafter material; and finally mark the edges indicated by the information panels on the blade, for the required cuts.

Note that this sequence of operations is carried out on each relevant arc to obtain the different cut-angles for all of the roof members.

12.10.7 Rafter Lengths

Figure 12.30: The common rafter and hip or valley rafter lengths per metre of run are engraved on each side of the blade, for roof pitches of 16–45° on one side and 45–75° on the other. To obtain the true or basic length of one of these rafters, the adjustable arm must be set to the correct pitch angle figure on the common-rafter arc. The two sets of table numbers now displayed – and indicated by reference headings – along the setting edge of the arm will be the common rafter length per metre of run and the hip or valley length per metre of run. When the figures shown are multiplied by the common rafter run, the true or basic lengths of rafters will be obtained.

12.10.8 Example Working Out

Figure 12.30: Assume a roof pitch of 35° with a span of 6.486 m.

1. Set the arm to number 35 on the common rafter arc.
2. Record the table numbers displayed under reference headings on the adjustable arm, thus:

   common rafter = 1.221; hip or valley rafter = 1.578
3. Divide the span by 2 to find common rafter run:

   6.486 ÷ 2 = 3.243 m

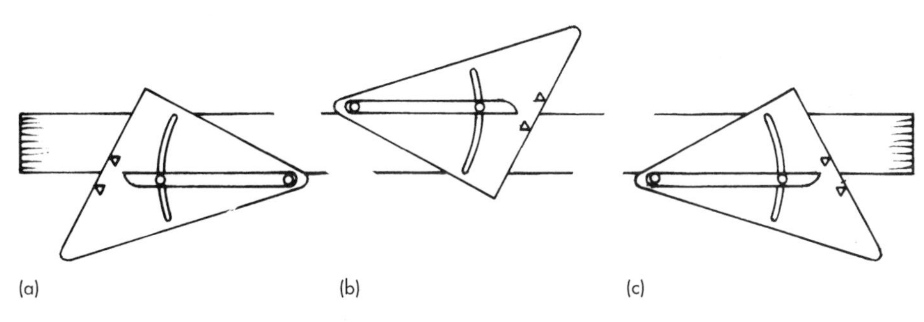

Figure 12.29 Reverse and opposite marking-positions

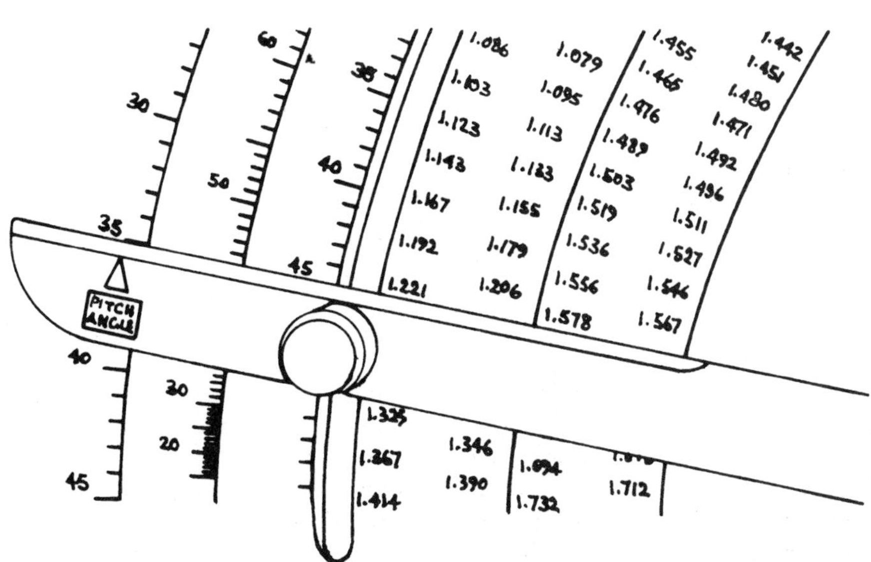

Figure 12.30 Rafter lengths per metre of run

4. True length of common rafter is

   1.221 × 3.243 = 3.959 m

   True length of hip or valley rafters is

   1.578 × 3.243 = 5.117 m

12.11 SETTING OUT A COMMON (PATTERN) RAFTER

12.11.1 Technique (for any Roofing Tool or Bevel)

Figure 12.31(a): Select a piece of rafter material which is straight and without twists and mark a common-rafter plumb cut A at the top of the rafter's face side, representing the centre of the ridge board. Next, mark another plumb cut B into the body of the rafter, at half the ridge-board thickness measured at right-angles to plumb line A, representing the actual plumb cut against the ridge board. Transfer these lines squarely across the top edge and make a shallow saw cut on the waste side of the edge-line (Figure 12.31(b)) to provide an anchorage for the hook of a tape rule.

12.11.2 Marking the True or Basic Length

Figure 12.31(c): Place the hook of the tape rule into the saw cut on the rafter's top edge, run the rule down to the eaves' area and mark the common rafter length – after having worked this out from one of the methods previously covered. Square this mark across the top edge and, from this point, mark another plumb cut C. This important plumb cut is known as the *backing line*, which is now measured for depth, divided by three and marked down two-thirds to indicate the corner of the birdsmouth D, from which point the common rafter seat-cut is marked.

12.11.3 Working out the Eaves' Projection

Figure 12.31(d): Finally, the rafter must carry on down, from the backing line, across the outer skin of brickwork E, to a distance measured horizontally from the face of the wall and known as the eaves' projection F. As referred to earlier (Figure 12.8, showing basic setting-out terms), the total projection horizontally equals E + F minus the thickness of the fascia board. The concealed

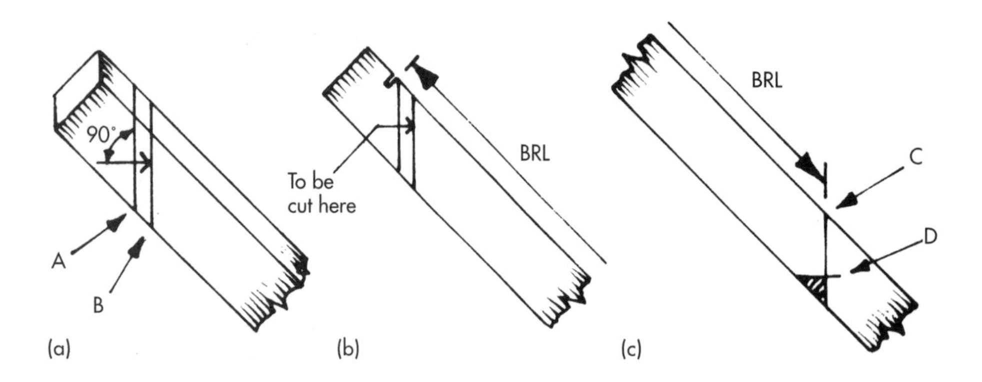

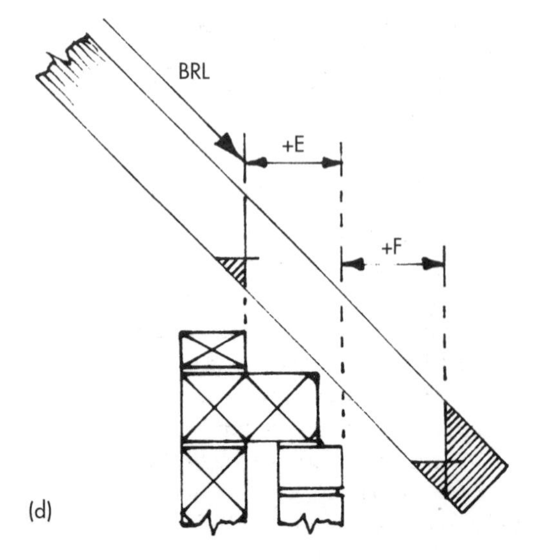

Figure 12.31(a) First step. **(b)** Second step. **(c)** Third step. **(d)** Fourth step

projection E is equal to the wall-thickness, minus wall plate. The visible eaves' projection F can be scaled from the elevational drawing showing the roof. For example, assume a wall of 275 mm, a visible eaves' projection of 200 mm, a wall plate 100 mm wide and a fascia board 20 mm thick. The sum would be

$$E = 275 - 100 = 175\,\text{mm}$$
$$F = 200 - 20 = 180\,\text{mm}.$$

Total eaves' projection = 175 + 180 = 355 mm.

12.11.4 Adding the Eaves' Projection

With the metric rafter square, this value (355 mm) can be set on the tongue while the blade is lined up to the backing-line plumb mark, the square being slid up or down until the blade lines up precisely and the 355 mm point on the tongue can be marked at the end of the rafter to become the plumb cut for the fascia board.

With the Roofmaster or the *Roofing Ready Reckoner*, the sum of 355 mm can be used as the figure for the run of the total projection, multiplied by the common-rafter table figure for, say 35°, i.e. 0.355 × 1.221 = 0.433 m (433 mm), to be measured down the top edge of the rafter, from the backing line to the top of the fascia-board plumb cut. If a soffit board is to be used, this plumb cut is usually marked down a half to two-thirds its depth and a seat cut established at this point.

12.11.5 Cutting and Checking the First Pair

Figure 12.32: The pattern rafter just marked should be double-checked (a wise trade saying is 'check twice, cut once'), as this first rafter is used for marking out the rest. Next, square the face marks across any edges not yet done. Cut all the bevels carefully, including the birdsmouth and write the word 'PATTERN' boldly on the face of the rafter. Next, only one common rafter should be marked and cut from the pattern, the pair laid on the ground with an offcut of ridge board between the plumb cuts and a tape rule stretched across the birdsmouth-cuts to check the span. If only slightly out, adjust the rafters to the correct span, then check the plumb cuts for a good fit against the ridge-board offcut. If satisfactory, the rest of the common rafters may be cut, keeping any cambered or sprung edges on top. If unsatisfactory, go through a checking procedure, as suggested in Figure 12.32(b).

Note that some carpenters leave the marked fascia plumb-cuts to be checked with a string line after all of the rafters have been fixed and then they cut them off *in situ*, for better alignment. However, if a soffit seat-cut is required, this is very awkward to cut *in situ* and should be done at the initial cutting stage.

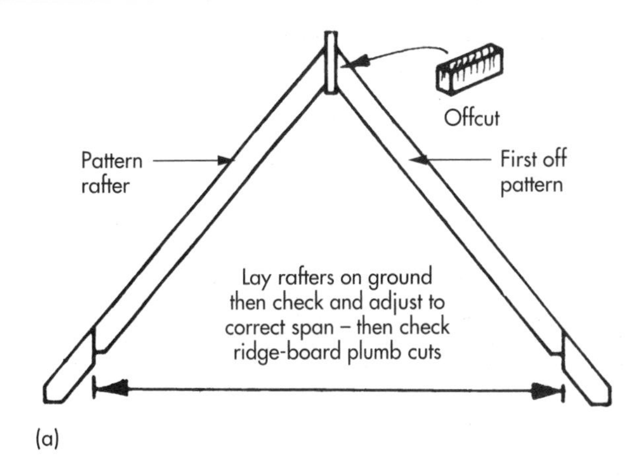

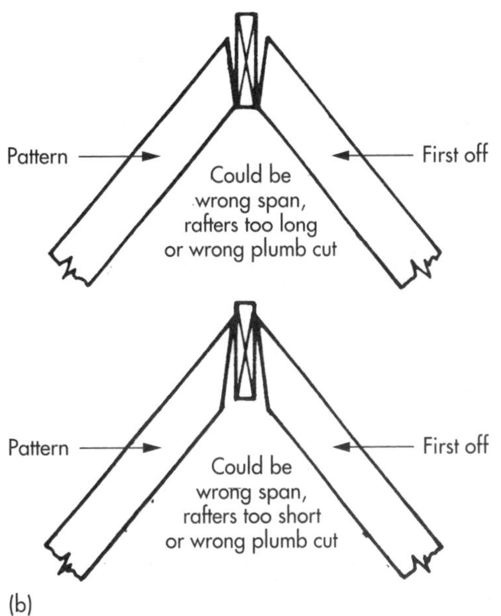

Figure 12.32(a) Cutting and checking first pair. **(b)** Possible error

12.12 SETTING OUT A CROWN (OR PIN) RAFTER

12.12.1 Slight Difference

The procedure for setting out a crown rafter is identical to that for setting out a common rafter, except that instead of deducting half the *ridge-board* thickness from the first plumb cut – as described in the common-rafter text to Figure 12.31(a) – half the *common-rafter* thickness is deducted. Therefore, the first plumb cut A represents the geometrical centre of the hip end and the second plumb cut B, again measured into the body of the rafter, at right-angles to the plumb line, represents the actual plumb cut against the saddle board.

12.13 SETTING OUT A HIP RAFTER

12.13.1 Technique

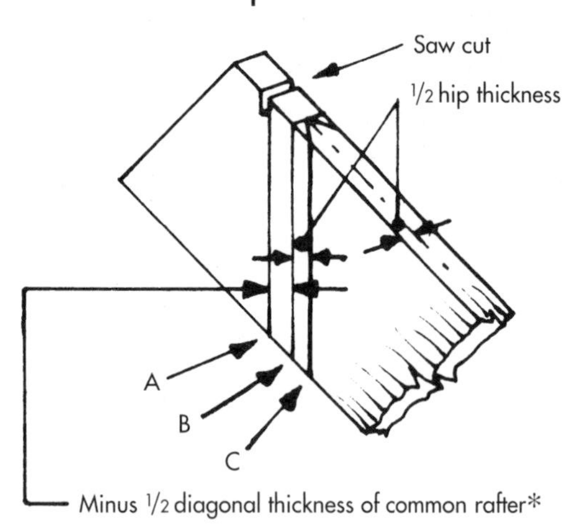

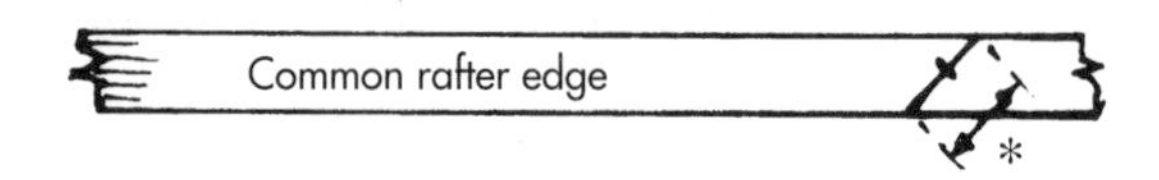

Figure 12.33(a) Setting out a hip rafter

Figure 12.33(a): This technique is applicable for any roofing tool or bevel. First, mark a hip-rafter plumb cut A at the top of the rafter's face side, representing the geometrical centre-point intersection of the hip end. Next, mark another plumb cut B into the body of the rafter, measured at right-angles to plumb line A and equal to half the common rafter thickness measured diagonally at 45°, as illustrated. This represents the arris or sharp edge of the compound-angled plumb cut. Transfer these lines squarely across the top edge and make a shallow saw cut on the waste side of edge-line A, to provide an anchorage for the hook of a tape rule – and mark the centre of edge-line B to provide the central point for the opposing hip edge cuts.

The simple way of finding this bevel, as explained previously in the text to Figure 12.21 (bevel formulas), is to mark another plumb cut C, again into the body of the rafter on each side, measured at right angles to line B and equal to half the hip's thickness. Where these plumb lines meet the top edge opposite each other, together with the centre-mark of edge-line B, they present the three points for marking the hip edge cut bevels, as illustrated.

12.13.2 Marking the True or Basic Length

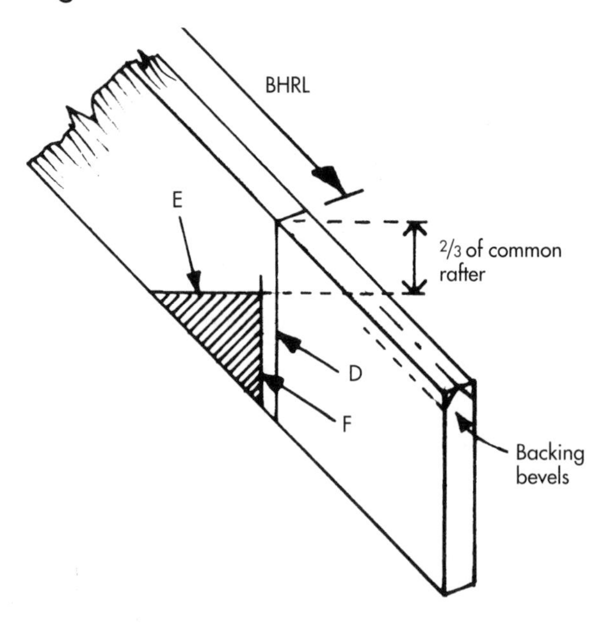

Figure 12.33(b) Hip with backing bevels

Figure 12.33(b): Place the hook of the tape rule into the saw cut on the rafter's top edge, run the rule down to the eaves' area and mark the hip rafter length – after having worked this out from one of the methods previously covered. Square this mark across the top edge and, from this point, mark a hip rafter plumb-cut D, representing the external corner of the half-lapped wall plate. If the hip rafters were receiving backing bevels (dihedral bevels) on their top edges, plumb line D – the important backing line reference for rafter-length – would also be the line to measure down and mark the same two-thirds measurement worked out for the common rafter (as illustrated) to form the birdsmouth after marking a hip rafter seat cut E at this depth.

12.13.3 Addition for Loss of Corner

Figure 12.33(c): Now, as the external corner of the bedded wall-plate is usually removed, to allow a square abutment of the hip plumb-cut against the wall plate, it follows that the amount removed has to be added to the hip plumb-cut in the birdsmouth. As illustrated, the amount removed diagonally from the corner geometrically equals half the thickness of the hip rafter. Therefore, if the hip was 32 mm sawn thickness, 16 mm would be cut off the corner of the wall plate and 16 mm would be added back into the birdsmouth, as indicated by plumb line F in Figure 12.33(b).

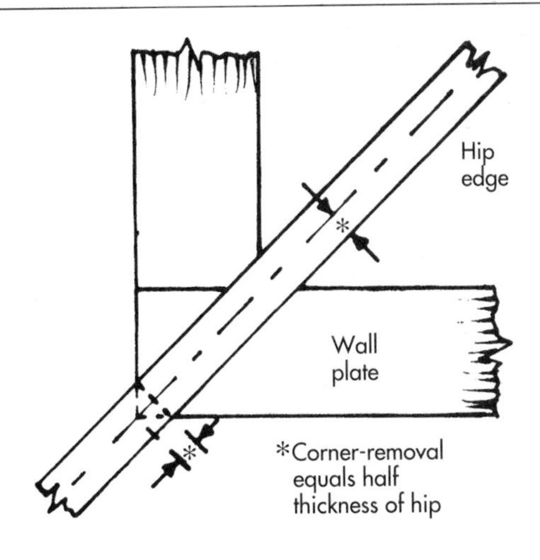

Figure 12.33(c) Addition for loss of corner

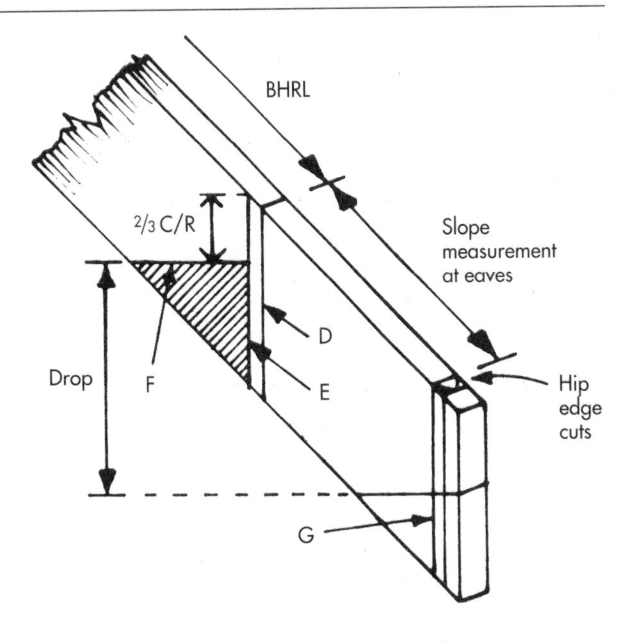

Figure 12.33(d) Hip without backing bevels

12.13.4 Hip Rafters Without Backing Bevels

Figure 12.33(d): This simplified hip rafter, left with a square top-edge, is more in evidence nowadays, since the economic omission of boarding (sarking) which was fixed over the whole roof surface, under the sarking felt, some years ago. Therefore, having marked out the top-end of the hip rafter, as described in the text to Figure 12.33(a), hook in and run the tape rule down to mark the length. As before, square this across the top edge and from this point, mark a hip plumb cut D, representing the external corner of the wall plate. Next, the amount to be removed diagonally from the wall-plate corner (half the thickness of the hip rafter), say 16 mm, is measured into the body of the rafter, at right-angles to plumb line D, and marked to become hip plumb cut E. This time, it is this *inner plumb line* which, as illustrated, is marked down with the same two-thirds measurement worked out originally for the common rafter. At this depth, a hip seat cut F is marked to form the birdsmouth.

12.13.5 Hip Rafter Eaves' Projection

Figure 12.33(e): The total eaves' projection for the common rafters, as described in the text to Figure 12.31(d), was worked out to be – as an example – 355 mm from the plumb cut backing-line to the plumb cut for the fascia board, measured horizontally (or 433 mm measured down the slope of the rafter, as worked out by the *Roofing Ready Reckoner* or the Roofmaster for a 35° pitch). Using the common-rafter working out of 355 mm again for the hip-rafter eaves' projection, bearing in mind that the hip is at 45° in plan and is therefore diagonally longer, you can find the horizontal diagonal projection with the metric rafter square. Do this by setting 355 mm on the blade and 355 mm on the tongue and by measuring the diagonal, which, in this example, produces a figure of 502 mm. A figure, fractionally more accurate than this, can be found by the other

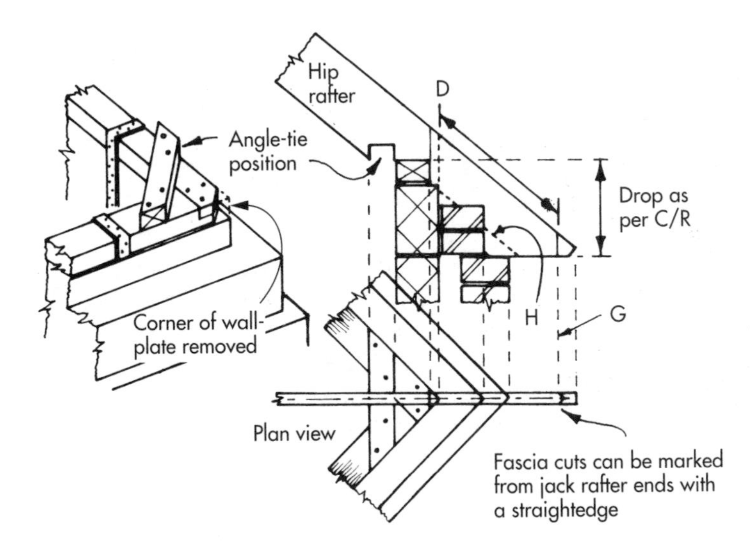

Figure 12.33(e) Hip eaves' detail

two methods; by using 45° as the pitch and the common-rafter-table figure per metre of run, i.e.

$$0.355 \times 1.414 = 0.50197\,\text{m (say 502 mm)}$$

12.13.6 Down the Hip-edge

Figures 12.33(d) and (e): Using the *Ready Reckoner* or the Roofmaster to find an alternative measurement for applying down the slope of the rafter, for the same 35° pitch, the sum would be the run of the common-rafter eaves' projection, multiplied by the hip table-figure:

$$0.355 \times 1.578 = 0.560\,\text{m (560 mm)}$$

This would be applied from the top of the backing line D to the top of another hip plumb cut to be marked at G, from which hip edge cuts are marked, as illustrated, against which the mitred fascia boards will be fixed.

12.13.7 Additional Notches

Figure 12.33(e): As illustrated, additional birdsmouth notches might be necessary to clear wall projections which clash with the hip's extra depth. Alternatively, the hip may be reduced in depth on the underside, from a point starting from and equal to the bottom outer-edge of the corner wall-plate, indicated by the dotted line H. Also, if corner angle-ties are to be used, the notches for them should now be marked ready for cutting on the inner edges of the extended birdsmouth seat cuts.

12.13.8 Practical Considerations

Figures 12.33(f) and (g): Finally, it must be mentioned that the length of the hip and its eaves' projection, rather than relying completely on geometrical principles, is often determined by practical methods, especially if the walls at the hipped end are out of square or the wall plates are out of level. One of many practical methods is to reduce the width of a truly square-ended rafter offcut to equal two-thirds the common-rafter backing-line height (height above the birdsmouth), fix it temporarily as illustrated in Figure 12.33(f), to be flush to the diagonally offcut wall plate and rest a length of hip rafter on it edgeways and on the crown/common rafter intersection at the top, as indicated in Figure 12.33(g).

12.13.9 Use for Marking or Checking

Where the hip rafter rests at the top, in a self-centralizing position, the sides of the actual hip-edge plumb cuts can be marked and where it rests at the bottom, the actual plumb cut for the birdsmouth can be marked. Then the seat cut for the birdsmouth can be marked down (in the case of hips without backing bevels) at two-thirds the common-rafter plumb-cut depth. Alternatively, hip rafters that have already been marked out by calculation can still be checked by this technique, prior to being irreversibly cut.

12.13.10 Variation of Method

Figures 12.33(g) and (h): An interesting variation on the above method is to omit the rafter offcut which equalled two-thirds' backing-line height, rest the length of hip rafter on the actual diagonally offcut wall-plate edge instead (Figure 12.33(h)), and again on the crown/common rafter intersection at the top, mark these two extreme resting points on the bottom-face of the rafter and square them across the edge. This length of hip material can now be used as a rod, to lay diagonally on the face of a hip rafter previously set out by calculation, to check that the rod-marks relate exactly to the marked corner of the birdsmouth at one end while relating to the top edge of the hip edge cuts at the

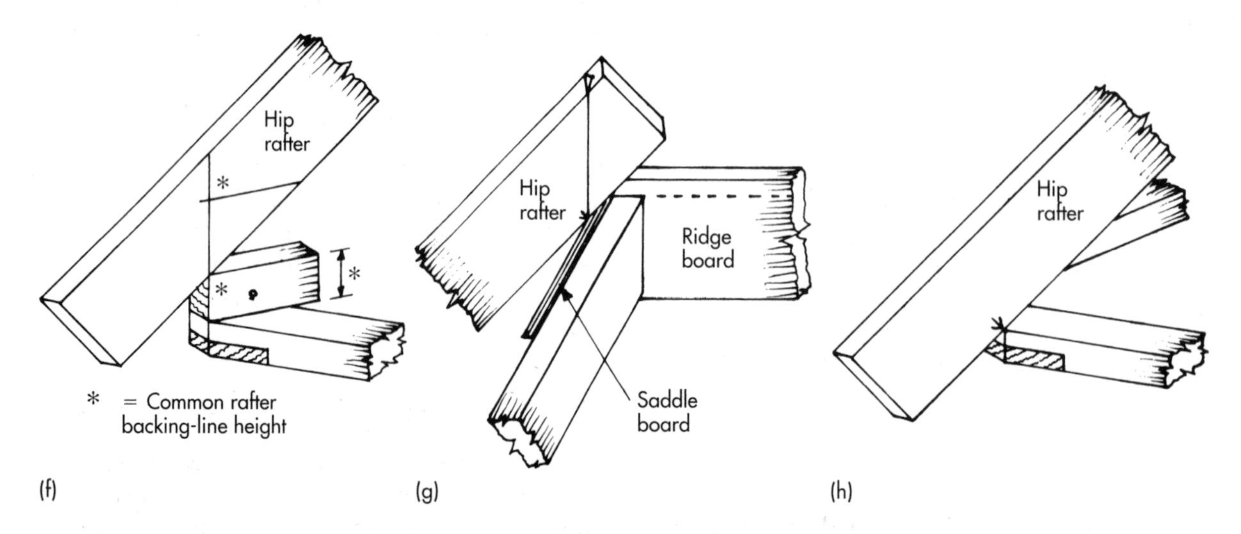

Figure 12.33(f)–(h)

other. Alternatively, hip rafters can be set out this way, without length-calculation, by first marking out the eaves' projection and birdsmouth and then by laying the rod to run up diagonally from the birdsmouth corner, pivoting as necessary at the top of the hip until the mark on the rod relates to the top edge, where it is marked to become the inner plumb cut of the hip edge cuts.

12.13.11 Eaves' Drop

Figure 12.33(e): The practical way to deal with the hip eaves' projection, is to mark out and cut the soffit seat-cut only, and leave the hip-edge plumb cuts for the fascia board, to be marked and cut *in situ* by string-line or straightedge, once the hips, commons and jack rafters are fixed. The position of the hip rafter seat cut for the soffit board is measured down vertically from the birdsmouth seat cut, at a distance known as *the drop*. This drop measurement is taken from the pattern common rafter.

12.14 SETTING OUT JACK RAFTERS

12.14.1 Calculated Diminish

As mentioned earlier in the text to Figure 12.11, jack rafters, being equally spaced, should diminish in length by a constant amount in relation to the length of the common rafter. From information given in the *Roofing Ready Reckoner*, the accompanying booklet to the metric rafter square and the booklet that comes with the Roofmaster, working out and applying the diminish, to find the length of each pair of diminishing jack rafters, is not difficult to understand, but again there is a practical method that can be used.

12.14.2 Practical Method

Figure 12.34(a): After the skeletal structure of a hip-ended roof has been pitched (the sequence yet to come), the next operation is to fix the jack rafters in pairs on each side of the hip rafters; start by working away from one of the first pair of common rafters. Ideally, you will need a steel roofing square, but an improvised wooden square will do and can be quickly and easily made out of, say 50 × 25 mm battens, to resemble a miniature builder's-square or, as illustrated, to resemble a tee square, simply constructed with a lapped, clench-nailed joint.

12.14.3 Technique

Figures 12.34(b) and (c): As illustrated, the tee square or the roofing square is held firmly and squarely against the common rafter, with the tongue resting on the hip-rafter edge, while being slid up or down until the rafter-spacing mark or measurement relates exactly to the inner edge of the hip. At this point a mark is made and squared across to the other edge on the crown-rafter side of the hip. Edges A and B of this squaring, establish the highest points of the positions of the first pair of jack rafters.

12.14.4 Centres of Rafters

Figure 12.34(d): Next, mark the spacing-distance of the rafters (usually 400 mm centres on traditional roofs) on top of the wall plate, away from the vertical face of the common rafter. In practical terms, this will be 350 mm, the distance *between* the common and the first jack rafter. The spacing for the first jack rafters on either side of the crown rafter, will be minus the thickness of the saddle board, so the distance between these rafters will be 350 − 18 = 332 mm. This dimension can also be found by measuring squarely across from edge mark B (Figure 12.34(c)) to the face-side of the crown, ready to be reproduced on the wall plate. All following pairs of jack rafters will conform to normal rafter spacing.

12.14.5 Marking the Length

Figure 12.34(e): Having gathered a selection of varied offcuts and lengths of rafter material to be used up as jacks, mark out and cut the birdsmouth on each (if not the plumb cut for fascia) and lay aside. Next, take a tape rule or, preferably, a measuring batten (rod), and carefully check the distance by laying the rod in a pitched position, from the edge of the wall plate E (Figure 12.34(d)), up to the hip-edge mark A. Then, mark these points onto the rod and transfer the rod to the prepared jack rafter, to lay in a diagonal position, relating to the corner of the birdsmouth E and being marked at A, on the top edge of the jack rafter. From this point, the jack edge cut and jack side cut are marked, ready for cutting.

12.14.6 Fixing Pair After Pair

Figure 12.34(c): To avoid overloading one side of the hip rafter, jacks should always be fixed in diminishing pairs, exactly opposite each other. Once the first pair are fixed, they can be used for relating to and continuing the squaring technique for the next pair, as indicated at C and D.

12.14.7 Setting Out Valley Rafters

Figure 12.35(a)–(c): Geometrically, a valley rafter is identical to a hip rafter, but inverted. The plumb cut, the seat cut and the calculated length are the same. However, there are a few variations. First, because the

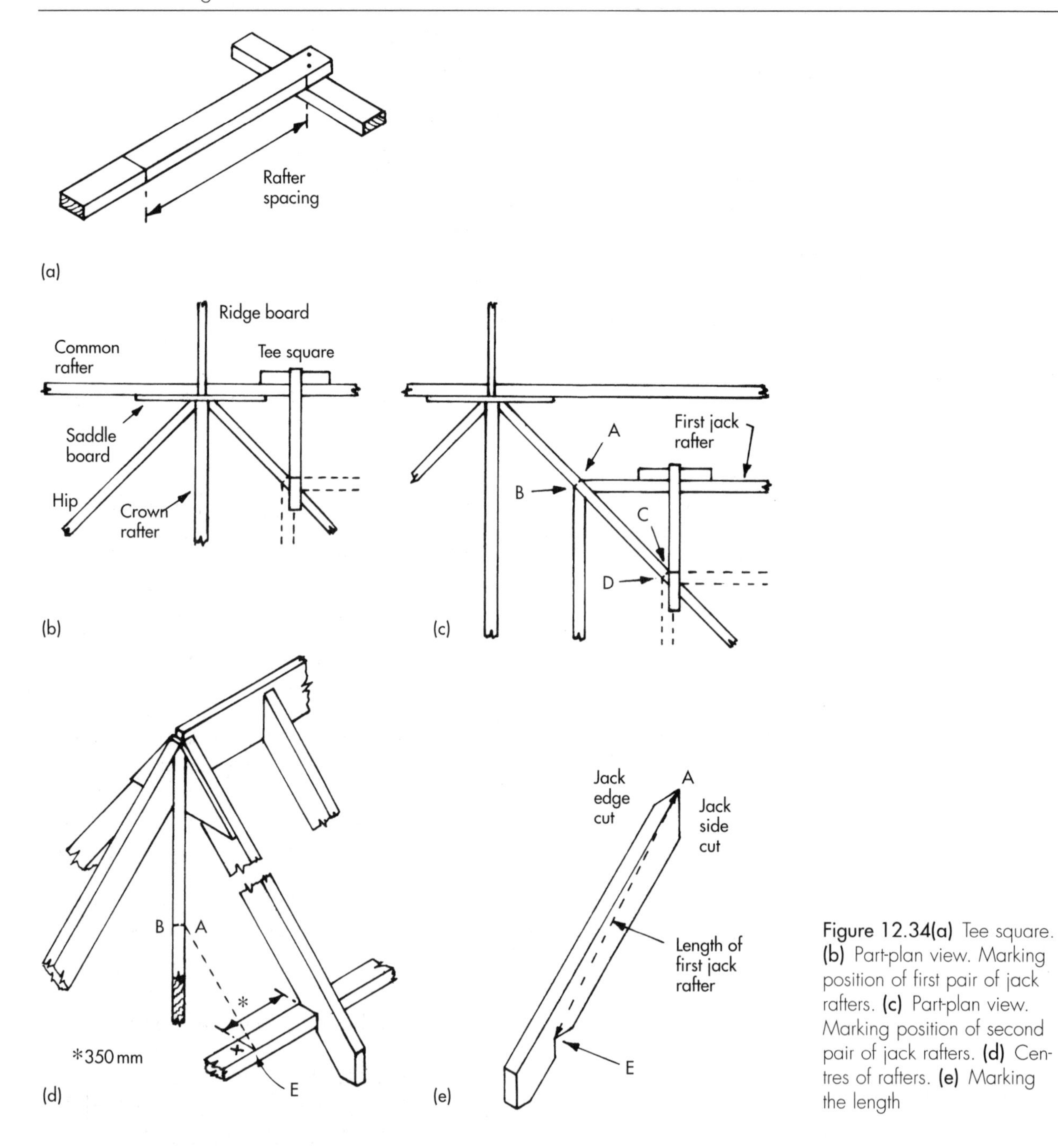

Figure 12.34(a) Tee square. **(b)** Part-plan view. Marking position of first pair of jack rafters. **(c)** Part-plan view. Marking position of second pair of jack rafters. **(d)** Centres of rafters. **(e)** Marking the length

valley rafter usually stems from a ridge-board junction, the reduction at the top from F to G, would be set in squarely at half the diagonal thickness of the *ridge board* (not the common rafter). Second, the plumb cut of the birdsmouth can be shaped with hip-edge cuts to allow a proper abutment against the internal angle of the wall plate. This is shown here, marked out between plumb lines H and I, equalling half the valley-rafter thickness, measured out squarely. The third and final variation is that when the feet of the cripple rafters are fixed, unlike jack-edges on hips, their top edges should protrude above the valley-rafter edge by the backing-bevel depth, as indicated in Figure 12.35(c).

12.15 PITCHING DETAILS AND SEQUENCE

12.15.1 Jointing the Wall Plates

Figure 12.36: Wall plates for gable-ended roofs are only required on the two side walls from which the roof pitches. If the walls exceed standard timber lengths of 6.3 m, the wall plates would have to be joined in length with half-lap joints, whereby the lap equals the width of the timber. In the case of a double hip-ended roof, the wall plates would be required on all four sides. These would be half-lap jointed on the four corners, as well as

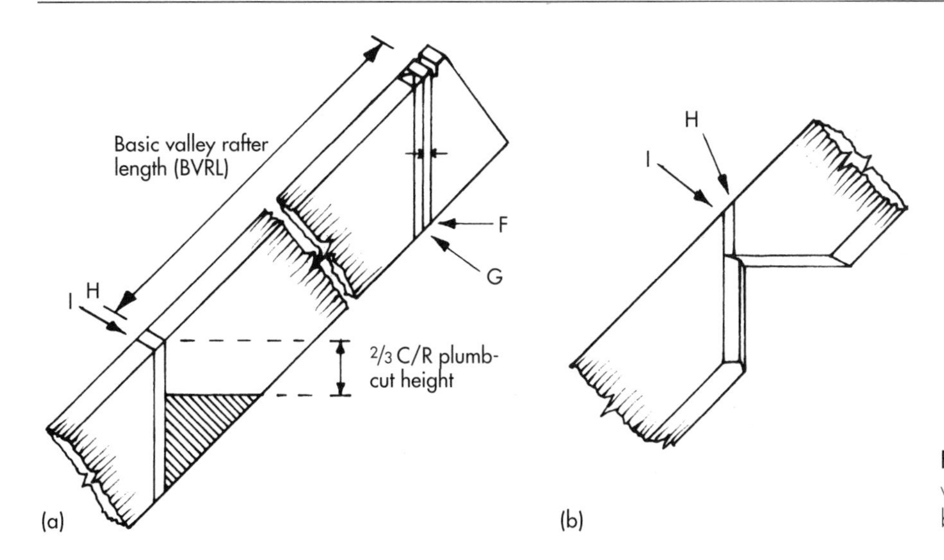

Figure 12.35(a) Setting out of valley rafters; **(b)** valley birdsmouth

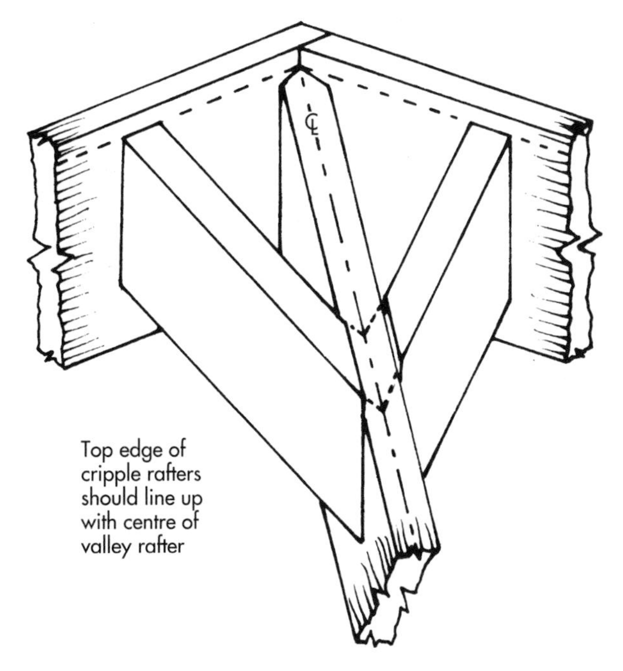

Figure 12.35(c) Protruding cripple edges

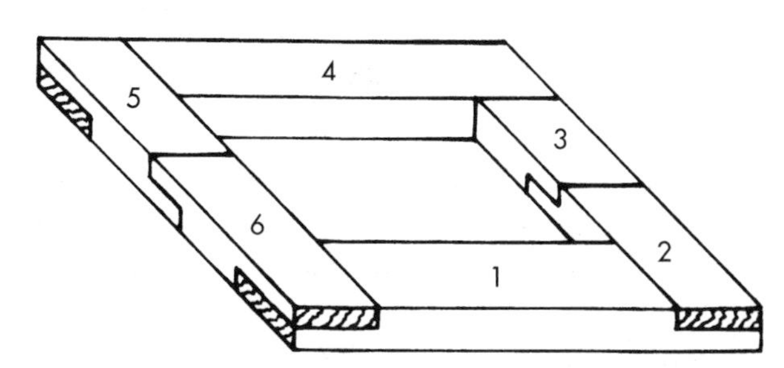

Figure 12.36 Sequence of lap-jointing for wall plates

where intermediate lengthening joints may be necessary. As illustrated, a sequence of lap-jointing should be worked out to allow successive pieces of plate to be dropped onto the open-joint and the mortar, rather than be pushed under a wrong-sided lap joint – which tends to trap the mortar in the under-joint.

12.15.2 Bedding the Wall Plates

Bedding the wall plates is best done as a team effort between carpenter and bricklayer. The bricklayer lays and spreads the mortar, beds and levels the plates and, in the case of a hip-ended roof, the carpenter checks that they are square (by using the 3–4–5 method) and parallel to each other across the span. To achieve this, in the case of a gable-ended roof, slight lateral adjustments of the plates may be made – within reason. The 3–4–5 method referred to, is a practical application of Pythagoras' theorem of the square on the hypotenuse of a right-angled triangle being equal to the sum of the squares on the other two sides. When used for squaring the wall plates, using metres as units, mark 3 m along the edge of one plate, 4 m along the edge of the other and, to prove a true right-angle, the diagonal measurement between these points should be 5 m.

12.15.3 Pitching a Gable Roof

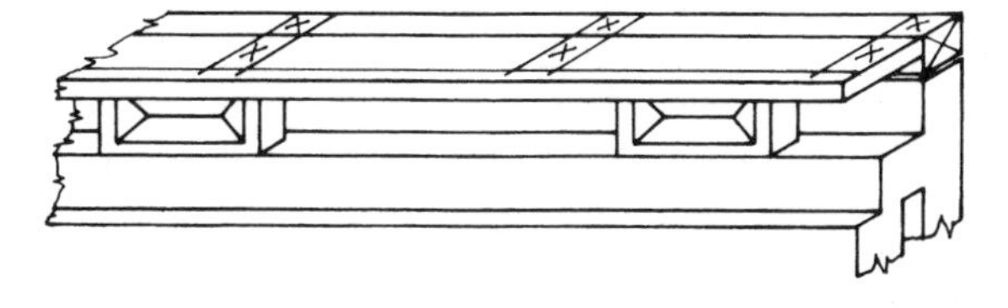

Figure 12.37(a) Marking rafter positions on plate and ridge

Figure 12.37(a): When the bedded plates are set, the ridge board may be laid out against the wall plate, if possible, and the positions of the common rafters spaced out to the required centres (usually 400 mm) and marked across both members. Next, in any of the clear areas indicated by the marking out, the vertical restraint straps are fixed over the plates, close to each end, one on each side of a lap-joint, and not more than 2 m apart

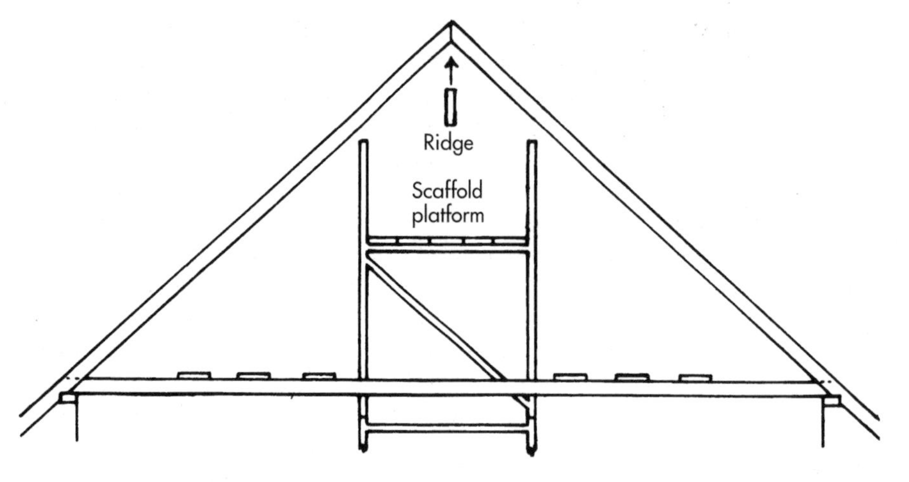

Figure 12.37(b) Inserting ridge board

– and then the ceiling joists, with any sprung edges kept uppermost, are fixed adjacent to the rafter marks, either by skew-nailing with 75 or 100 mm round-head wire nails or by being fixed into pre-fixed, shoe-type metal framing anchors. The ceiling joists, which are also fixed to the plates of any internal cross-walls, act as a working platform and should be close or open-boarded with an area of scaffold boards.

12.15.4 Technique and Sequence

Figure 12.37(b): At each end of the roof, a pair of rafters is pitched and fixed to the wall plates *and* the ceiling joists, their plumb cuts supporting each other at the apex. This is a two-man job, with one man at the foot of each rafter. An interlocking scaffold can be erected through the ceiling-joist area and, from this, the ridge board is pushed up between the rafter plumb-cuts and fixed into position. On each fixing, one 75 mm round-head wire nail is driven through the top edge of the rafter and two – one on each side – are skew-nailed through the sides into the ridge board. The nails first driven into the rafters' top edges usually pierce through the ridge board on each side, causing problems unless both opposite rafter-heads are fixed at the same time. To manage this single-handedly, leave the first nail-head protruding until the nail on the other side is partly or fully driven in.

12.15.5 Important Foot-fixings

The foot of each rafter must be fixed to the wall plate and the ceiling joist (acting as a structural tie) with at least three 100 mm round-head wire nails – one skew-nailed above the birdsmouth into the plate, which, as well as fixing, also tightens the rafter against the ceiling joist, and the other two driven in squarely through the side of the ceiling joist, into the rafter, as previously indicated in Figure 12.7(a).

12.15.6 Adding Purlins and Struts

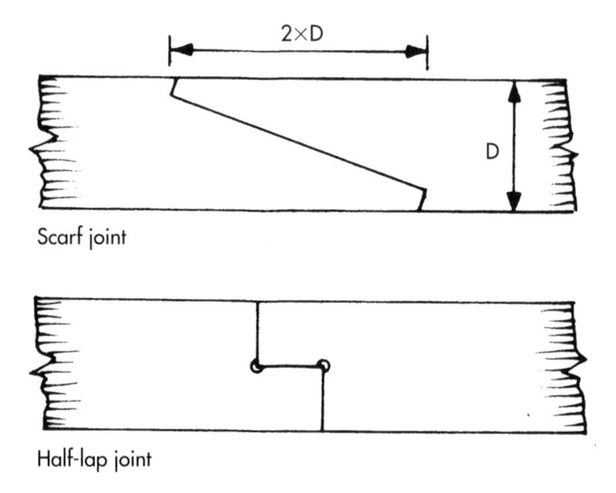

Figure 12.37(c) Purlin half-lap joint and scarf joint

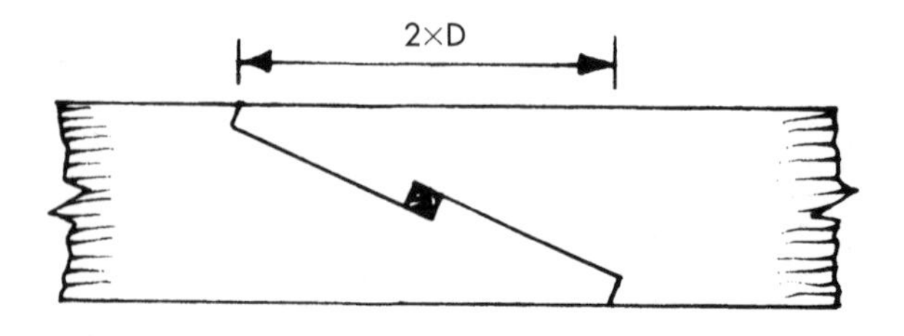

Figure 12.37(d) Ridgeboard scarf joint

Figures 12.37(c) and (d): According to the length of the roof (from gable to gable) and whether the purlins or ridge board need to be extended in length by jointing – as illustrated – a few more pairs of rafters may need to be fixed at strategic positions before the purlins are offered up and fixed by skew-nailing from above, through the available rafters, into the purlin edges. The struts, set out by using a rod and/or rise-and-run principle for lengths and bevels, are then fixed into position.

Next – to complete the main structure of the roof – the remainder of the rafters are filled in.

12.15.7 Restraint Straps

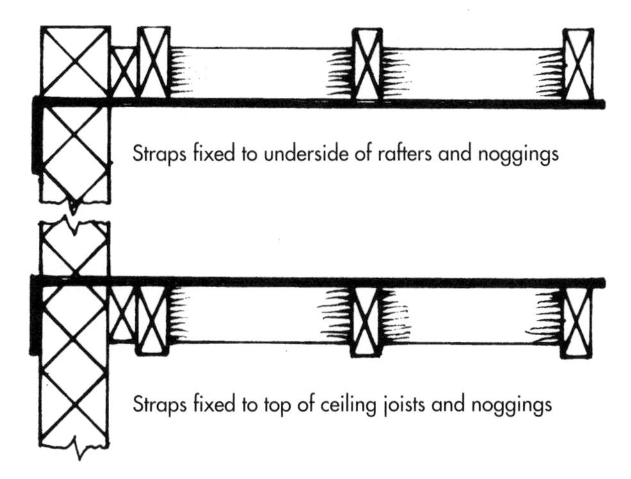

Figure 12.37(e) Restraint straps

Figure 12.37(e): Whether the gable walls are to be built before or after the roof is pitched, it must be remembered that horizontal restraint straps of 5 mm thick galvanized steel must be fixed at maximum 2 m centres across the ceiling joists and rafters, to be built into (to help stabilize) the gable-end walls.

12.16 PITCHING A HIPPED ROOF (DOUBLE-ENDED)

12.16.1 Setting out Wall Plates

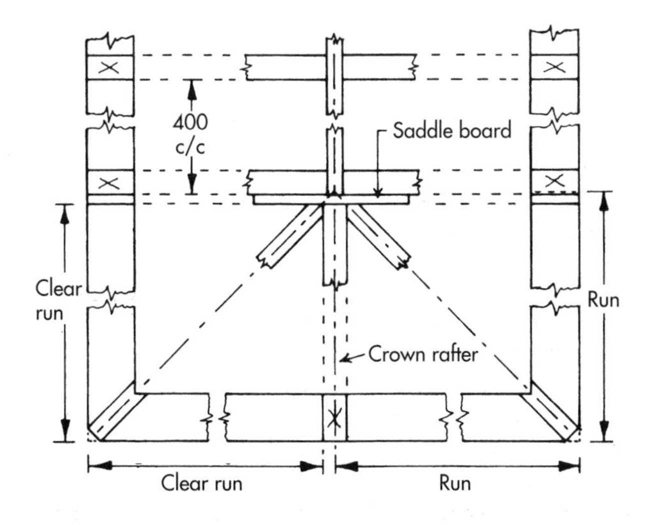

Figure 12.38(a) Setting out a hipped end

Figure 12.38(a): After the wall plates have been half-lapped (see Figure 12.36), bedded into position, checked for being level, parallel and square-ended and the mortar has set, the rafter positions may be set out to allow the vertical restraint straps to be fixed in any of the clear areas – as described for the gable-roof wall plates. To set out, check the actual span across the wall plates and divide by two to find the run. Mark this as a centre line on the wall plate at each hip-end and split the thickness of the crown rafter on each side, to be squared across the plates. The clear run is now measured and should equal the run minus half the crown-rafter thickness. This measurement is now marked in from each side of each hip-end and represents the face of the saddle board. The thickness of the saddle board – usually 18 mm WBP plywood – is now marked across the plates and this line is equal to the face of the first pair of common rafters at each end. The other rafters are spaced out between these pairs, at specified centres from one end, regardless of an odd spacing at the other end.

12.16.2 Marking and Cutting

As mentioned earlier in this chapter, most – if not all – of the marking and cutting of the various components in the roof should be done on the ground, laid out on pairs of saw stools, then hoisted or man-handled up to the roof.

12.16.3 Pitching Technique and Sequence

First, the ceiling joists are fixed, but only those in the middle area that attach to common rafters – not those at each end that attach to jack rafters. As before, a boarded area is laid with scaffold in position. The first pair of rafters at each end are pitched and fixed. The marked ridge board is inserted and fixed, the rafters being braced diagonally down to the wall plates. Then, the saddle boards are fixed at each end with 50–63 mm round-head wire nails (at least six in each board), followed by the crown rafters and then the hips – or the hips and then the crown rafters, if using the alternative hip-arrangement illustrated in Figure 12.4(a).

12.16.4 Adding Purlins and Temporary Struts

After fixing a few more pairs of common rafters, strategically placed to relate to any lengthening joints in the purlins or ridge board, it will be found to be advantageous to fix the purlins, after checking that the hips are not bowed and, if necessary, bracing them straight with diagonal battens down to the wall plates. Fixing the purlins at this stage reduces the struggle against a full complement of sagging rafters and provides an intermediate ledge upon which to manoeuvre the rafters on their way up to the ridge. At this stage, the purlins should be supported, if only with temporary struts until the joists at each end are complete and the binders have been added.

12.16.5 Finishing Sequence

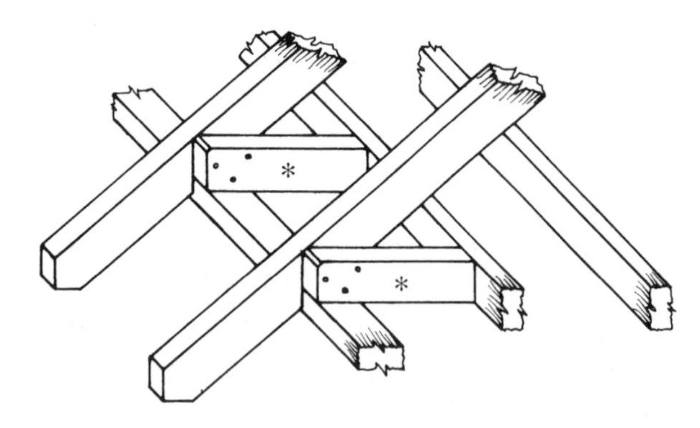

Figure 12.38(b) Return-joists on hipped end*

Figure 12.38(b): Finally, cut and fix the jack rafters, the remaining ceiling joists each end, the binders (skew-nailed from alternate sides into the top of each ceiling joist), the struts and straining pieces, the remaining common rafters, the hangers and, if required, collars to complete the main structure of the roof. On hipped roofs, as illustrated, short return-joists are fixed to the feet of the crown and jack rafters at the hipped ends. These return-joists only provide fixing points for the plasterboard ceiling at the wall-plate edge and, unlike ceiling joists, do not act as structural ties. For this reason, the *New Build Policy Guidance Notes BGN 9B*, recommend that horizontal restraint straps or timber ties (like binders) should be fixed across the tops of a minimum of three ceiling joists and be attached to the side of the crown and every second jack rafter.

12.17 FLAT ROOFS

12.17.1 Introduction

The structure of flat roofs is very similar to that of suspended timber floors, especially when kept level for the provision of a ceiling on the underside. If the ceiling is unimportant, or not required, the joists may be set up out of level to create the necessary *fall* (roof slope). Like floors, the roof joists span across the shortest distance between the load-bearing walls – or the longest distance when intermediate cross-beams are used – at similar centres to floor joists. They are subject to the same rules as floors regarding strutting being required when the span exceeds 2.5 m. Suitable joist sizes can be obtained either by structural design or by reference to Tables A17 to A22 for roof joists in *The Building Regulations' Approved Document A, 1992 Edition.*

12.17.2 Anchoring

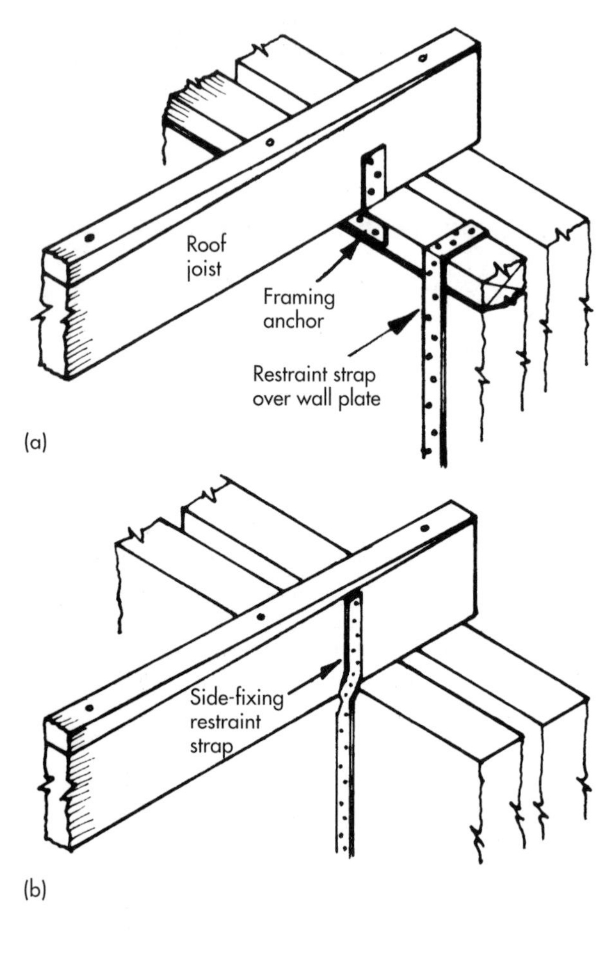

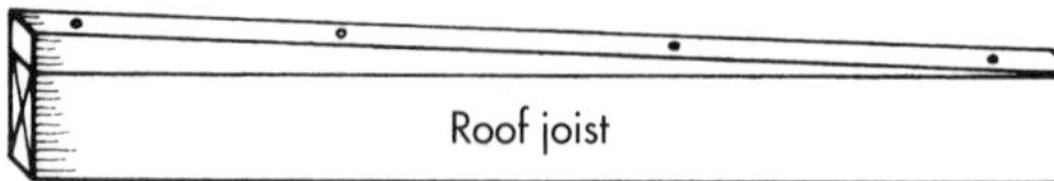

Figure 12.39 Flat roof details. (a) Anchoring. (b) Side-fixing restraint strap. (c) Firrings

Figure 12.39(a) and (b): As illustrated, the roof joists are either fixed to the wall plates by skew-nailing with round-head wire nails or by framing anchors. Galvanized restraint straps are fixed over the plates and to the inside face of the wall, as described for pitched roofs, at maximum 2.0 m centres – or, without wall plates, the joists may be anchored to the top of the walls with twisted side-fixing restraint straps.

12.17.3 Creating a Fall

Figure 12.39(c): Wedge-shaped timber fillets, or diminishing parallel-fillets, known as firring pieces, are fixed to the upper joist edges to create the necessary *fall* or roof slope. The recommendations for these falls vary between as low as 1 in 80 (25 mm in 2.0 m) and 1 in 40 (25 mm in 1.0 m). The *New Build Policy Guidance Notes BGN 9A* recommend the latter as a minimum fall for flat roofs. An alternative method of creating a fall, available nowadays, is to keep the structural roof truly flat and create the fall *above* the roof deck by using a system of pre-designed tapered roof insulation boards.

12.17.4 Flat Roofs Against Buildings

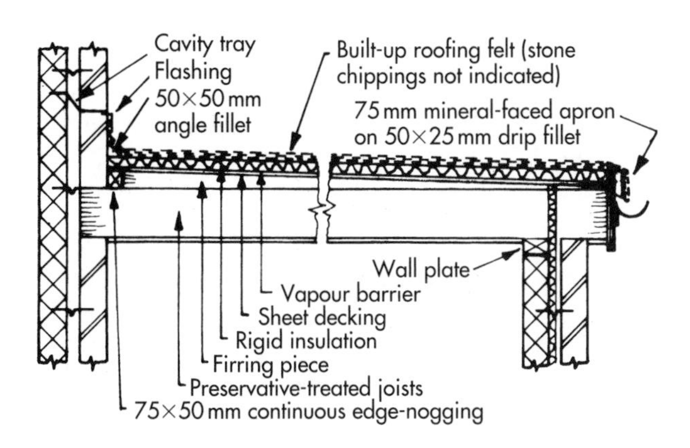

Figure 12.40(a) Flat roof butted to adjacent walls

Figure 12.40(a): Flat roofs may be independent (as on a detached garage, Figure 12.40(c)), or have one or more edges butted up to the face of an adjacent building. Sometimes, on single-storey buildings such as bungalows, the abutment is weathered-in with the pitch of the main roof at eaves' level. In the case of abutments to walls, the roofing material is turned up the wall and is covered by a lead flashing chased into the wall at a minimum height of 150 mm above the roof.

12.17.5 Cavity Trays

If the wall is of cavity construction, a cavity tray must be built in or be inserted to lap onto the top edge of the flashing, as illustrated in Figure 12.40(a). However, this creates a problem in conversion work, where a lean-to or flat roof is butted-up to a wall as an afterthought, and involves the bricklayer in cutting away a brick-course of the existing wall in a piecemeal operation to allow for building in a cavity-tray system.

12.17.6 Joists Into or Against a Wall

Figure 12.40(a) shows the joists at right angles to the adjacent wall, built into the outer skin of brickwork. They may also be carried on galvanized, type TW (timber to wall) joist hangers – a popular method in conversion work, as the top flange of the hanger is easily cut into a mortar-bed joint in the brickwork. This can be done with a disc angle-grinder and/or a traditional plugging chisel. When the hangers are inserted, any gaps above the flange should be caulked with bits of slate or strips of sheet lead.

12.17.7 Roof Fabric

The most common covering to flat roofs for some years now, has been bituminous roofing felt, although it has a limited life-span up against more traditonal coverings such as mastic asphalt, sheet lead or copper. However, high-performance roofing felts are now available, built-up as before in three layers with limestone chippings on top.

12.17.8 Decking Material

Sheet decking is laid and fitted on the roof joists, firrings or counter-battens, with cross-joints staggered in a similar way to flooring panels, fixed down with 50–56 mm by 10 gauge annular-ring shank nails at recommended 100 mm centres. Types of decking material include WBP exterior grade plywood, Sterling (strand-) board, moisture-resistant or bitumen-coated chipboard and pre-felted chipboard with roofing felt bonded onto the top surface.

12.17.9 Board Thicknesses

Thickness of decking is related to the joist-spacing – plywood can be 12 or 15 mm for joists at 400 mm centres, Sterling board 15 mm and chipboard 18 mm. When joists are at 600 mm centres, the thickness should increase to 18 mm for plywood, 18 mm for Sterling board and 22 mm for chipboard. Nails should be about $2\frac{1}{2}$ times the thickness of the board, so only the 22 mm chipboard qualifies for the 56 mm nails mentioned above. Decking should be laid with a 3 mm expansion joint between boards and board-joints should be covered with 100 mm wide bitumen felt strip to protect the edges of the boards while awaiting the arrival of the roofing specialist.

12.17.10 Projecting Eaves and Verges

Figures 12.40(b) and (c): When projecting eaves and/or verges are required for design purposes or to allow for

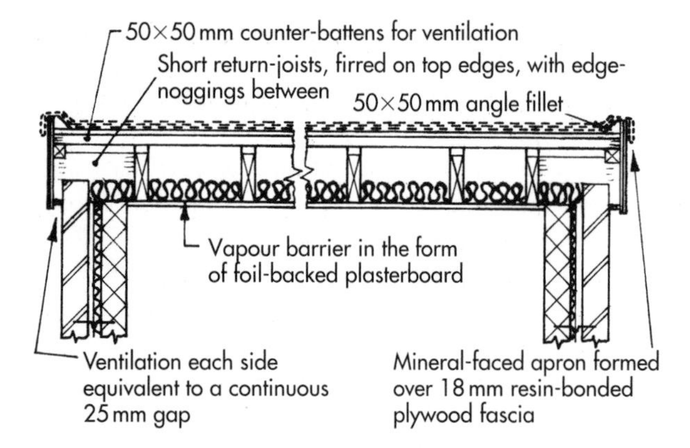

Figure 12.40(b) Verge details

ventilated soffits (as is necessary for *cold roofs*), the joists can be extended in length across the wall and made to extend on the sides. As illustrated three-dimensionally in Figure 12.40(c), this is achieved by the formation of short, projecting return-joists fixed at right-angles to the sides of the outer joists by using type TT (timber to timber) joist hangers, framing anchors or simply by skew-nailing. Even if the fascia boards are to be kept close to the face brickwork, these short return-joists will be needed for fascia-fixings and for better continuity of the whole roof structure.

12.17.11 Insulation to Flat Roofs

The Building Regulations' *Approved Document F2* states that excessive condensation in roof voids over insulated ceilings must be prevented, otherwise the thermal insulation will be affected and there will be an increased risk of fungal attack to the roof structure. This applies only to roofs where the insulation material is at ceiling level (cold roofs). Where the insulation is kept out of the roof void, placed on the deck (warm roofs), the risks of excessive condensation developing are not present and these roofs, therefore, are not covered.

12.17.12 Categories of Flat-roof Construction

There are three to be mentioned. The first, which is preferred nowadays because a better balance between heat-loss and condensation-control can be achieved, is known as a *warm deck flat roof*, which does not require ventilation of the roof void. The second, which *does* require ventilation and is more susceptible to condensation, is known as a *cold deck flat roof*. This type of roof, although covered by The Building Regulations of England and Wales, is not recommended in The Building Standards Regulations of Scotland (in which warm roof constructions are recommended). The third category of roof is known as a *hybrid flat roof*, covering certain roofs which do not fall within the warm or cold category. This is because some structural decks are themselves composed of insulating materials, such as woodwool or, in other cases, insulation is added above the deck in addition to insulation at ceiling level.

12.17.13 Warm Deck Flat Roof

Figure 12.40(d): The main feature of this roof is that the insulation is placed above the structural deck in the form of rigid boards, such as Thermazone polyurethane foam roofboards. These particular boards are 600 × 1200 × 50 or 80 mm thick; their edges are rebated for

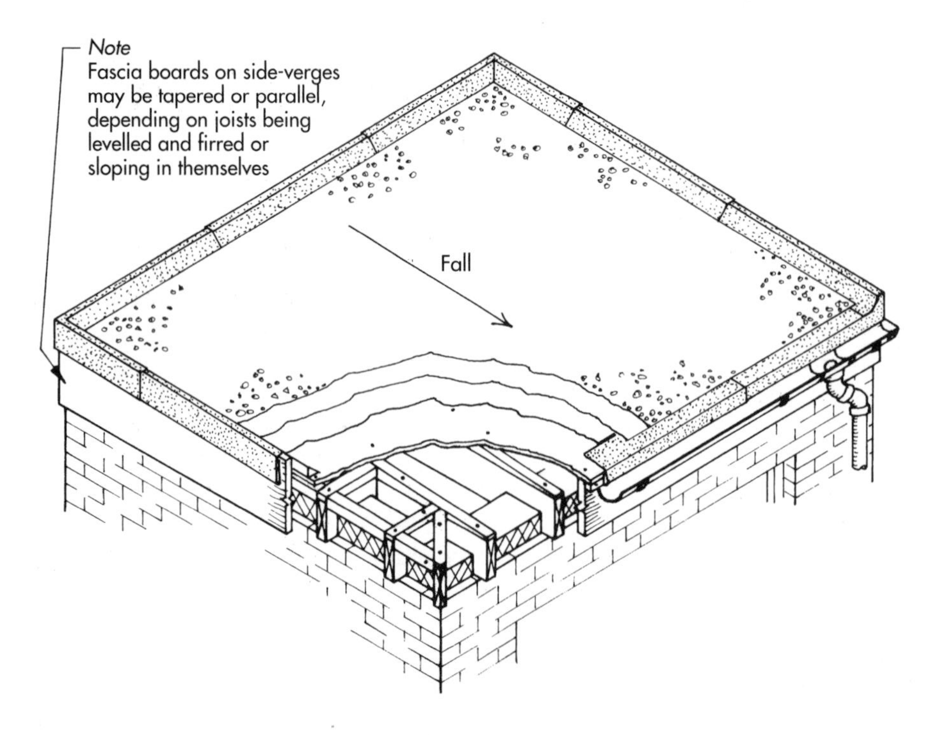

Figure 12.40(c) Partly-exposed view of independent flat roof

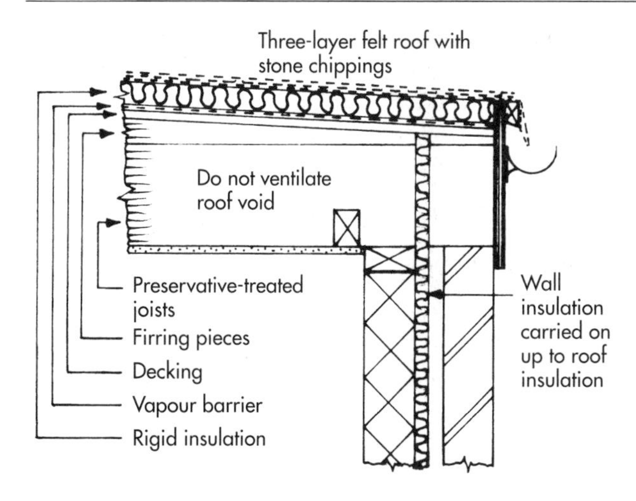

Figure 12.40(d) Warm deck flat roof (eaves' detail)

interlocking to avoid cold bridging and they have bitumen glass fibre facings on each side.

12.17.14 Construction

Figure 12.40(d): The rigid insulation is bonded or fixed with mechanical fastenings to a layer of felt vapour barrier, which has been fully bonded to a sheet decking. The decking is fixed to tapered firring pieces which are fixed to the joists. To avoid cold bridging around the perimeter of the roof, any cavity insulation, as illustrated, should be carried on up to meet the deck insulation. In this type of construction, the roof-void is not to be ventilated. All roof timbers should be preservative-treated. The waterproof covering to the roof is recommended to be of three-layer high performance built-up felt and limestone chippings.

12.17.15 Cold Deck Flat Roof

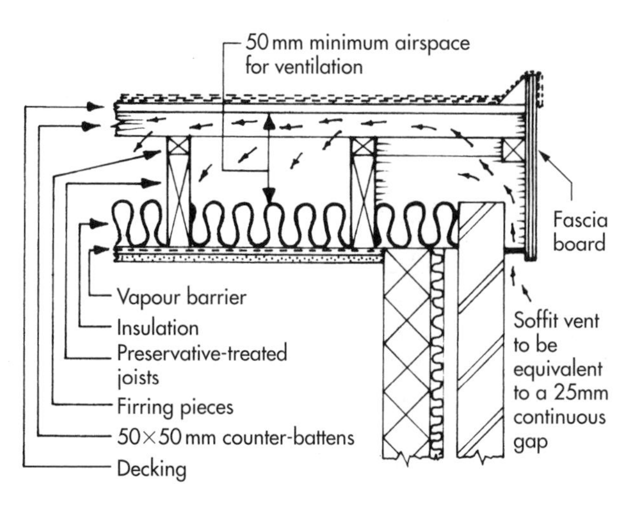

Figure 12.40(e) Cold deck flat roof (verge detail)

Figure 12.40(e): The main feature of this roof is that the insulation is placed *below* the structural deck, between the joists at ceiling level. To avoid cold bridging, the ceiling insulation should join up to the cavity insulation, as illustrated, with care being taken not to block the perimeter air vents. These vents, positioned in the soffit area, can be continuous or intermittent and must be on opposite sides of the roof for cross-ventilation. Where this is not possible, then this type of roof should not be used. The opening of the vents, incorporating an insect screen mesh, should be equivalent to a continuous 25 mm gap. Additional, intermediate roof vents are recommended for spans over 5 m and for roofs with an irregular plan-shape.

12.17.15 Construction

Figure 12.40(e): *Approved Document F2* also recommends that there should be a free airspace of at least 50 mm between the insulation and the underside of the roof deck. This will not normally cause a problem if the continuous air vents can be placed at right-angles to the joists on opposite sides of the roof, but in situations where the air vents are parallel to the joists, as illustrated, airflow across the joists can be achieved by positioning 50 × 50 mm counter-battens at joist-spacings across the roof joists, fixed on top of the firring pieces and to the short, projecting return-joists. Under the insulation (which can be of a rigid or flexible, but not loose type), a vapour barrier should be placed at ceiling level, using minimum 500 g polythene or metallized polyester-backed plasterboard. All roof timbers, as before, should be preservative-treated and the waterproof roof-covering should be as recommended for the warm deck flat roof.

12.18 DORMER WINDOWS AND SKYLIGHTS

12.18.1 Introduction

Roof lights in the form of dormer windows and skylights are usually found in roof spaces used for storage or habitation. Both of these windows involve a trimmed opening in the roof slope and the use of thicker trimming and trimmer rafters, according to the size of opening and the amount of trimmed rafters to be carried. The trimming rafters that carry the trimmers and their load of trimmed rafters, can also be – and usually are nowadays – formed by fixing two common rafters together, as indicated in the illustration.

12.18.2 Dormer Windows

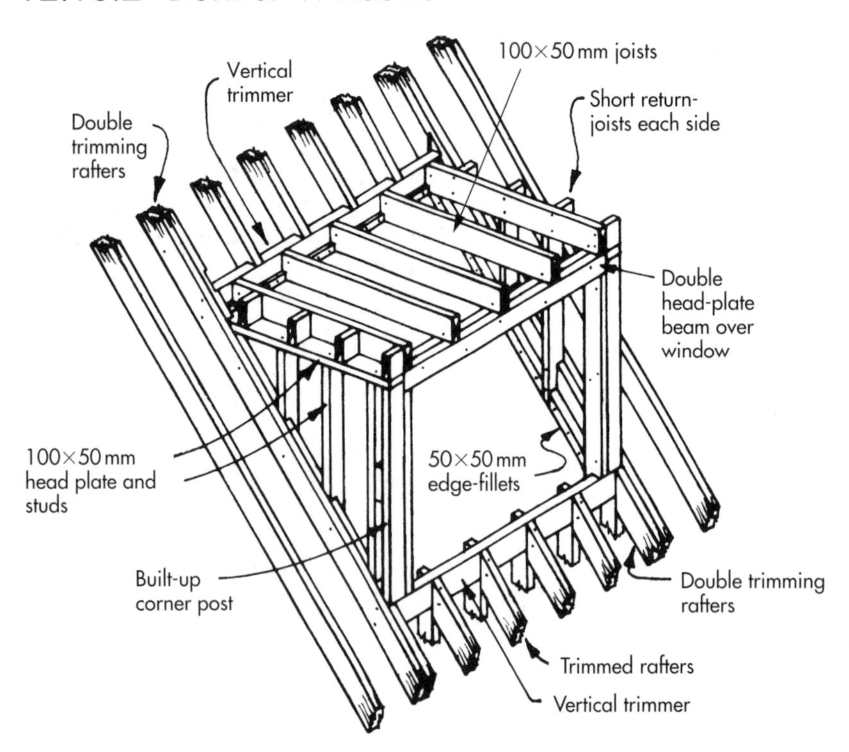

Figure 12.41 Skeleton dormer with window omitted

Figure 12.41: These traditional constructions protrude vertically from the eaves or middle area of the roof and have triangular sides known as cheeks, framed up from minimum 100 × 50 mm sawn studs, sheathed externally with ex. 25 mm diagonal boarding (parallel to the roof slope) or 15 mm WBP plywood or Sterling board. If the cheeks are to receive tile or slate cladding, a breather membrane of building paper (not a roof underlay) should be fixed to the sheathing behind the preservative-treated tile battens. The windows can be uPVC from the outset, but are usually specified initially (for economy) to be the wooden stormproof casement type, with wide rebates for 14 mm double-glazed sealed units.

12.18.2 Dormer Roof

Figure 12.41: The roof may be flat and the 100 × 50 mm minimum joists firred to slope backwards to the main roof or to fall to a front gutter and corner downpipe which discharges onto the main roof. Dormer roofs may also be segmental, semi-circular, etc., or pitched with a gable end or hipped end and tiled or slated in keeping with the main roof. Typical construction details of the skeletal dormer are shown in the illustration. All timbers not rated according to BS 5268: Part 5, should be preservative-treated and the waterproof roof-covering should be as recommended for the warm-deck and cold-deck flat roofs.

12.18.3 Ventilation and Condensation Control

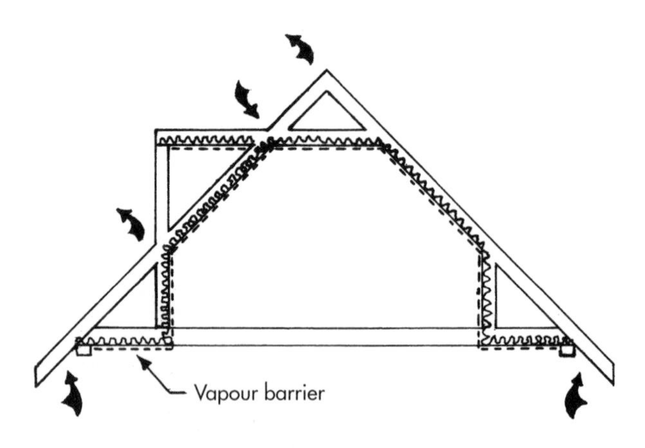

Figure 12.42(a) Ventilation and continuous insulation

Figure 12.42(a): Where there is a well-ventilated space within a pitched roof, above the insulation at ceiling level, a vapour control layer is not normally required; but where there is limited space above the insulation, making it difficult to ventilate effectively, as in the case of a loft room (attic), a vapour control layer *would* be necessary, as and where indicated in the illustration. The whole loft area, including the dormer, should be insulated up against the dwelling-side of the construction, against the vapour barrier, between the timbers of the dormer cheeks, the joisted roof, the rafters where affected, the ceiling joists, the limited areas of floor joists behind the ashlaring and between the ashlar studs as well.

12.18.3 Detailed Recommendations

Figure 12.42(a): In recommending the above, BS 5250 also recommends that the insulation on inclined and vertical faces must be held firmly in place to prevent slipping and that where there is limited space for the insulation, such as above the ceiling to the sloping rafters, a minimum 50 mm clear airspace between the insulation and the sarking should be provided. To achieve this, it may be necessary to increase the size of the roof and dormer timbers to accommodate the correct thickness of insulation.

12.18.4 Ventilation Requirement

Figures 12.42(a)–(c): To ensure the 50 mm clear airspace is maintained, BS 5250 suggests that consideration be given to installing some form of inert baffle boarding between the roof timbers to restrict the insulation. This recommendation could be met by using strips of 12 mm thick soft-fibre insulation board, easily cut on site, or pre-cut by the supplier's mill from the imperial-sized, metricated boards of 2.440 × 1.220 m. The strips could be cut slightly oversize and pushed into a friction fit and/or positioned against small protruding pre-set nails, as indicated in Figure 12.42(b). A separate baffle board, about 200/250 mm wide, is recommended between each rafter at the eaves, above the structural wall, again to restrict the insulation and so ensure that a minimum 25 mm clear airspace is maintained at this point, as indicated in Figure 12.42(c).

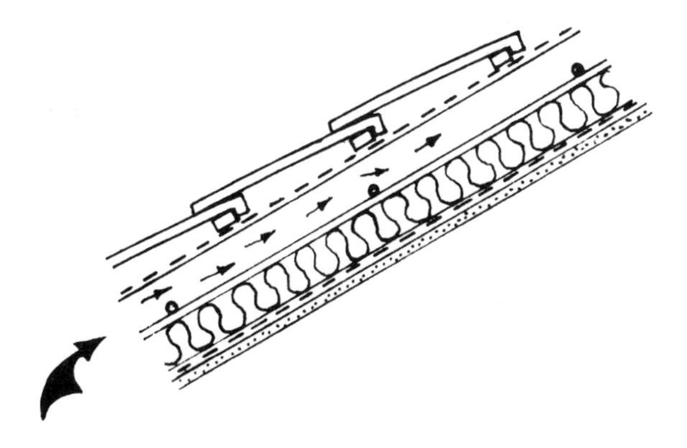

Figure 12.42(b) Maintaining minimum 50 mm airspace by using push-fit baffle boards

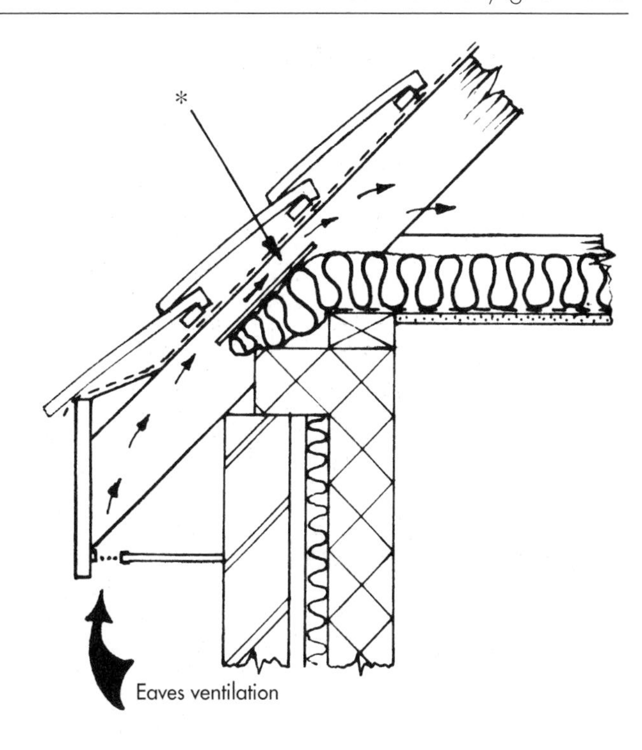

Figure 12.42(c) Maintaining minimum 25 mm airspace by using baffle boards*

12.18.5 Low-Level and High-Level Ventilation

Figures 12.42(a) and (c): With roofs pitched above 15°, low-level eaves' ventilation should normally be equivalent to a continuous 10 mm gap, but where there are areas of limited space above the insulation, as with an attic or loft room, the opening area of the vents – incorporating an insect screen mesh – should be increased to be equivalent to a continuous 25 mm gap, as indicated in Figure 12.42(c), and high-level ventilation openings at the ridge should be provided, equivalent to a continuous 5 mm gap.

12.18.6 Mid-roof Ventilation

Figure 12.42(a): Additional ventilation openings should be provided when there are obstructions in the ventilation path, such as a roof light or dormer window. These vents should be placed immediately below the obstruction, equivalent to 5 mm × obstruction-length and immediately above the obstruction, equivalent to 10 mm × obstruction-length. Finally, to achieve cross-ventilation within the dormer's cold-deck flat roof, counter-battens, as described earlier, will have to be used.

12.19 SKYLIGHTS

12.19.1 Introduction

These lay in the same plane as the roof slope and traditionally consisted of a glazed skylight window, hinged on the underside at the top and fixed to a raised curb or lining. Although metal aprons of lead, zinc, etc. were fixed to the back and side edges of the skylight, these windows were not always weathertight.

12.19.2 Roof Windows

Figure 12.43 Roof window (skylight)

Figure 12.43: Modern skylights, referred to as roof windows by the manufacturers, are a different proposition. These skylights are very sophisticated and reliable. They are made from preservative-impregnated Swedish pine, clad on the exterior with aluminium. The casements are of the horizontal pivot type with patent espagnolette locks, seals and draught excluders, and are double glazed with sealed units. The windows, suitable for roof pitches between 20° and 85°, are easily fixed to the rafters with metal L-shaped ties provided. Metal flashings and full fitting and fixing instructions are also supplied with each unit.

12.19.3 Loft Conversions

If these windows are to be installed in a roof already tiled or slated, as in a loft conversion, depending on the size of the window, it may be necessary to shore up the roof temporarily within the loft, to enable the opening to be made. This can be as simple as fixing two 100 × 50 mm timber plates, flat-faced across the rafters, one slightly higher than the proposed opening, the other slightly lower; 100 × 50 mm sole plates are fixed across the ceiling joists and similar-sized struts are fitted between the high and low plates. If possible, the struts should be at right-angles to the roof slope. The affected rafters can now be cut and removed safely to enable a top and bottom trimmer to be fixed in position. The faces of the trimmers should also be at right-angles to the roof slope. When the trimmers are fixed to the trimming rafters and to the ends of the newly trimmed rafters, the shoring can be released.

12.20 EYEBROW WINDOWS

12.20.1 Introduction

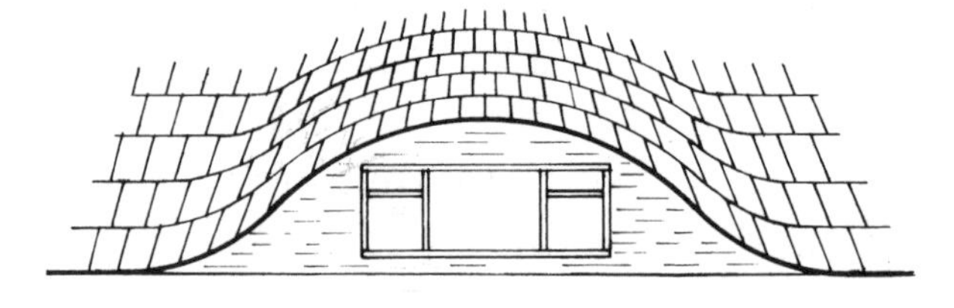

Figure 12.44(a) Eyebrow window

Figure 12.44(a): The eyebrow window is another form of roof light, serving roof spaces used for storage or habitation. This type of window is similar in most respects to a dormer window, whereby a trimmed opening in the roof slope will be required with triangular studded cheeks to the sides of the opening. These cheeks, which emanate from the trimming rafters, are not seen externally – unlike the cheeks of a dormer, which are seen. Vertical timbers known as ashlar studs are fixed to the underside of the window trimmer, the top edge of which should protrude about 100 mm above the roof slope to form the apron below the window sill. As with the dormer, the window, in a present-day construction, would initially be of the wooden stormproof casement type. To keep the window independent, to allow for future replacement, the old practice of resting the ceiling joists on the head of the window frame should be avoided. Instead, they should span from the upper trimmer in the roof to the double-plated beam at the head of the structural window opening.

12.20.2 Eyebrows and Steep Pitches

Figure 12.44(b): Roofs with tiled eyebrow windows need to be of a steep pitch of about 50–60° to accommodate the shallower pitch of the curved roof to the raised eyebrow, which should not be less than 35°. This is because the only tiles that can be used on roofs with eyebrows are non-interlocking 165 mm × 265 mm ($6\frac{1}{2}$in × $10\frac{1}{2}$ in) plain tiles, which are not recommended for roofs below 35° pitch. As illustrated, the geometry for the serpentine shape of the eyebrow is

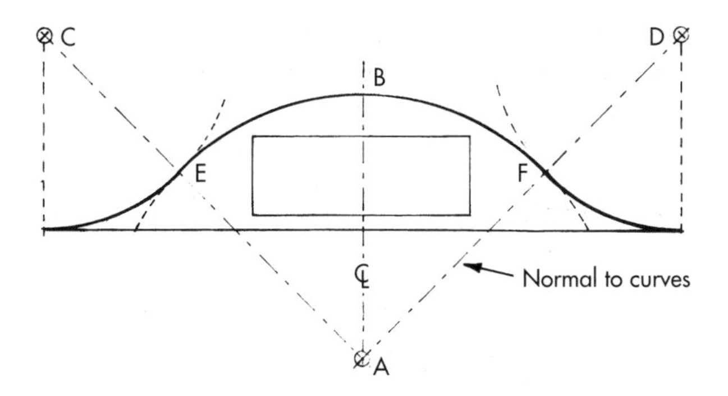

Figure 12.44(b) Setting out the eyebrow shape

best established to suit the predetermined window height and width.

12.20.3 Setting out the Eyebrow Shape

Figure 22.44(b): First, in scaled form (usually done by the architect), draw the outline of the window frame with a centre line A–B drawn vertically through it. By trial and error, try the compass or trammel point at different positions on the centre line until a suitable segment appears to fit above the frame. Now determine the outer limits of the segment by judging a suitable point E, near the base of the window frame and draw a line through it, radiating from the centre A. Measure the straight distance between E and B and set the same distance from B to mark point F. Establish another line radiating from centre A through F. These lines are geometrical normals to the curves required in reverse positions (cyma reversa) and at judged points along these normals, as at C and D, the centres can be established to complete the eyebrow (serpentine) shape.

12.20.4 Method of Construction

Figure 12.44(c): The construction details of traditional eyebrow windows can be found to vary to some extent, thereby creating variations in the methods of construction that are used. However, the following method is recommended. Assuming that the skeletal structure of the main roof is nearing completion and a free space between the double trimming rafters is already established to accommodate the eyebrow window, the first step is to position and fix the 100 × 50 mm floor plate A to the floor joists, ready to carry the ashlaring. Then, against the inside face of the double trimming rafters, fix a vertical stud B on each side of the opening, extended in height to run up about 50 mm above the estimated roof curve. Now fix the trimmer C between the vertical studs B, by forming housing joints, nailing or with modern fasteners such as Mafco CL-type framing anchors.

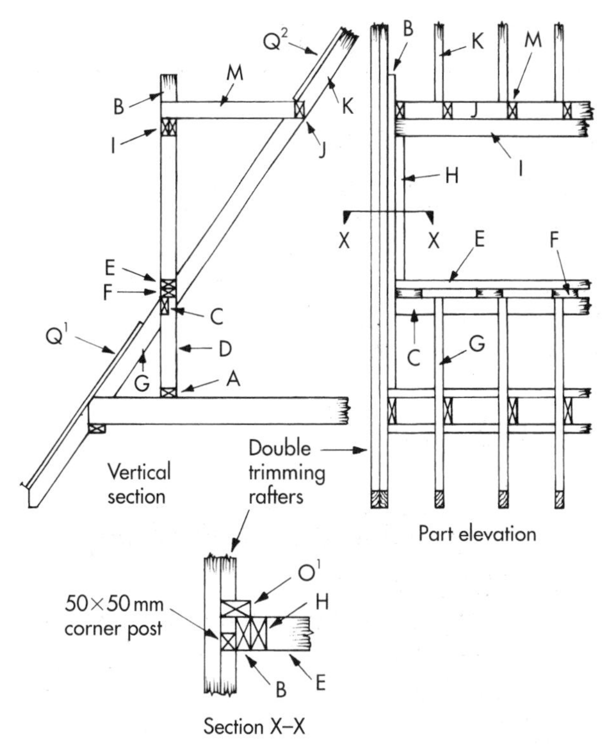

Figure 12.44(c) Method of construction

12.20.5 Sub-sill/Apron to Window

Next, cut and fix the ashlaring D to the floor plate and trimmer, at centres equal to rafter-spacings. Then cut and fix the sub-sill E between the vertical studs B, with the top edge protruding at least 100 mm above the roof to act as an apron below the window; the sub-sill is also fixed to 100 × 50 mm offcuts F, pre-fixed to the trimmer. Now complete the bottom section by cutting and fixing the trimmed rafters G to the wall plate at the bottom and the trimmer C at the top. Above the sub-sill now cut and fix the vertical sub-frame stud H to the face of stud B on each side of the opening, equal in height to the window frame plus a minimum tolerance of 7 mm for fitting and for the lead apron to be dressed under the sill. Studs H act as bearings for the double head-plate beam I over the window, which should now be cut to length, fixed together by stagger-nailing through the face side at about 400 mm centres, then fixed in position at each end with 100 mm round-head wire nails.

12.20.6 Upper Trimmer and Trimmed Rafters

Now start the ceiling structure by levelling across from the top of the head-plate beam I and marking the side of the trimming rafters on each side of the opening. This establishes the underside of the upper trimmer J, which should now be cut and fixed in a similar way to that described for the lower trimmer. Next, with birdsmouth cuts at the foot and plumb cuts at the head, cut and fix the trimmed rafters K from trimmer J to the ridge board.

12.20.7 Ceiling above the Window

The 100 × 50 mm ceiling joists M can now be fitted and fixed. They rest on the window-head beam I and are skew-nailed flush to its front edge, while at the other end they are fixed to the face of the trimmer J in various ways similar to fixing the ends of the trimmers C and J. As illustrated in the elevational view in Figure 12.44(c), the ceiling joists M should be positioned and fixed immediately to the side of an alignment with the trimmed rafters K.

12.20.8 Studded Side Cheeks

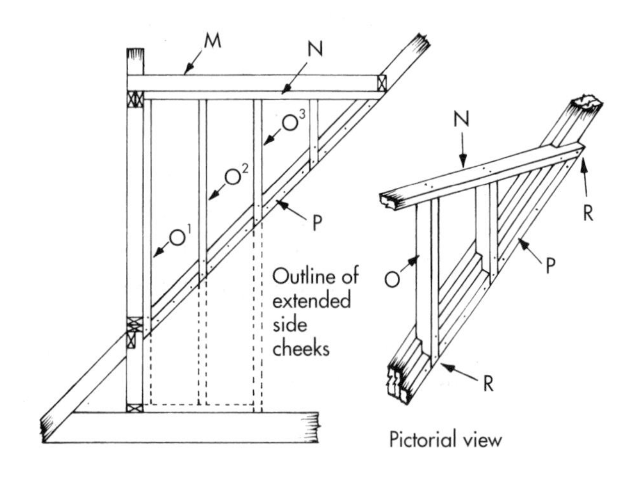

Figure 12.44(d) Studded side-cheeks

Figure 12.44(d): Sometimes the side cheeks of dormer and eyebrow windows – or a part of them, as illustrated by the broken lines – carries on down to the floor and the line of ashlaring each side of the window takes up more of the available floor space, giving an increased vertical face to the dwarf walls of the room. However, the following notes assume that the cheeks are to be triangular.

12.20.9 Triangular Cheeks

First, notch the bottom of stud O^1, as seen at R in the pictorial illustration, to lap onto the trimming rafter, then cut squarely to length to finish 50 mm below the top of the window-head beam. Fix to the side of stud B and through the half-lap notch to the trimming rafter with 100 mm wire nails. Next, notch the head-plate N, in a similar way to O^1, to fit against the trimming rafter near trimmer J and cut to length to fit squarely on top of stud O^1 and fix in position. Now, at 400 mm centres, fit and fix the remaining vertical studs O^2, O^3 etc., similar to O^1, fixed at the head and through the half-lap notch. Finally, cut and fix the 50 × 50 mm edge fillets P as shown.

12.20.10 Making the Full-size Template/Sheathing

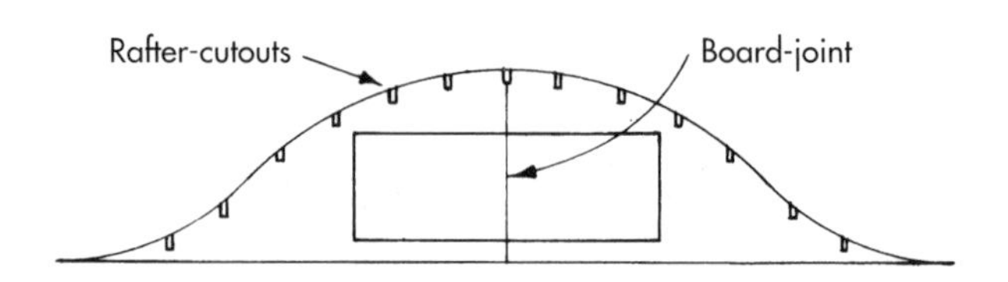

Figure 12.44(e) Making the template/sheathing

Figure 12.44(e): The next important step is to lay out one or two sheets of 12–18 mm WBP plywood or Sterling board and by scaling measurements from the working drawing, set out the eyebrow shape full size by using the geometrical method already described. The idea is that these boards should be set out to include the whole vertical face of the eyebrow window. This will give the dual advantage of the board(s) being fixed against the front of the studded structure already erected, to act as a template for the formation of the eyebrow shape, while also becoming a permanent sheathing.

12.20.11 Marking the Rafter Cutouts

Once the eyebrow shape is marked, it should be cut out with a jigsaw. Next, the eyebrow-rafter positions – equal to the main roof common-rafter centres – must be marked around the curved edge of the template and set down for depth, as indicated in Figure 12.44(e), giving the appearance of a castellated edge. To simplify this, the one- or two-piece template can be laid out flat on the roof slope, either immediately below the window opening, with the template's base resting against temporary nails driven in near the fascia plumb-cut, as indicated at Q^1 (Figure 12.44(c)), or on the roof slope above the window, with the base resting against the ceiling joists M, as indicated at Q^2. Either way, the template must be in an exact position laterally, i.e. the centre of the template must be equal to the centre of the window-width. Resting against the rafters like this, their positions can be easily marked onto the curved-template edge and lines drawn down at right-angles to the base, equal in depth to the eyebrow-rafter plumb cut, as illustrated in Figure 12.44(f).

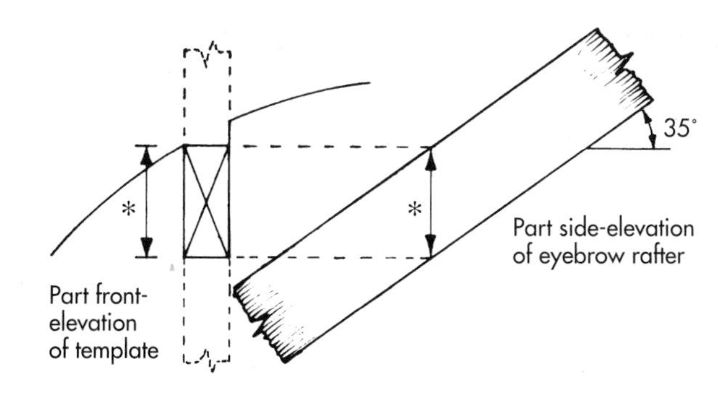

Figure 12.44(f) Cutouts equal plumb-cut depth*

12.20.12 Cutting the Window Opening

Before removing the template/sheathing from the roof to make the cutouts, it should be offered up and adjusted for its true position against the face of the studded structure, then marked around the inside edge to outline the window opening. This, then, can be cut when making the cutouts for the eyebrow rafters – which should now be done, again with the aid of a jigsaw.

12.20.13 Pitching the Eyebrow Roof

Figure 12.44(g): It must be appreciated that, because the eyebrow rafters are pitched at the same angle, say 35°, around the serpentine profile, the surface shape and its effect on the eaves is geometrically similar to a cylinder resting at 35° to the horizontal plane, whereby its base, representing the eaves, would be at 55° to the horizontal plane. From a side elevation, this would be at right angles (35° + 55° = 90°) to the cylinder/roof slope, as indicated between points A, B and C in the illustration. This can be proved in a practical way by bending a strip of 6 mm plywood, say 265 mm wide, equal to the tile depth, over the eyebrow shape near the edge and noting that it would follow the same path indicated at A–B. If you can imagine this plywood strip being equal to a line of tiles laid side by side, you should then appreciate that the unequal projection of the eaves' edge must be like this for the tiles.

12.20.14 Fixing the Template

The template/sheathing board(s) should now be fixed lightly in position, to be stable enough to work against, but able to be removed easily if necessary. This is because the studwork yet to be built-up behind the template might be awkward to fix properly in certain places with the template in position.

12.20.15 Eyebrow Rafters

The eyebrow rafters, seen in their diminishing lengths in Figure 12.44(g), have an acute-angled cut at one end for fixing to the top edges of the main roof's rafters, like sprocket pieces, and are either finished with a plumb cut at the other, front end, or with a shallow plumb cut and

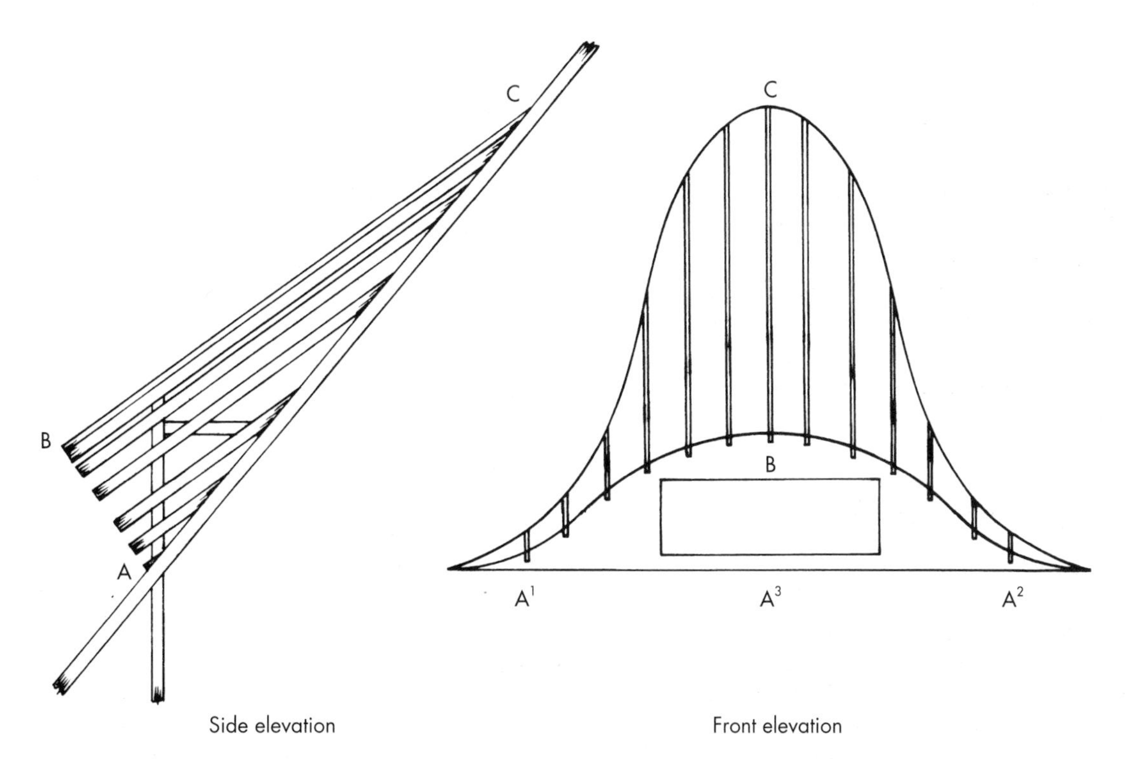

Figure 12.44(g) Formation of eyebrow rafters: left, side elevation; right, front elevation

a seat or soffit cut. Mostly, eyebrow eaves' rafters are left visible, but can be found covered up with a soffit lining. Whether visible or not, a strip of 6 mm WBP plywood will be required, either for fixing on top of the rafters in the area of the projecting eaves, if the rafter-ends are left visible, or for shaping to a diminish and fixing on the underside of the rafters, if it is to be lined with a soffit.

12.20.16 Finding the Angles

Figure 12.44(h): The illustrations used here assume a main roof pitch of 50° and a pitch of 35° for the eyebrow rafters. The main angle required is for the sprocket cut. Thinking of this, as illustrated, in the form of right-angled triangles, the cut for this very acute angle would be

$$a - b = 50° - 35° = 15°$$

Also illustrated is the setting out from a roofing instrument, to provide an alternative way of visualizing the angle required. If a soffit seat-cut is needed, we already know this to be 35° and, therefore, a plumb cut, if required, would be 90° − 35° = 55°.

12.20.17 Finding the Lengths

On this type of roof, a practical approach to finding the diminishing rafter-lengths will be more expedient. One way of doing this, is to establish the longest rafter in the crown position and then use it as a guide to finding the length of its neighbour on each side. First, cut the 15° sprocket angle on one end and lay the rafter in position, resting in the profile cutout and on the main rafter. Because the sprocket cut is shallow and therefore long, small adjustments up or down until the cut is properly seated will ensure that the rafter is at its required angle of 35°. Now fix this position with a temporary nail. At the eaves' end, judge the approximate amount of projection required, with a margin for error of, say 100 mm, and mark for the initial cut. If considered necessary, check this by squaring down to a point near the apron upstand, as indicated at A–B–C in Figure 12.44(g).

12.20.18 Cutting and Fixing the Eyebrow Rafters

The first rafter is now removed and tried in the cutouts on either side and, again, a judgement is made regarding the approximate amount of projection required. The side of the rafter is marked accordingly and the marked length is transferred to other timbers to make two more rafters. Rafter number one can now be fixed back in place with the temporary nail through the sprocket-cut again. Newly cut rafters two and three can now be used to determine the diminished length of the next pair, rafters four and five. Then rafters two and three can be fixed like rafter number one with a temporary nail through the sprocket-cut into the main rafters. This sequence is carried on down the eyebrow shape in pairs of rafters, one on each side of the crown position, until the ends are reached at main-roof level. Only the sprocket-cut ends are nailed, the other ends should be unfixed for now, resting in the cutouts.

12.20.19 Marking the Eaves' Edge

The precise eaves' edge can now be determined, as mentioned earlier, by pinning down a strip of 6 mm plywood over the rafter-ends of the eyebrow and marking the line on each rafter-top. This task can be made easier by using the three points A^1, A^2 and B, referred to in Figure 12.44(g). First, on each extreme of the eyebrow, at the lowest and smallest sprocket, mark points A^1 and A^2 and projecting, say, 75 mm from the vertical face of the

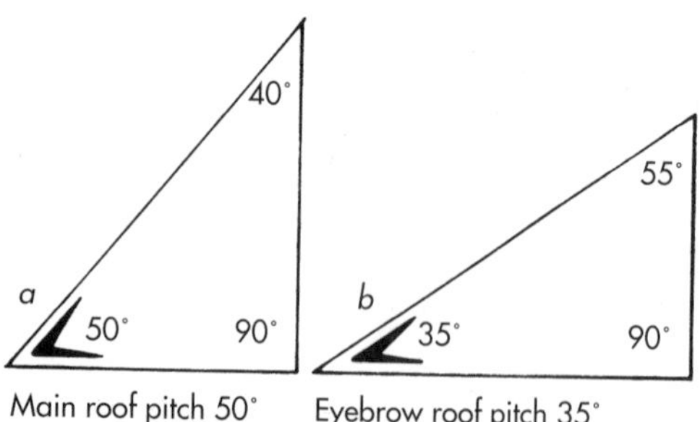

Main roof pitch 50°

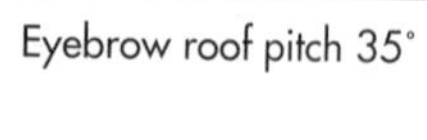

Eyebrow roof pitch 35°

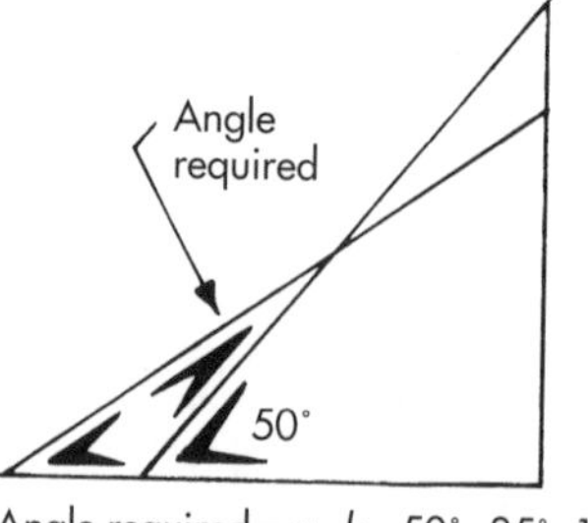

Angle required = a − b = 50° − 35° = 15°

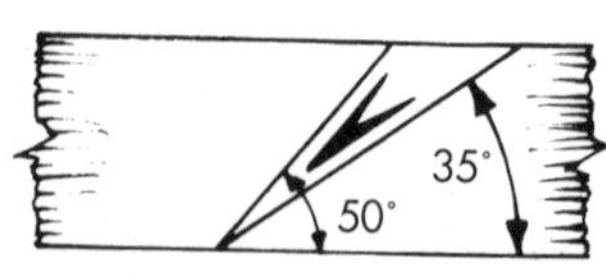

50° − 35° = 15°
Setting out from roofing instrument

Figure 12.44(h) Finding the angles

sheathed window. Strike a line across the tops of the trimmed rafters from point A^1 on one side to point A^2 on the other and mark the line midway at A^3. Now square up from this middle point A^3 to mark point B on the eyebrow rafter in the crown position. This can be done with a straightedge, with one end resting on mid-point A^3 and the other being adjusted at point B up against a roofing or try-square. These three points, marked on the rafter-tops, greatly assist in positioning the ply and establishing the eaves' edge.

12.20.20 Eaves' Finish

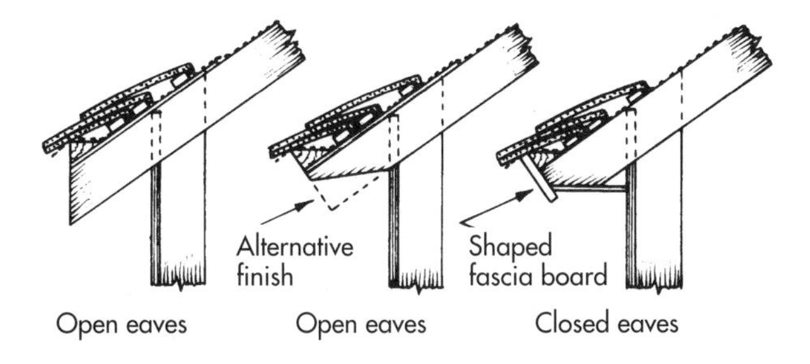

Figure 12.44(i) Eaves' finish

Figure 12.44(i): Next, remove the temporarily nailed rafters and mark and cut the required eaves' finish, related to the eaves'-edge line just established. Three optional finishes are shown in Figure 12.44(i). The two open-eaves details are most common and seem to be more in keeping with the aesthetics of eyebrow windows. The 6 mm WBP plywood is shown fixed on top of the rafters, to give a better visual finish on the exposed underside. This has to be wide enough to mask the greatest projection at the crown. A tilting fillet is also fixed at the eaves' edge, to give the required uplift to the eaves' tiles. This may need a few strategic half-depth saw cuts to ease it around the serpentine shape, on top for hollows and on the underside when going over the crown.

12.20.21 Closed Eaves

The closed-eaves detail in Figure 12.44(i) requires a narrow fascia board, backed up by a tilting fillet, as shown. The fascia board, of about 75 mm width and normal thickness, must be shaped to follow the eyebrow curvature. The soffit lining, of 6 mm WBP plywood, needs to be fixed to a level seat cut – to be less problematic geometrically – and is best fixed before the fascia to allow the outer edge to be trimmed against the rafter-ends, if necessary. When the eaves' cuts are known or have been decided, marked out, cut and completed, refix the eyebrow rafters in their previous positions, adding a few more 100 mm wire nails as final fixings.

12.20.22 Fixing Side-support Studs

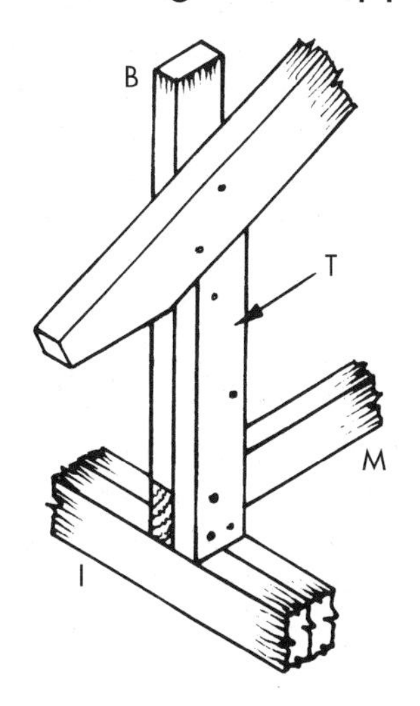

Figure 12.44(j) Studding on window beam

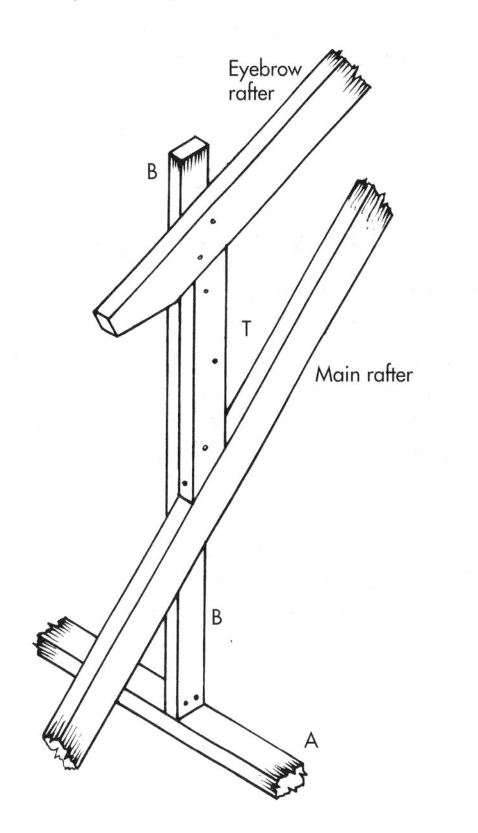

Figure 12.44(k) Studding on rafter and floor plate

Figures 12.44(j) and (k): The final studwork directly behind the sheathing consists of two vertical studs per eyebrow rafter and a row of staggered noggings. The first, vertical stud B, is a side-support stud which, as shown, either rests squarely on the ceiling joist, when the studding is on the window beam, or rests squarely on the floor plate which continues through on each side of the window opening. With the sheathing/template still in position, studs B should now be cut and fixed.

They are cut with a square-ended allowance, to project at the top, skew-nailed at the base and side-fixed to the eyebrow rafters. Side-fixings should also be made where studs B rest against main-roof rafters (Figure 12.44(k)).

12.20.23 Fixing Load-bearing Studs

Now that the eyebrow rafters are supported by the side-support studs B, the sheathing/template can be removed, if considered necessary, to facilitate the fixing of studs T, which tend to get more awkward to fix as they diminish in height. These studs are load-bearing studs which, as shown, are cut with an acute angle at the top to fit under the eyebrow rafter and either rest squarely at the bottom, on the window beam, or are cut with another acute angle to rest on the main-roof rafter. With an eyebrow pitch of 35°, the top angle would be 90° − 35° = 55°; and with a main roof pitch of 50°, the bottom angle would be 90° − 50° = 40°. Whether removing the sheathing/template or not, studs T, when cut and fitted, should be side-fixed to studs B, skew-nailed to the window beam, where applicable, and an edge-fixing nail should be driven through the sharp point of the acute angles. 100 mm round-head wire nails should be used throughout.

12.20.24 Fixing Final Noggings and Sheathing

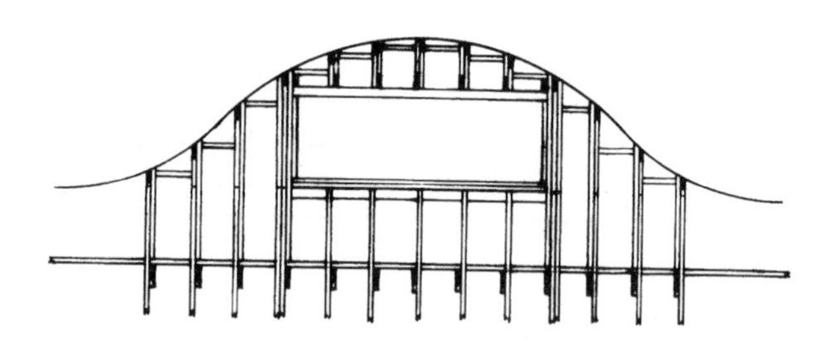

Figure 12.44(l) Staggered noggings

Figure 22.44(l): To give the studded eyebrow structure more rigidity, 100 × 50 mm noggings should be cut in and fixed as near to the top as possible, as indicated in the exposed elevational view. Because the noggings should be kept in a horizontal position, they will take on a staggered appearance. Finally, fix the sheathing/template to the whole studded structure with 50 mm round-head wire nails at approximately 200 mm centres and cut off the square-ended projections of studs B, flush to the top of the eyebrow rafters, following the top edge of the sheathing/template.

12.21 LEAN-TO ROOFS

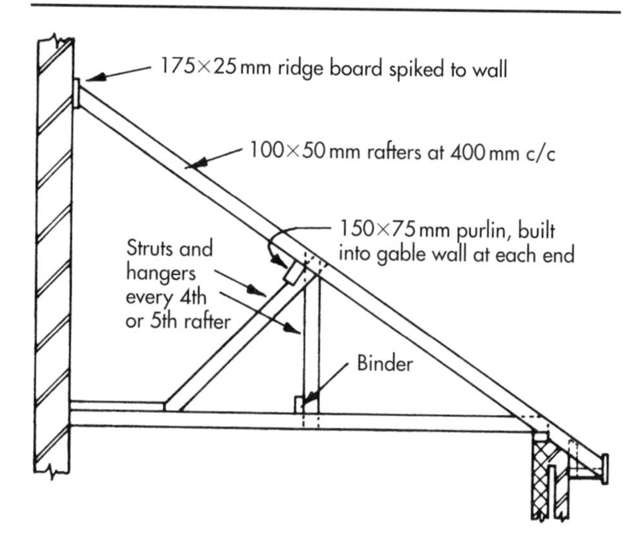

Figure 12.45 Traditional lean-to (double) roof

Figure 12.45: This type of roof, usually found on parts of the building that extend beyond the main structure, comprises mono-pitched rafters *leaning* on the structural wall in various ways. This connection to the wall was usually in the form of a wall plate resting on wrought-iron corbels built into the wall at about 1 m centres, or a wall plate bedded on continuous brick corbelling projecting from the face of the wall. If the potential thrust of a particular lean-to roof can be discounted, the connection may be simply a ridge board fixed to the wall to take the plumb cuts of the rafters.

12.21.1 Ceiling-joists Connection

Traditionally, ceiling joists were built in or cut into the main wall. Nowadays, TW type joist hangers could be used. Technically, without purlins, this roof would be termed a *single roof* and would be restricted to a span of about 2.4m.

12.22 CHIMNEY-TRIMMING AND BACK GUTTERS

Figure 12.46: When a chimney stack passes through a roof, the rafters are trimmed around it in a similar way to trimmed openings for dormer or eyebrow windows and skylights. The trimmer rafters can be stop-housed, fixed with TT type joist hangers, CL type framing anchors, or simply butt-jointed against the trimming rafters – and may be vertical or leaning to the roof pitch. The former position is preferred, as this allows the trimmed rafters a birdsmouth bearing on the trimmers. Triangular blocks, boarding or sheeting material, as illustrated, form the usual back gutter to the stack.

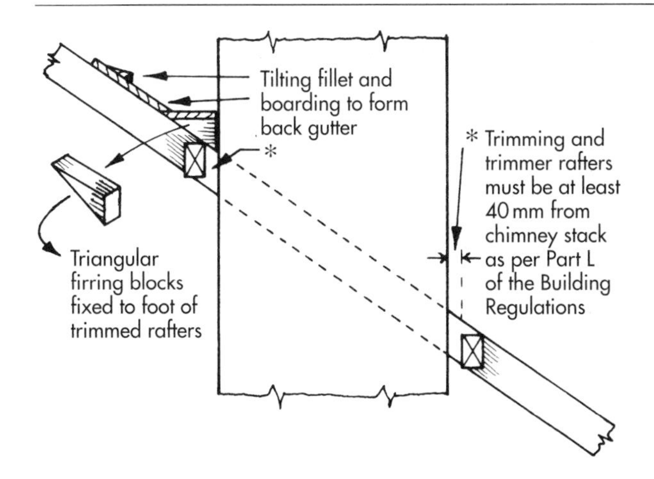

Figure 12.46 Chimney trimming and back gutter

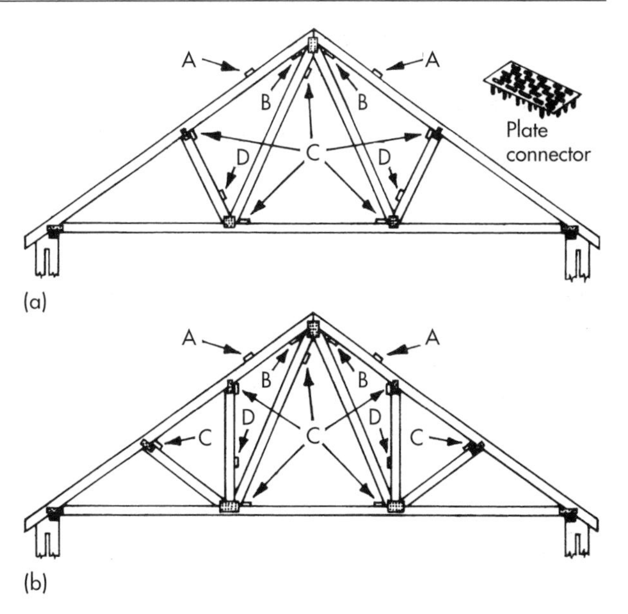

Figure 12.47(a) The Fink or W truss. (b) The fan truss

12.23 TRUSSED RAFTERS

12.23.1 Introduction

As mentioned briefly in the beginning of this chapter, roofing on domestic dwellings now predominantly comprises factory-made units in the form of triangulated frames referred to as trussed rafters. These assemblies are made from stress-graded, prepared timber, to a wide variety of configurations according to requirements. Most shapes have a named reference and the two most common designs used in domestic roofing are the Fink or W truss, and the Fan truss. All joints are butt-jointed and sandwiched within face-fixing plates on each side. These plates are usually of galvanized steel with integral, punched-out spikes for machine-pressing onto the joints. After initial positioning of the trusses, they must be permanently braced.

12.23.2 Bracing

Figures 12.47(a) and (b): The bracing-arrangement details given here and in the illustrations – as well as the other references to truss rafters – are based on current information given in the technical manuals obtained from Gang-Nail Systems Ltd, a member company of International Truss Plate Association.

- *Braces A*: 75 × 25 mm or 100 × 25 mm temporary longitudinal bracing, used to stabilize the position of the trusses during erection.
- *Braces B*: 97 × 22 mm (minimum size) permanent diagonal bracing, forming 45° angles to the rafters, running from the highest point on the underside of a truss to overlap and fix to the wall plate, starting at a gable end and running zigzag throughout the length of the roof, fixed to every truss with two 3.35 mm × 65 mm long galvanized wire nails. There should be not less than four braces (two on each slope) of any short-length duopitched roof. All joins of incomplete bracing-lengths should be overlapped by at least two trussed rafters. The angle of bracing given above as ideally 45°, should not be less than 35° or more than 50°.
- *Braces C*: 97 × 22 mm (minimum size) permanent longitudinal bracing, with fixings and overlap allowances as for diagonal bracing, positioned at all *node points* (points on a truss where members intersect), with a 25 mm offset from the underside of the rafters (top chords) to clear the diagonal bracing, extended through the whole length of the roof and butting tight against party or gable walls.
- *Braces D*: 97 × 22 mm (minimum size) permanent diagonal web chevron bracing, each diagonal extended over at least three trusses, required for duopitched spans over 8 m and monopitched spans over 5 m.

12.23.3 Advantages

One of the advantages of trussed-rafter roofs is the clear span achieved, without the traditional need for load-bearing partitions or walls in the mid-span area. Another advantage, to the building designer, is that the specialist truss-fabricator will only need basic architectural information to plan the truss layout in detail for the building designer's approval.

12.23.4 Site Storage and Handling

It is important to realize that although the trusses are strong enough to resist the eventual load of the roofing materials, etc., they are not strong enough to resist certain pressures applied by severe lateral bending. These pressures can have a delaminating effect on the metal-plated joints and are most likely to occur during truss delivery, movement across the site, site storage and lifting up into position – especially the motion of see-sawing over the top edges of walls when the truss is laying on its side face.

Recommendations

Storage on site should be planned to be as short a time as possible, preferably not more than 2 weeks. In bad weather, stored trusses should be protected by a waterproof cover, arranged to allow open sides for air ventilation. The trusses should be stored on raised, level bearers to avoid distortion and contact with the ground.

Vertical Storage

This is the preferred method. The trussed rafters are stored in an upright position, stacked close together against a firm, lean-to support at each end, resting on bearers at the position where the wall plates would normally occur, built up to ensure that any eaves' projection clears the ground and any vegetation present.

Horizontal Storage

This alternative method, where trussed rafters are laid flat, stacked up upon each other, requires a greater number of bearers which should be carefully arranged to give level support at close centres and be directly under every truss joint. This is to reduce the risk of joint-damage and long-term deformation of the trusses. If subsequent sets of bearers are used higher up in the stack to take another load of trusses, the bearers must be placed vertically in line with those below.

Inverted Storage

A third method of storage, preferred by some manufacturers and builders, is to invert the trussed rafters and support them on built-up side frames. These frames, resembling braced stud-partitions, are secured with raking struts and are set up parallel to each other and at the same span as the roof's wall plates (the length of the bottom chord or ceiling tie). The height of the two side frames must be more than the rise of the roof, to ensure that the apex of the upside-down trusses clears the ground.

Manhandling

On wide-span trusses – which are more liable to joint-damage from sideways-bending – it may be necessary to use additional labour to provide support at intermediate positions. When carrying trusses across the site, it may be safer and more manageable for a truss to be reversed so that the apex hangs down. On the other hand, when being offered up into its roof position, the truss should be upright and the eaves' joints should be the main lifting points. Laying trusses on their sides and pushing/pulling/see-sawing them across walls and scaffolding, etc., may make manhandling them easier, but is a completely unacceptable practice.

Mechanical Handling

When mechanical means are used, the trusses should be lifted in banded sets and lowered onto suitable supports. The recommended lifting points are at the rafter (top chord) or ceiling tie (bottom chord) *node* points (where the joints occur). Lifting single trusses should be avoided, but if unavoidable, a suitable spreader bar should be used to offset the sling-forces.

12.23.5 Providing Profiles for Gable Ends

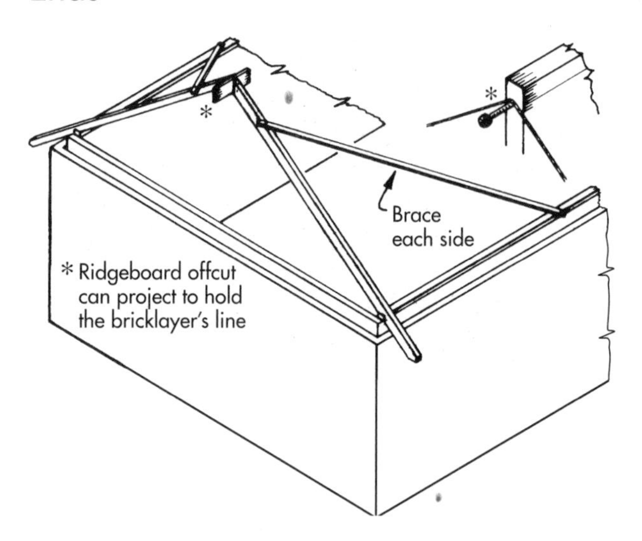

Figure 12.48 Profile erected for gable end

Figure 12.48: Gable-end walls are usually completed, or partially completed, before the trussed-rafter roof is erected. When this happens, a single trussed rafter frame or – if the trussed rafters are not yet on site – a pattern pair of common rafters, as illustrated, is fixed and braced up at each gable end to act as a profile for the bricklayer to use as a guide in shaping the top of the raking walls. If gable ladders (described later) are to project over the face of the wall, then the brickwork should be built up only to the approximate underside of the truss and completed after the ladders have been fixed in position.

12.24 ERECTION DETAILS AND SEQUENCE FOR GABLE ROOFS

12.24.1 Wall Plates and Restraint Straps

Wall plates for trussed rafters are jointed and bedded as already described for traditional pitched roofs. The next step is to mark the positions of the trusses at maximum 600 mm centres along each wall plate. This will indicate the clear areas for the positioning of the vertical restraint straps, which are now fixed.

12.24.2 Procedure

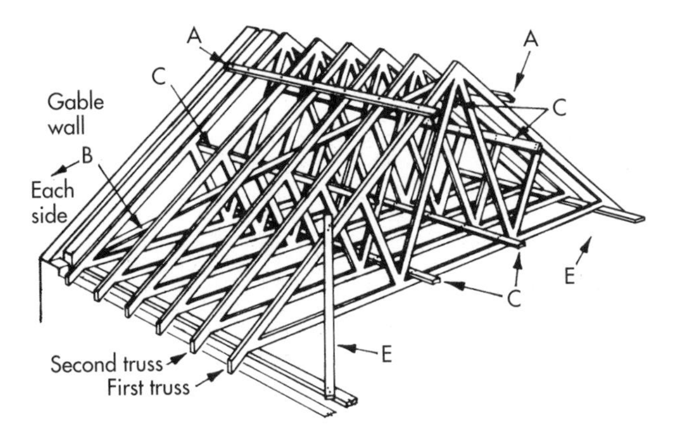

Figure 12.49 Erection of Fink trussed rafters

Figure 12.49: The erection procedure is open to a certain amount of variation, providing care is taken in handling and pre-positioning the trusses on the roof. Bearing that in mind, the following procedure is based on the notes given in the technical manual on trussed rafters, mentioned on page 153.

By using framing anchor truss clips (the recommended fixing, which does not rely on the high degree of skill required for successful skew-nailing) or by skew-nailing from each side of the truss with two 4.5 mm × 100 mm galvanized round wire nails per fixing, the first truss is fixed to the wall plates, in a position approximately equal to the first pair of common rafters in a hipped end. This determines the apex for the diagonal braces, marked B. Stabilize and plumb the truss by fixing temporary raking braces E on each side, down to the wall plates.

Fix temporary battens A, on each side of the ridge and resting on the gable wall. Position the second truss and fix to the marked wall plate and to the temporary battens A, after measuring or gauging to the correct spacing. Proceed until the last truss is fixed near the gable wall. Now fix the diagonal braces B with two 3.35 mm × 65 mm galvanized round wire nails per fixing and then continue placing and fixing trusses in the opposite direction, braced back to the first established end. Finally, fix the braces marked C throughout the roof's length and fix horizontal restraint straps at maximum 2 m centres across the trusses onto the inner leaf of the gable walls, as illustrated previously in Figure 12.37(e).

12.25 HIPPED ROOFS UNDER 6 m SPAN

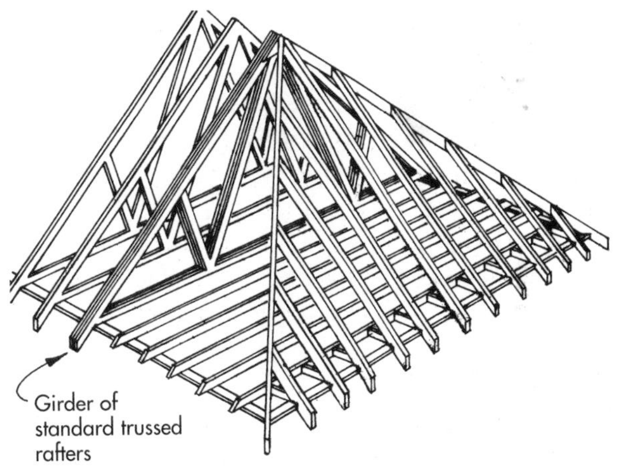

Figure 12.50 Hip-end construction for roof under 6.0 m span

Figure 12.50: As illustrated, the recommended hip-end for a roof of this relatively small span is of traditional construction. The main difference being that instead of the saddle board and hips being fixed to a ridge board and the first pair of common rafters, they are fixed onto a *girder truss* made up of manufactured truss rafters. This consists of two or three standard trussed rafters securely nailed together by the supplier (preferably), or fixed on site to a nailing pattern stipulated by the supplier.

12.25.1 Procedure

After the wall plates are jointed, bedded and set, the hipped ends are set out as already described for traditional roofing, the positions of the standard trusses marked and the vertical restraint straps positioned and fixed. The erection sequence starts with the fixing and temporary bracing of the girder truss, followed by the infill of the standard trusses and bracing. The hip ends are then constructed, using hip rafters and jack rafters of at least 25 mm deeper section than the truss rafters to allow for birdsmouthing to the wall plates. Ceiling joists – unlike those illustrated – may also run at right-angles to the multiple girder truss, supported on the girder truss by minihangers.

12.26 HIPPED ROOFS OVER 6 m SPAN

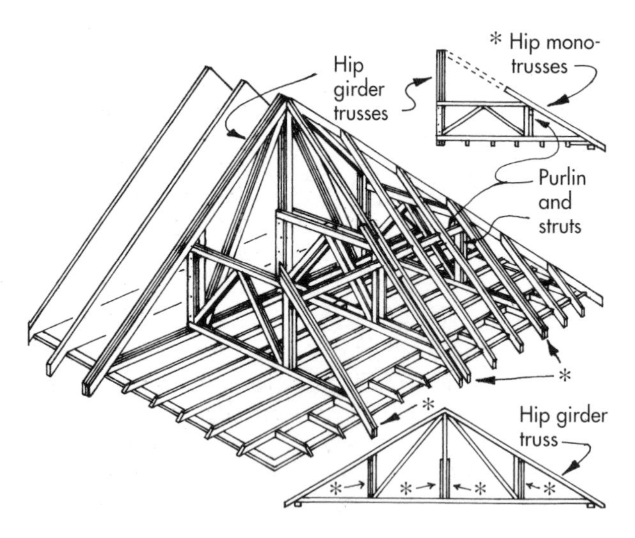

Figure 12.51 Hip-ended construction for roof over 6.0 m span

Figure 12.51: There are various alternative methods of forming hip ends in trussed rafter roofs. The example given here, therefore, is just one of the methods that deal with large spans. Assuming that the wall plates have been jointed and bedded and are now set, the erection procedure is as follows.

12.26.1 Procedure

Three special trusses, known as *hip girder trusses*, are – as mentioned for the previous hip end – securely nailed together by the supplier (preferably), or fixed on site to a nailing pattern stipulated by the supplier. The girder is fixed at the half-span (run) position and infill ceiling joists are laid and fixed. In three positions, as illustrated, other special trusses rest across the ceiling joists and are fixed to them and to the vertical members of the girder truss. These secondary trusses are known as *hip mono trusses* and six are required at each hip end. Two in the centre are nailed together with rafter-thickness packings in between and fixed in position to straddle the central vertical girder member – and house a half-length *flying* crown rafter. The other hip mono trusses, nailed directly together in pairs, without packings, are fixed at the quarter-span position on each side of the girder truss and are splay-cut like jack rafters to fit the hip rafters. A short purlin and vertical struts, as illustrated, are fixed to the mono trusses. Hips and jack rafters, as before, must be at least 25 mm deeper to allow for birdsmouthing to the purlin and wall plates.

12.27 ALTERNATIVE HIPPED ROOF UP TO 11 m SPAN

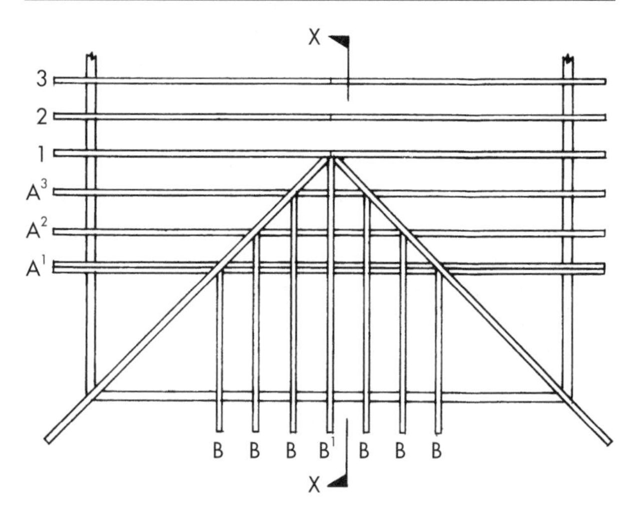

Figure 12.52(a) Standard centres hip system

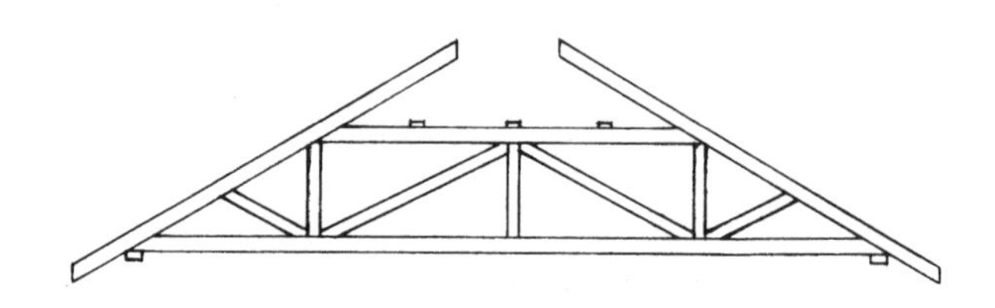

Figure 12.52(b) Flat-top hip truss (in position A)

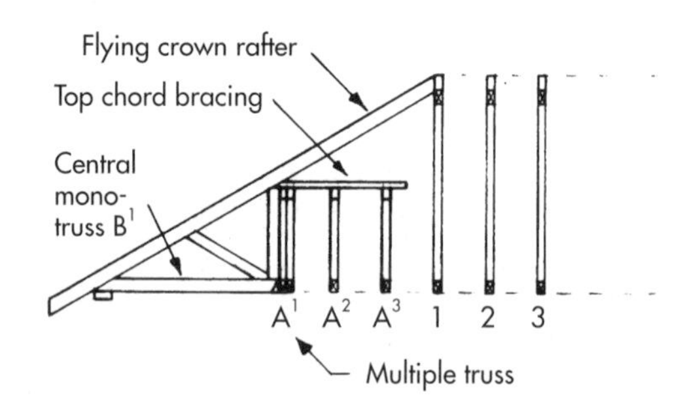

Figure 12.52(c) Section X–X

Figures 12.52(a)–(c): The most common construction for a hipped roof up to this span is referred to as a *standard centres hip system*. It is made up of a number of identical flat-top hip trusses A (Figure 12.52(b)) spaced at the same centres as the standard trusses, and a multiple girder truss of the same profile which supports a set of monopitch trusses B, set at right-angles to the girder. The corner areas of each hip are made up of site-cut jack rafters and infill ceiling joists attached to the hip girder. The flat tops (chords) of the hip trusses require lateral bracing back to the multiple hip girder.

12.27.1 Procedure

Assuming that the wall plates have been marked out for the hip end(s) and the truss positions, and that the vertical restraint straps have been fixed, a standard truss is first fixed at the half-span (run) position, labelled 1 in the illustration. The remaining standard trusses (2, 3 etc.) are then erected and braced. A ledger rail of 35 × 120 mm section – instead of a saddle board – is fixed at the apex of truss 1, at a height to suit the hip rafter's depth.

Next, the multiple girder truss A^1 is fixed, set in from the hip end by the span (length of bottom chord) of the mono truss B; then the two intermediate flat-top hip trusses, A^2 and A^3, are fixed and braced. Now a string line is set up, representing the *in situ* position of the hip rafters, and the flying rafters of the flat-top hip trusses are marked and cut back, allowing for the hip rafter's thickness. Next, the central mono truss B^1 is fixed, after its flying rafter has been trimmed and notched onto the ledger rail and its bottom end has been fitted into a pre-fixed truss shoe attached to the bottom chord of the girder. Now the hip rafters can be cut to length, birdsmouthed, notched and fixed in position. The remaining mono trusses B are then fitted and fixed into pre-fixed bottom-chord truss shoes, after trimming the flying rafter of each, ready for fixing to the hip rafter. Finally, the corner areas of each hip are completed with loose ceiling joists attached to the girder truss with mini hangers and loose infill jack rafters.

12.28 VALLEY JUNCTIONS

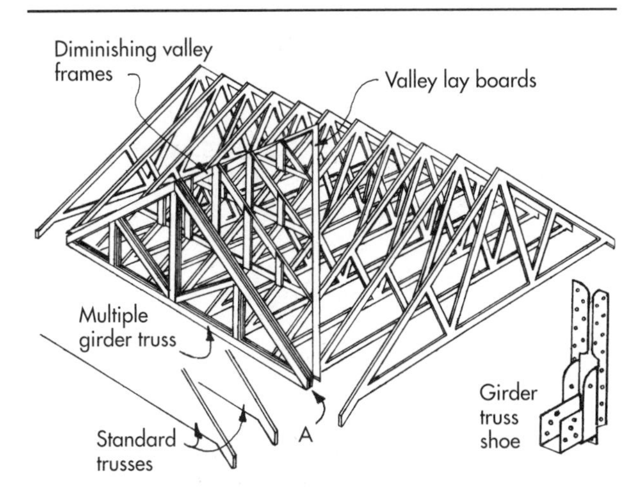

Figure 12.53 Valley junctions

Figure 12.53: Where a roof is so designed as to form the letter T in plan, a *valley set*, known as *diminishing valley frames* (as illustrated), can be fixed and braced directly onto the main trussed rafters in relation to lay boards or inset noggings. The noggings are required when the ends of the valley frames do not coincide with an underlying truss position. Lay boards or battens, running along the edges of the valley, are normally required by the roofer as a means of attaching the splay-cut ends of the tiling battens that are fixed at right-angles to the trusses.

12.28.1 Intersecting Girder Truss

At the intersecting point marked A, where the offshoot roof meets the trimmed eaves of the main roof – if no load-bearing wall or beam exists at this position – it will be necessary to have a multiple girder truss. This is to carry the ends of the trimmed standard trusses via hangers known as *girder truss shoes* (Figure 12.53). The girder is formed, either in the factory or on site, by nailing three intersection girder trusses together to a stipulated nailing pattern. Because of the heavy loads being carried, it may be necessary for the girder truss to have larger bearings in the form of concrete padstones.

12.29 GABLE LADDERS

Traditionally, when a verge projection was required on a gable end, this was achieved by letting the ridge board, purlins and wall plates project through the wall by the required amount – usually about 200 mm – to act as fixings for an outer pair of common rafters and the barge boards.

12.29.1 Modern Method

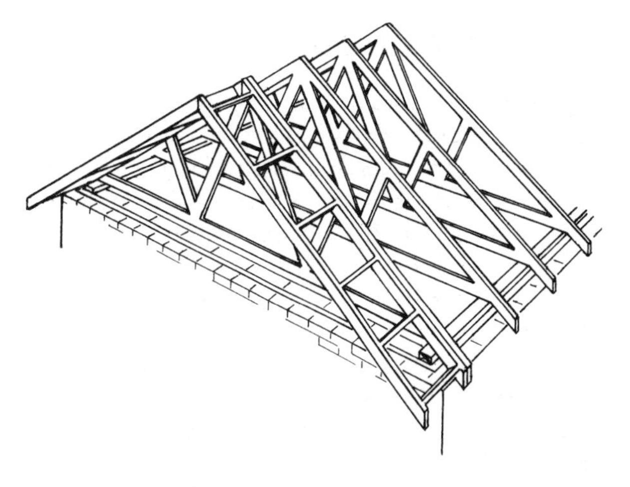

Figure 12.54 Gable ladders

Figure 12.54: As trussed-rafter roofs do not have purlins or ridge boards, the verge-projection is achieved by fixing framed-up assemblies, known as gable ladders, directly onto the first truss on the inside of the gable wall, as illustrated. The ladders, supplied by the truss manufacturer, are fixed on site by nailing at 400 mm centres and are subsequently lined with a soffit board on the underside and barge boards on the face side, after being built-in by the bricklayer. The usual practice is for the gable wall to be built-up to the approximate underside of the end truss, then completed after the ladders have been fixed.

12.30 ROOF HATCH (TRAP)

When trusses are spaced at 600 mm centres, it should be possible to simply fix trimmer noggings between the bottom chords of the trusses to form the required roof-trap hatch. In other cases, when the trusses are closer together or a bigger hatch is required, it will be necessary – but not desirable – to cut the bottom chord of one of the trusses, as illustrated.

12.30.1 Procedure

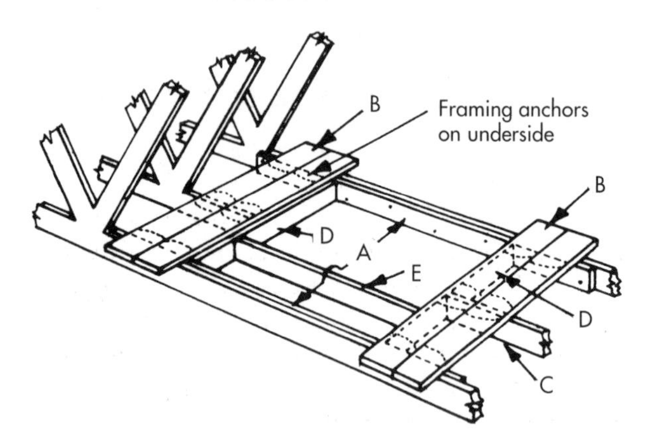

Figure 12.55 Forming roof trap

Figure 12.55: Typical details are shown for forming a trap hatch in the central bay of the trussed rafters. First, 35 mm thick framing timbers A are fixed to the inside faces of the ceiling joists that will be acting as trimming joists. Next, two 150 × 38 mm boards B, spanning three trusses, must be fixed in position on each side of the proposed trap hatch before the central ceiling-joist chord is cut. The boards are fixed on the underside to the side of the ceiling-joist ties with bracket-type framing anchors and 31 mm × 9 gauge square-twisted sherardized nails, as per the manufacturer's instructions. The central chord C is then cut, and the opening trimmed with two trimmers D and, if required, an infill joist E.

12.31 CHIMNEY TRIMMING

Figure 12.56: Where the width of a chimney is greater than the normal spacing between trussed rafters, the trusses may have a greater spacing between them in the area of the chimney stack, providing the increased

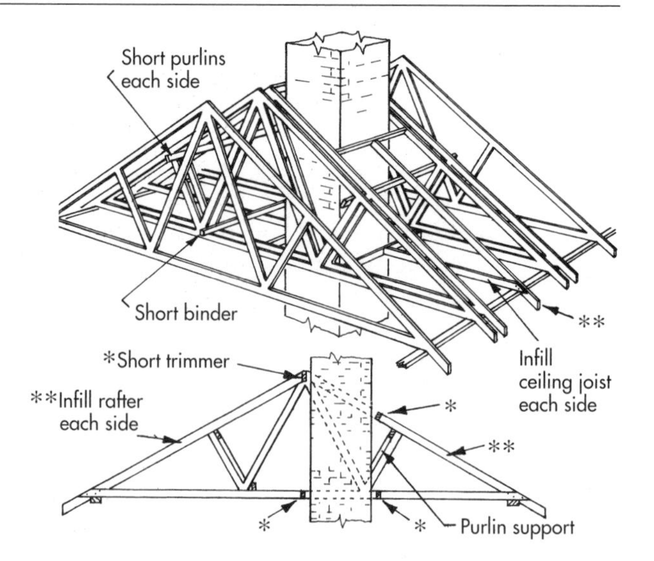

Figure 12.56 Chimney trimming

spacing is not more than twice the normal truss spacing. As illustrated, the non-truss open areas remaining on each side of the chimney stack are filled in with loose infill rafters, ceiling joists, trimmers, binders and short purlins strutted against the webs of the nearest standard trusses on each side.

As per the Building Regulations, the timbers should be at least 40 mm clear of the chimney stack. The infill rafters, which are nailed to the side of the infill joists and to the wall plate, should be at least 25 mm deeper than the trussed rafters to allow for a birdsmouth to be formed at the wall plate.

12.32 WATER-TANK SUPPORTS

Figure 12.57: For domestic storage of 230 or 300 litre capacity, the load is usually spread over three or four trusses. The details illustrated here are those recommended for a 300 litre tank within a Fink or W truss roof of up to 12 m span.

12.32.1 Spreader Beams

Two spreader beams of 47 × 72 mm section extend in length over the bottom chords of four trusses, as illustrated, up against the web on each side (close to the node points), sitting vertically on edge, not flat. The permanent longitudinal bracing, which is normally in this position, is offset in this area and fixed at the sides, up against the spreader beams.

12.32.2 Cross-bearers

Two cross-bearers of 35 × 145 mm section are now skew-nailed to the spreader beams, positioned at one-sixth the distance of the beam's bearing-length from

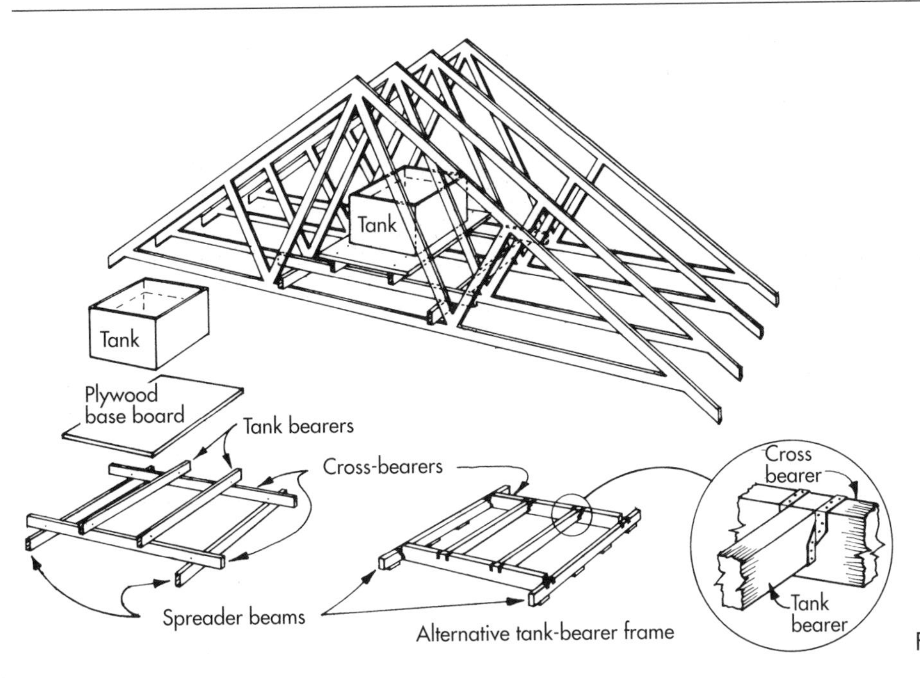

Figure 12.57 Water-tank supports

each end – which is midway between the spacings of the first and second truss and the third and fourth truss in contact with the spreader beam.

12.32.3 Tank Bearers and Base Board

Next, two tank bearers of 47 × 72 mm section are skew-nailed across the first bearers, relative to the tank width, and a WBP plywood base board – not chipboard – is fixed to these. Like the spreader beams, the cross-bearers and tank bearers must sit vertically on edge, not flat.

12.32.4 Alternative Tank Support

If more headroom is needed above the tank platform, an alternative tank-bearer frame, the same or similar to that illustrated, may be used. This is made up of joist hangers and/or truss shoes, so the deeper cross-bearers, being parallel and between the trusses, can be dropped lower if required, on pre-positioned joist hangers. To allow for long-term deflection, there should be at least 25 mm between cross-bearers and the ceiling and the same between tank bearers and the ceiling ties of the trusses.

13

Erecting Timber Stud Partitions

13.1 INTRODUCTION

Partitions are secondary walls used to divide the internal areas of buildings. Although often built of aerated building blocks, other materials, including timber, are frequently used – especially above ground-floor level, on suspended timber floors, where block partitions would add too much weight unless supported from below by a beam or a wall.

13.2 TRADITIONAL BRACED PARTITION

Figure 13.1: This is only shown for reference and comparison with the modern stud partition illustrated in Figure 13.3 and the trussed partition seen in Figure 13.2. The partition was made up of 100 × 75 mm head, sill(s), door studs and braces – and 100 × 50 mm intermediate studs and noggings. The diagonal braces, which were bridle jointed to the door studs and sill, were included partly to give the partition greater rigidity against sideways movement; and partly to carry some of the weight from the centre of the partition down to the sill-plate ends, which were housed into the walls.

13.2.1 Jointing Arrangement

The main frame of this type of partition was through-morticed, tenoned and pinned (wooden pins or nails); the intermediate uprights (studs) were stub-tenoned to the head and sill; the door head was splay-housed and stub-tenoned to the door studs; the door studs were dovetailed and pinned to the sill; and the staggered noggings were butt-jointed.

13.3 TRADITIONAL TRUSSED PARTITION

Figure 13.2: As with the above braced partition, the trussed partition, with the advent of modern materials and methods, has been obsolete for many years. It was used for carrying its weight and the weight of the floor above. This should be taken into account before commencing any drastic alterations or removal on conversion works – as these partitions can still be found in older-type buildings.

Note that timber partitions are now referred to as *stud* partitions, or *studding* (derived from old English *studu*, meaning post).

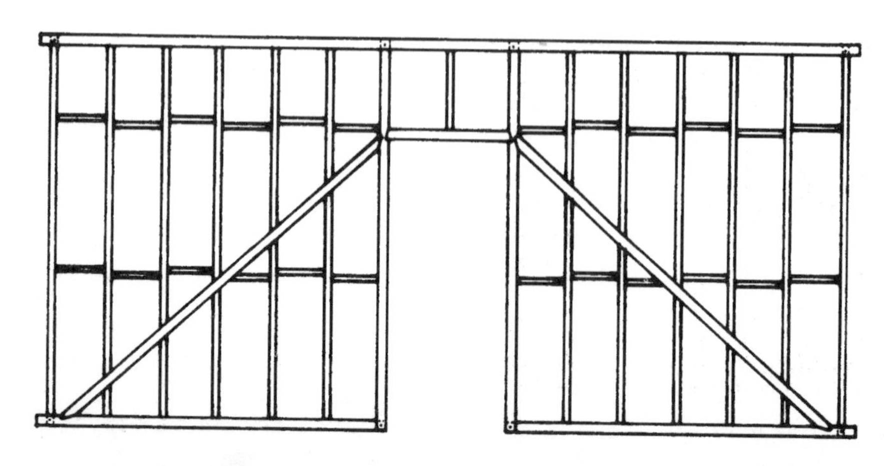

Figure 13.1 Traditional braced partition

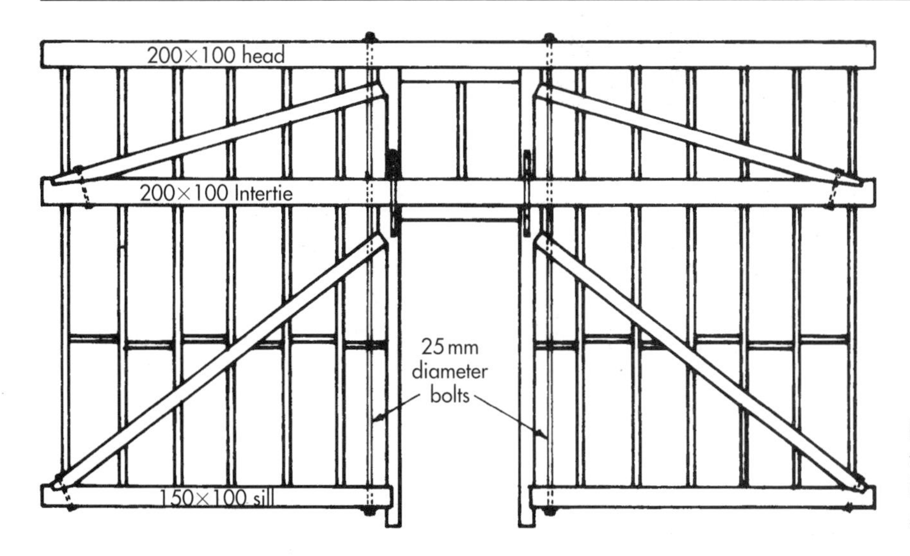

Figure 13.2 Traditional trussed partition (100 × 100 mm door posts, braces and straining heads; 100 × 50 studs and noggings)

13.4 MODERN STUD PARTITION

Figure 13.3: Although these partitions can be made in a joinery shop and re-assembled on site, with certain tolerances made for the practicalities involved, the common practice nowadays is to cut and build them up on site (*in situ*), piecemeal fashion. The detailed sequence of doing this is given below.

13.4.1 Sill (or Floor) Plate

Figure 13.3: The sill, labelled A on the illustration is cut to length and, if straight, can be used for setting out the floor position. Alternatively, its position can be snapped on the floor with a chalk line related to tape-rule measurements. The position of any door opening must be deducted from the sill plate setting-out. This deduction is an accumulation of

door width (say 762 mm)
thickness of door linings (say 28 mm × 2)
a fitting tolerance (say 6 mm)
door studs (50 mm × 2)

to give a total of 924 mm.

The plate is fixed to joists with 100 mm round-head wire nails, or to floor boards or panels with 75 mm nails or screws, or to concrete or screeded floors with nylon-sleeved Frame-fix or Hammer-fix screws, or cartridge-fired masonry nails or bolts, at approximately 900 mm centres.

13.4.2 Head Plate

Figure 13.3: The position for the head plate B can be fixed by plumbing up to the ceiling from the side of the sill at each end, either with a pre-cut wall stud and spirit level, a straightedge and spirit level, or a plumb bob and line, then by snapping a chalk line across the ceiling. The head plate is then cut to length, set out with intermediate stud-positions and propped up with two or three temporary uprights – purposely oversize in length – as illustrated in Figure 13.4. The plate is fixed to the ceiling joists with 100 mm round-head wire nails.

13.4.3 Wall Studs

Figure 13.3: The wall studs C should be marked to length either by a pinch-stick method (Figure 13.5(a)) or by offering up a slightly oversize stud, resting on the

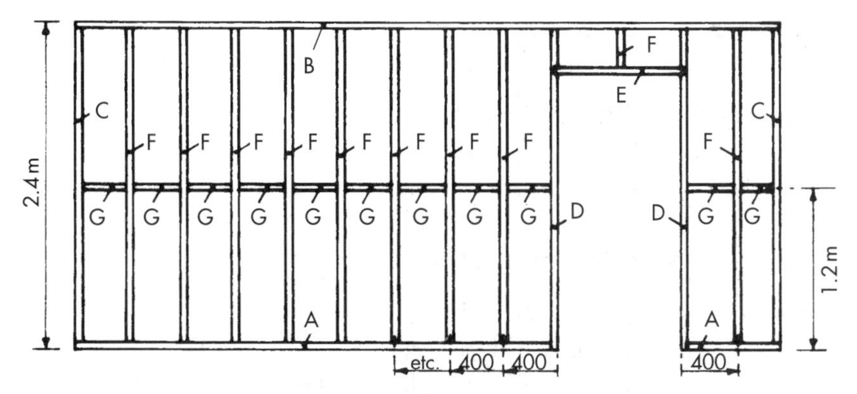

Figure 13.3 Modern stud partition

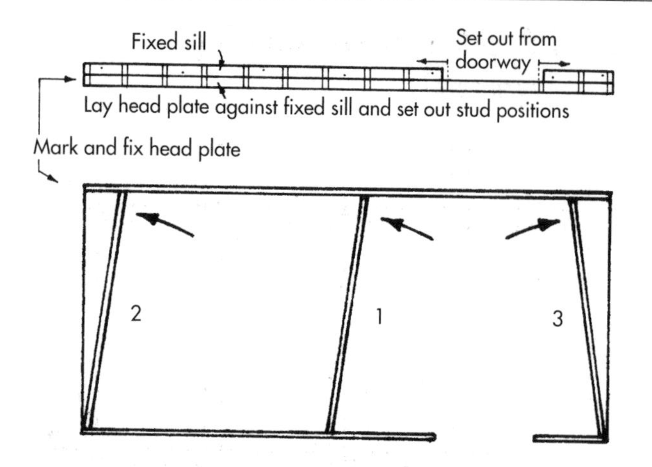

Figure 13.4 Marking and fixing the head plate

floor plate and marked at the top, as indicated in Figure 13.5(b). Add 1–2mm for a tight fit, then cut to length and fix to the block-wall with 100 mm cut clasp nails or with Frame-fix or Hammer-fix screws. There should be at least three wall fixings and the ends of the stud should be skew-nailed to the plates, as shown in Figure 13.5(c).

13.4.4 Door Studs

Figure 13.3: The door studs D are now marked, as above, from floor surface to the underside of the head, tightening-allowance added, then cut, carefully plumbed and fixed. The base of each stud is fixed with two 100 mm round-head wire nails, driven slightly dovetail-fashion into the end of the floor plate and one central 75 mm nail skew-nailed into the floor. At the top, each stud is skew-nailed with three 75 mm wire nails, using the skew-nail technique indicated in Figure 13.6.

13.4.5 Door Head

Figure 13.3: Next, the position of the door head E is marked on the door studs, as illustrated in Figure 13.7.

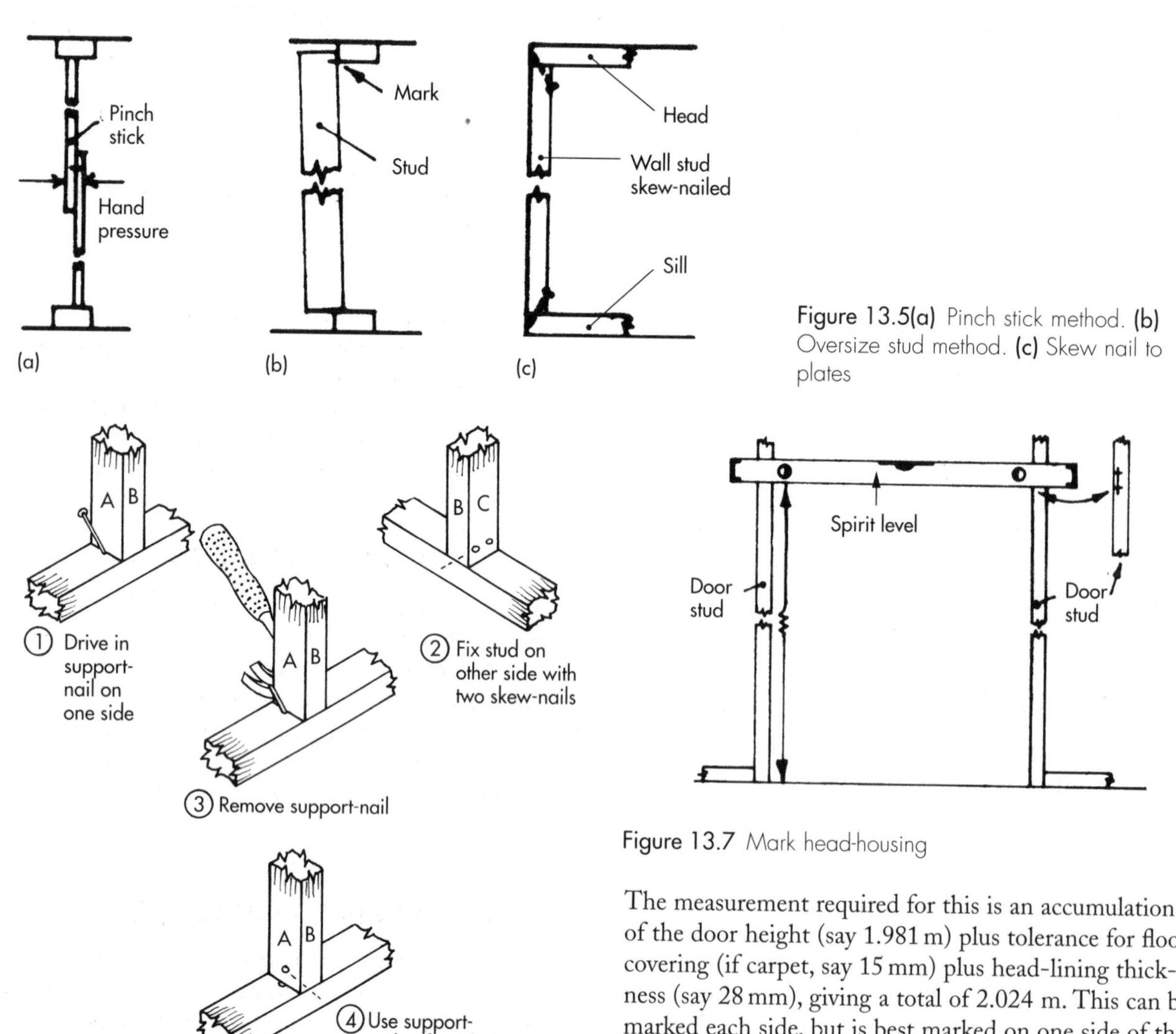

Figure 13.5(a) Pinch stick method. (b) Oversize stud method. (c) Skew nail to plates

Figure 13.6 Skew-nail technique

Figure 13.7 Mark head-housing

The measurement required for this is an accumulation of the door height (say 1.981 m) plus tolerance for floor covering (if carpet, say 15 mm) plus head-lining thickness (say 28 mm), giving a total of 2.024 m. This can be marked each side, but is best marked on one side of the door opening, then transferred across with a spirit level. If the head is to be butt-jointed and nailed (which is the most common trade-practice, see Figure 13.9(d)), cut to length to equal the width between studs at floor-plate level, or 924 mm (example measurement) as worked out in Section 13.4.1, and fix through the door studs with two 100 mm round-head wire nails each side. Alternatively, if to be housed (see Figure 13.9(b)) add 24 mm to the length, mark and cut the 12 mm deep door-stud housings each side *in situ*, working from a saw stool or steps, slide the head into the housings and fix through the door studs with two 75 mm round-head wire nails each side.

13.4.6 Intermediate Studs

Figure 13.3: These are next in sequence studs F, cut to length for a tight fit as described for wall studs in Section 13.4.3. If the plates are not already set out for the intermediate stud positions, as suggested earlier, then these positions should now be marked out on the plates to suit the plasterboard sizes. The boards used are usually 2.4 m × 1.2 m and either 9.5 mm or 12.5 mm thick. These sizes dictate the spacing of the studs at either 400 mm centres (6 × 400 = 2.4 m) or 600 mm centres (4 × 600 = 2.4 m). The 9.5 mm thickness should only be used on studding spaced at 400 mm centres. The studs are nailed into position with three 75 mm wire nails to each abutment, using the skew-nail technique shown in Figure 13.6.

13.4.7 Noggings

Figure 13.3: These final insertions are short struts, G, that stiffen up the whole partition. If they are centred at 1.2 m from the floor, as shown (by measuring up at each end and snapping a chalk line), the joint of the plasterboard will be reinforced against the noggings. The noggings are cut in snugly and fixed by skew-nailing or end-nailing as indicated in Figure 13.8(a). To lessen the risk of bulging the door studs, the fixing of the noggings should be started from the extreme ends, working towards the door opening and being extra careful with the final nogging insertions.

13.4.8 Studding Sizes

The timber used for studding is usually 100 × 50 mm sawn (unplaned) softwood, or ex. 100 × 50 mm prepared (planed to about 95 × 45 mm finish) softwood. For economy, 75 × 50 mm sawn, or ex. 75 × 50 mm prepared is sometimes used. Sawn timber is more common, but prepared timber is also used a lot nowadays to lessen the irregularities transferred to the surface material. To this end, *regularized* timber, machined to a reduced, more constant sectional size, is available and used for partitions nowadays.

13.4.9 Alternative Nogging Arrangements

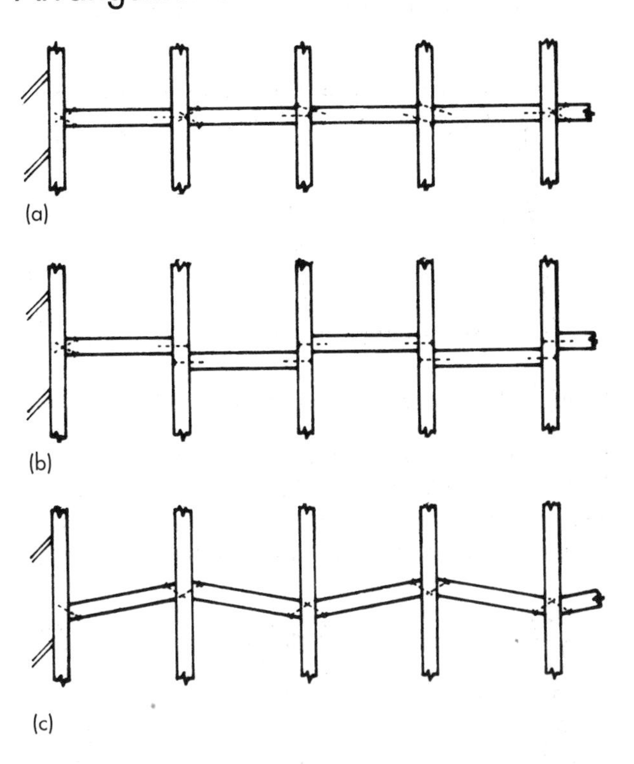

Figure 13.8 Alternative methods of nogging. **(a)** Straight noggings. **(b)** Staggered noggings. **(c)** Herringbone noggings

Figure 13.8: Ideally, successive rows of noggings at 600 mm centres vertically (between the 1.2 m spacings shown in Figure 13.3) should be used to give greater rigidity and support to the plasterboard, although one row is normally sufficient. Points for and against the three alternative nogging arrangements are given below.

13.4.10 Straight Noggings

Figure 13.8(a): These can be positioned to reinforce horizontal plasterboard joints, but are not the easiest to fix. Various alternative methods of fixing are indicated by dotted lines denoting the nailing technique. The technique for skew-nailing (Figure 13.6) can be used here, with the support nail positioned under the nogging. Another technique, using a temporary strut, is shown in Figure 13.19.

13.4.11 Staggered Noggings

Figure 13.8(b): These cannot effectively reinforce the plasterboard joints, but as indicated, are easier to fix by end-nailing. Of course, if the plasterboard was placed on end, with vertical joints being reinforced on the intermediate studs, there would be nothing against this method.

13.4.12 Herringbone Noggings

Figure 13.8(c): These are positioned at an angle of about 10°, are easy to fix and achieve a tight fit, even with inaccurate cutting. If correctly positioned, they give about 90 per cent reinforcement to the plasterboard joints. On the minus side, this method has a tendency to bulge the door studs.

13.5 DOOR-STUD AND DOOR-HEAD JOINTS

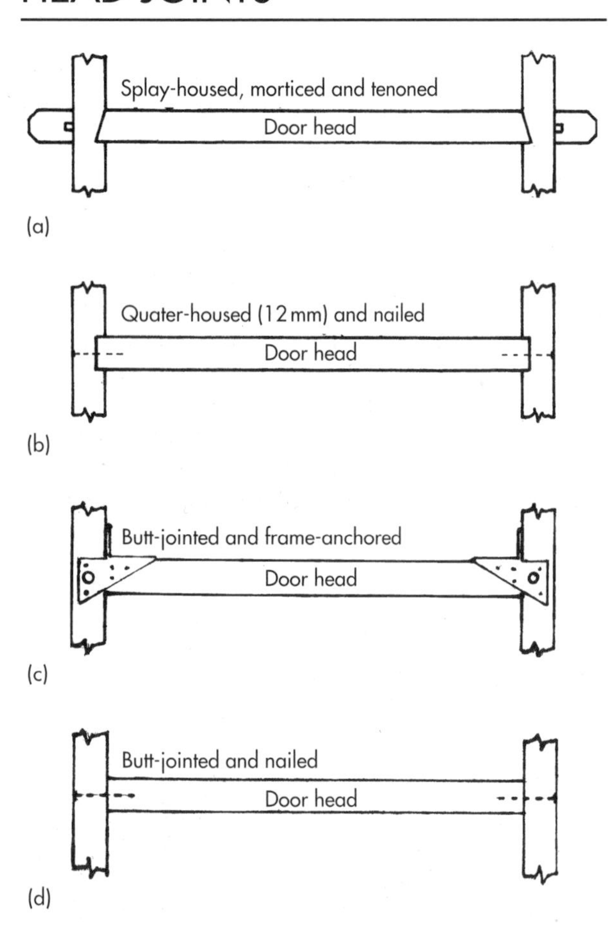

Figure 13.9 Door-stud and head joints. **(a)** Too elaborate. **(b)** Good compromise. **(c)** Modern method. **(d)** Common method

Figure 13.9: The strength of these joints is important, as any weakness, especially resulting in an upwards movement of the door-head stud, can affect the door-lining head. Problems like this usually occur for two reasons: if a tolerance gap exists between the door-head stud and the lining-head; and if the door-head/door-stud joint is not strong enough to resist hammer-blows when the door-lining legs are being fixed at the top, or – more likely – when the door stop is being fixed to the underside of the door-lining head. Evidence of movement will appear as unsightly gaps to the corner housing-joints of the door lining. The risk of this happening can be avoided by using the joints shown in Figures 13.9(b) or (c) and by inserting packing or wedges in the gap between door-head and lining-head, directly above each door-lining leg (see also Chapter 6 which covers the fixing of door frames, linings and doorsets).

13.5.1 Splay-housed, Morticed, Tenoned and Draw-bore Wedged

Figure 13.9(a): This traditional door-stud/head joint, although ideal for the job and strongly resistant to displacement from timber which might twist and from lateral or vertical hammer-blows, is too elaborate and time-consuming nowadays.

13.5.2 Quarter-housed and Nailed

Figure 13.9(b): This is one of the recommended methods and is a good compromise between the other two extremes. The door-stud quarter housings (i.e. a quarter of the thickness) restrict twisting and upwards movement of the door-head and if the partition is well strutted with noggings, lateral movement should not be a problem.

13.5.3 Butt-jointed and Frame-anchored

Figure 13.9(c): This modern method is also recommended. It has all the virtues of the quarter-housed joint in restricting movement and is less time-consuming. The butt-jointed head can be nailed through the door-studs initially, then the framing anchors fixed at each end, or the framing anchors can be fixed in position on the head, the head located and fixed into the door-studs via the remaining framing-anchor connections. The anchors are recommended to be fixed with 3 mm diameter × 30 mm-long sherardized clout nails.

13.5.4 Butt-jointed and Nailed

Figure 13.9(d): This method is the most common trade-practice for attaching door-head to door-studs but, for reasons already stated, it is not the best. In the past, it was the method used on cheap work, which has now become the norm.

13.6 STUD JOINTS TO SILL AND HEAD PLATE

Figure 13.10: There are four methods of jointing vertical, intermediate studs to the head plate and sill (floor) plate, as follows.

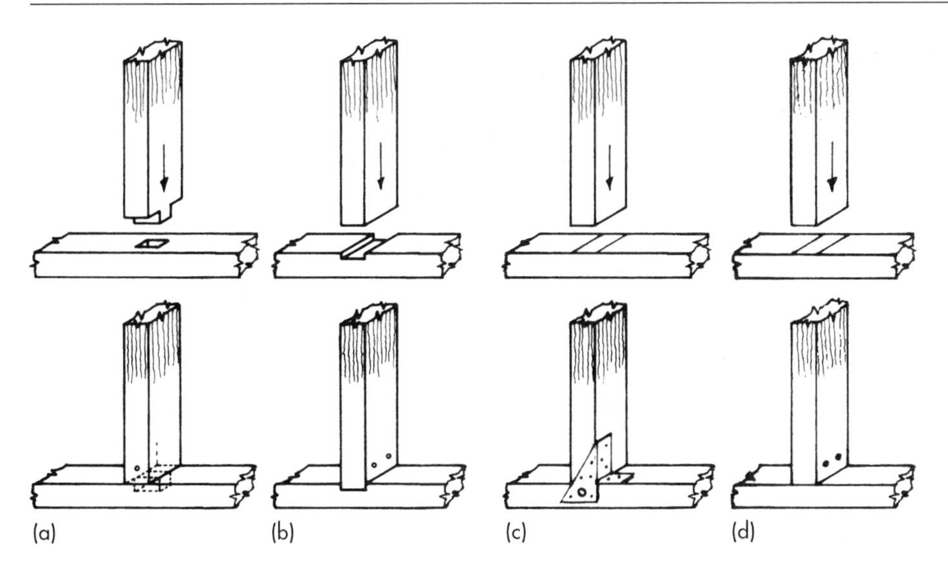

Figure 13.10 Stud joints to sill or head plate

13.6.1 Stub-tenoned

Figure 13.10(a): The short tenons are morticed to half depth into the head and sill. This method involves too much hand work on site and is best suited to preformed partitions being made in the joinery shop, where machinery is available. Such partitions would be sent to the site in pieces, designed with length and height tolerances, ready for assembly and erection. This is useful in occupied premises, such as offices and shops, where the site work would be reduced.

13.6.2 Housed or Trenched

Figure 13.10(b): The housings are cut into a quarter-depth of the plate thickness and the studs are skew-nailed at each joint with two 75 mm round-head wire nails. This method can be easily handled on site; however, although housings are ideal for easier nailing, straightening and retaining any twisted studs, the method is rarely used nowadays because of the added time element.

13.6.3 Butt-jointed and Frame-anchored

Figure 13.10(c): This modern method, already referred to in Section 13.5.3, has most of the benefits afforded by housing joints – with the exception of straightening out a stud already in a state of twist. The studs need to be fitted tightly and two framing anchors per joint – one on each opposite face-edge – are nailed into position with 3 mm × 30 mm sherardized clout nails.

13.6.4 Butt-jointed and Skew-nailed

Figure 13.10(d): Again, this is the most common trade-practice, not because it is the best, but because it is the quickest method of jointing. Originally used only on cheap work, this method is now widely used. The stud should be a tight fit, otherwise the relative strength of this joint is very much impaired. Three 75 mm wire nails should be used, two in one side, one in the other (as shown in Figure 13.6). This technique requires one support-nail to be partly driven into the plate, at about 45°, on one line of the stud-position, so that when the stud is against it, the nail-head protrudes to the side. The stud is then fixed against the support-nail with two fixings, the support-nail removed with the claw hammer and used as the central fixing on the other side.

13.7 DOOR-STUD AND SILL-PLATE JOINTS

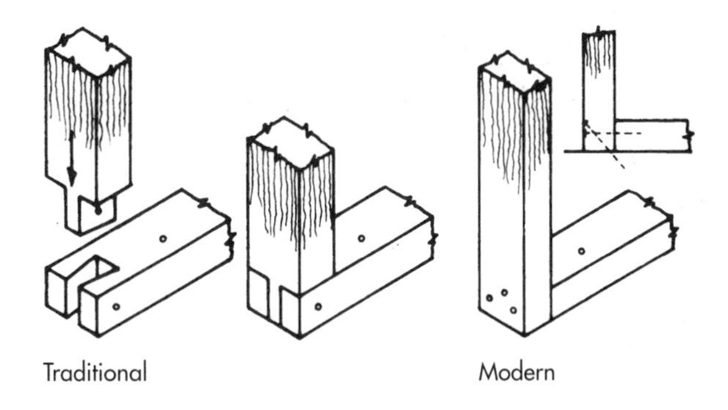

Figure 13.11 Door-stud/sill-plate joints

Figure 13.11: Traditionally, these joints were dovetailed and pinned (dowelled or nailed), as shown, to retain the base of the door-stud effectively. Studs of 63 or 75 mm thickness were used.

The present-day method uses door studs reduced to 50 mm thickness, resting on the floor and butted and nailed against the sill. As described previously, two 100 mm wire nails are driven slightly dovetail-fashion into the end of the plate and one centrally placed 75 mm wire nail is skew-nailed into the floor.

13.8 CORNER AND DOORWAY JUNCTIONS

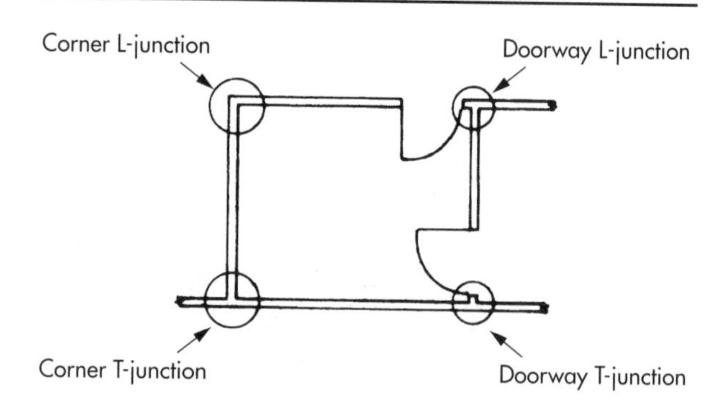

Figure 13.12 Corner and doorway junctions

Figure 13.12: This small-scale plan view of a room might be an uncommon layout for partitioning, but serves to illustrate the four junctions requiring different treatment. The best treatment each could receive would involve the use of three vertical studs, but other treatments shown here are sometimes used.

13.8.1 Corner L-junction

Figures 13.13(a)–(c): The plan view (a) shows the first choice of construction using three full-length corner studs to provide a fixing-surface on each side of the internal angle. As seen in the isometric view (b), short offcut blocks should be fixed in the gap existing between two of the studs, to add rigidity and give continuity to the rows of noggings. For possible economy of timber and to achieve a similar result, a practical method of using *vertical* noggings is shown (c). This could save a full-length vertical stud, as usually the noggings required can be cut from offcuts and waste material.

Alternative Methods

Figures 13.13(d) and (e): The first alternative (d) is reasonable, but would involve interrupting the partitioning operation to allow for plasterboard to be fixed to at least one side of the first partition, before the second partition could be built. The second alternative shown (e) allows the whole partition to be built by providing a fixing-surface to each side of the internal angle, in the form of 50 × 50 mm vertical noggings fixed between the plates and the normal horizontal noggings. However, unless these vertical noggings are housed-in, there is a risk of them being displaced when plasterboard fixings are driven in.

13.8.2 Doorway L-junction

Figure 13.14: Where a doorway meets a corner L-junction, the problem of providing fixing surfaces on the internal angle is the same as before, and a similar treatment (a), using vertical noggings between normal horizontal noggings, is used instead of a full-length stud. The other methods (b) and (c), are similar alternatives to those shown in Figures 13.13(d) and (e) and the same considerations and comments apply.

Provision for Architrave

In good building practice, another consideration at this doorway junction, is that there should be provision for a

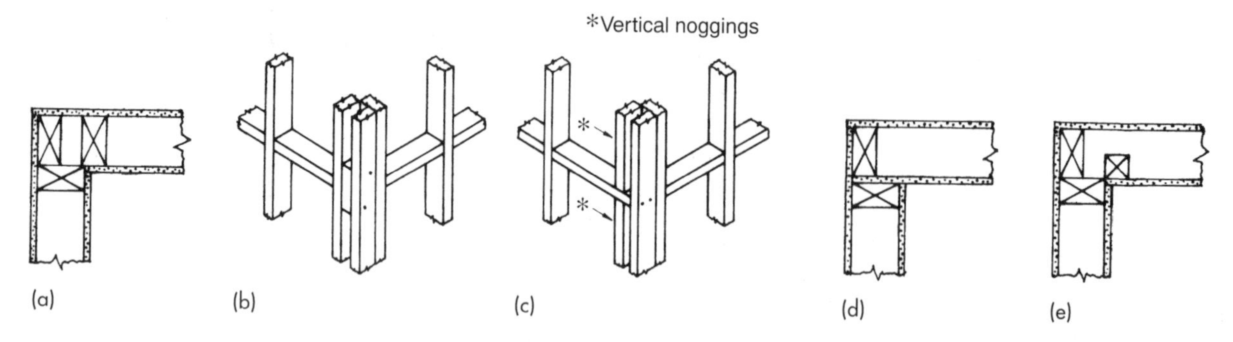

Figure 13.13 Corner L-junction

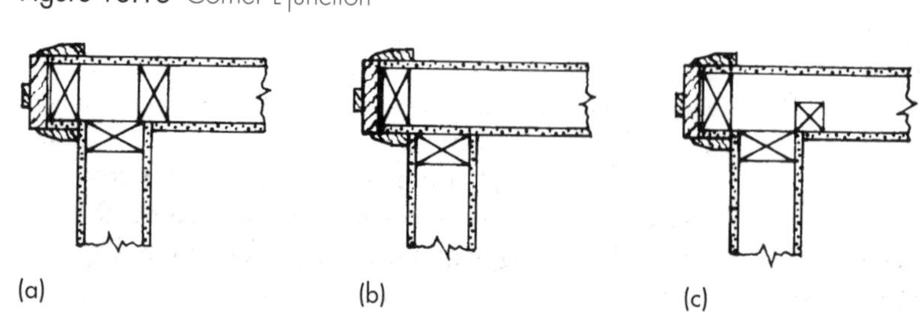

Figure 13.14 Doorway L-junction

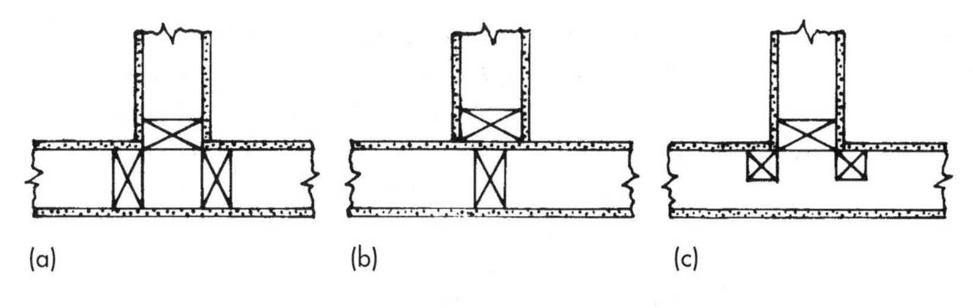

Figure 13.15 Corner T-junction

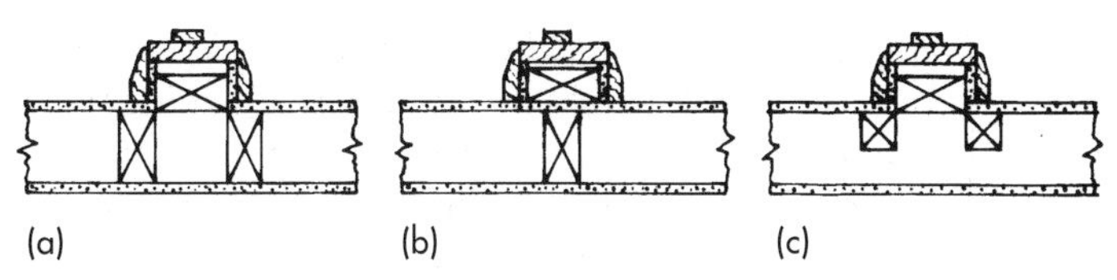

Figure 13.16 Doorway T-junction

full-width architrave on the side where the architrave is touching the adjacent wall. In practice, 3 or 4 mm less than the architrave width is provided on the inside angle to allow for eventual scribing of the architrave against an irregular plaster surface.

13.8.3 Corner T-junction

Figure 13.15: As illustrated in the plan-view details, this type of junction receives similar treatment in its three variations to those illustrated and described before.

13.8.4 Doorway T-junction

Figure 13.16: As seen in these illustrations, similar treatments apply in the same descending order. Another consideration in this situation is to pack out the lining (as shown), if necessary, to achieve full width architraves each side.

13.9 FLOOR AND CEILING JUNCTIONS

Figure 13.17(a): Although stud partitions do not normally present weight problems on suspended timber floors, certain points must be considered when a partition runs parallel to the joists and

- rests on the floor boards either in a different position to a joist below as at A, or
- rests in a position that coincides with a joist below, as at B; (both situations inhibiting floor-board removal for rewiring, etc.);
- misses a joist required for head-plate fixings, C;
- creates a problem in board-fixings on each side of the head, if erected before the ceiling is boarded, as at D.

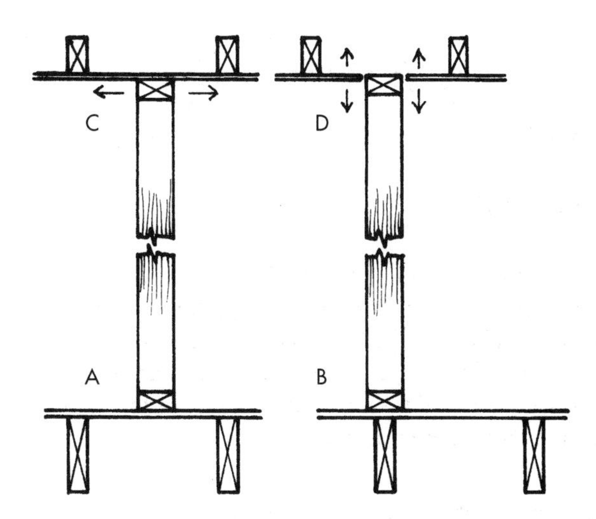

Figure 13.17(a) Vertical sections

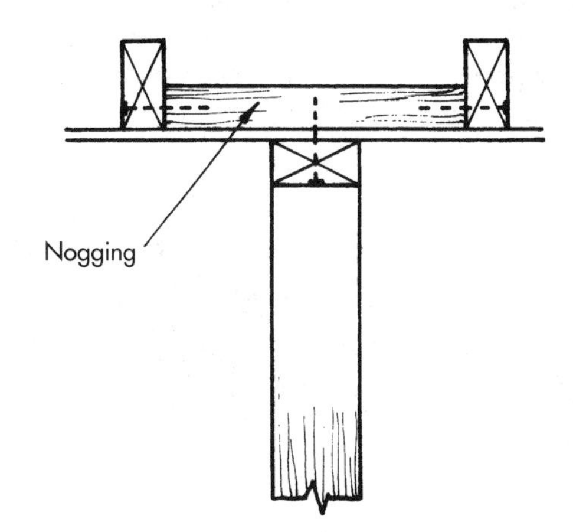

Figure 13.17(b) Head-fixings

13.9.1 Creating Head-fixings

Figure 13.17(b): This shows a method of overcoming the lack of head-fixings by inserting 100 × 50 mm (or less) noggings between the ceiling joists at about 1 m centres. Where there is access above the ceiling joists, as in a loft, these noggings can be fixed before or after the ceiling is boarded; however, where there is no access above, as with floor joists that have been floored, then obviously the noggings must go in before the ceiling is boarded.

13.9.2 Double Ceiling-and-floor Joists

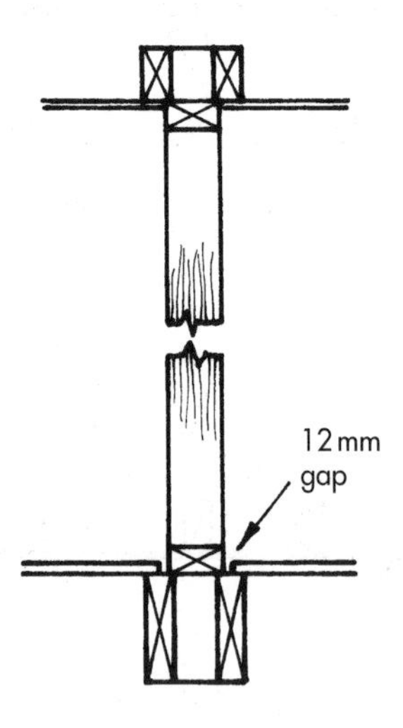

Figure 13.17(c) Double ceiling-and-floor joists

Figure 13.17(c): This shows a way of overcoming all the previous issues, but uses more timber and requires extra work and careful setting-out at the joisting stage. Arrangements of double ceiling-and-floor joists, with support-blocks between, are set up. The blocks should be inserted at a maximum of 1 m centres and fixed from each side with 100 mm round-head wire nails, staggered as in Figure 13.17(d).

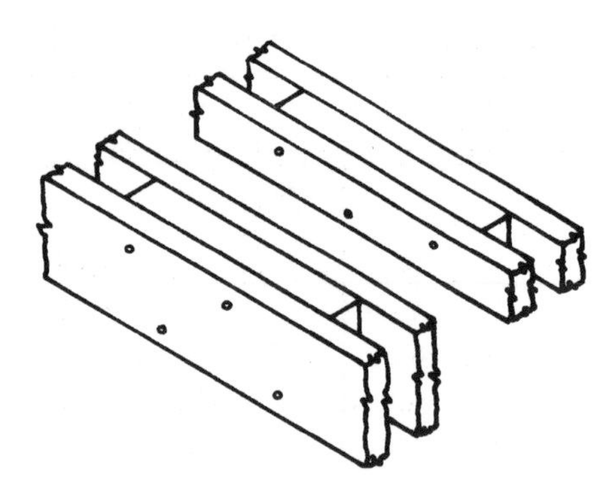

Figure 13.17(d) Support-blocks

13.9.3 Creating a Beam Effect

Figure 13.17(e): To achieve more of a beam effect with the double floor-joists – and so offset the disadvantage of a direct load – the support-blocks could be replaced by a continuous middle-joist, bolted into position with 12 mm diameter bolts, 50 mm round or square washers and 75 mm diameter toothed timber connectors at maximum 900 mm centres.

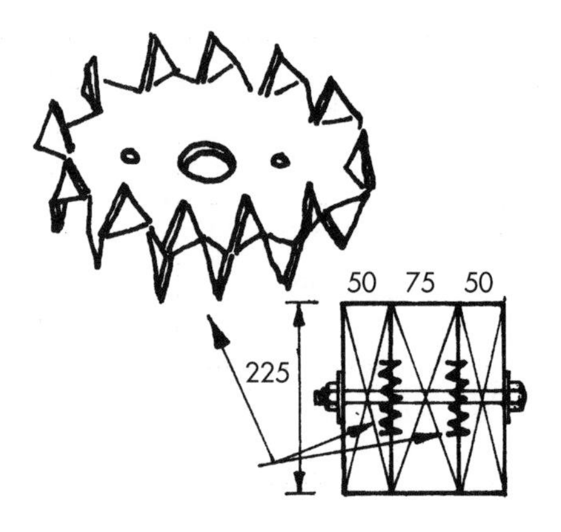

Figure 13.17(e) Timber-connectors between joists

13.9.4 Partition across Joists

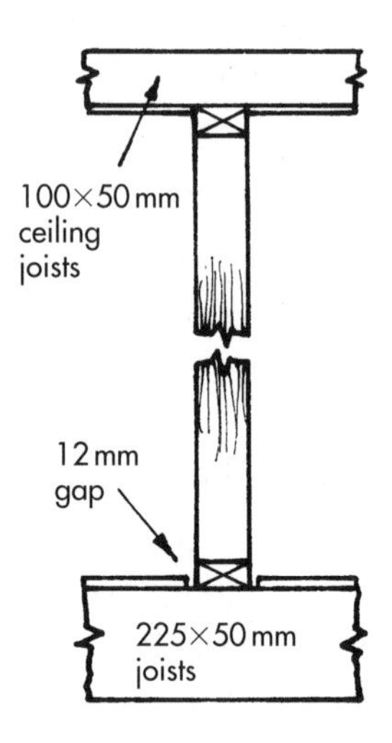

Figure 13.17(f) Partition at right-angles to joists

Figure 13.17(f): This illustrates a situation that presents no problems, when the partition runs at right-angles to the floor and ceiling joists. The sill or floor plate is best fixed on the joists, to allow for expansion and contraction of the floor membrane – hence the 12 mm gap each side – but can be fixed on the flooring. Likewise, the head plate can either be fixed directly to the joists or to the boarded ceiling.

13.9.5 Fixing Boards for Skimming

Figure 13.18(a): Plasterboards come in a variety of sizes, perhaps the most popular of these being 2.4 m × 1.2 m of 9.5 mm or 12.5 mm thickness. The illustration shows boards of 9.5 mm thickness, laid on edge, fixed to stud-spacings of 400 mm centres, in an arrangement suitable for skimming with finishing plaster, after reinforcing the joints with bandage or hessian skrim. Boards are fixed with 30 mm galvanized clout nails, at approximately 150 mm centres.

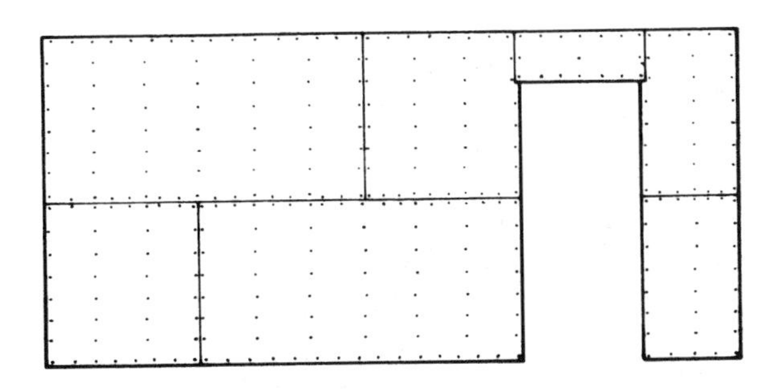

Figure 13.18(a) Fixing boards for skimming

13.9.6 Fixing Boards for Dry Finish

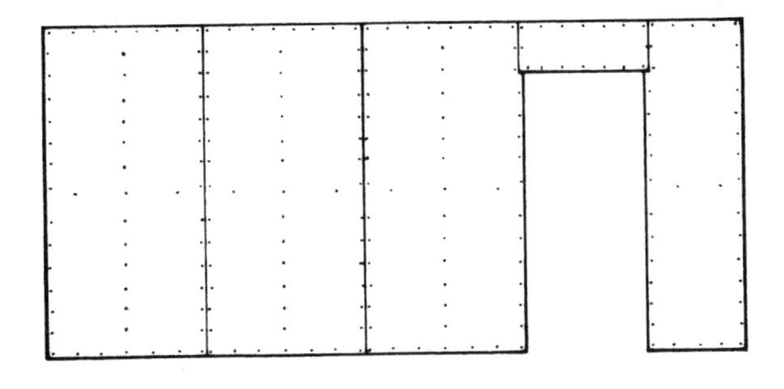

Figure 13.18(b) Fixing boards for dry finish

Figure 13.18(b): The illustration shows TE boards (tapered-edge plasterboards) of 12.5 mm thickness, laid on their ends to eliminate horizontal joints and fixed to stud-spacings of 600 mm centres, using either 38 mm galvanized clout nails, at approximately 200 mm centres. Alternatively, countersunk, sherardized screws of a similar length can be used to achieve a countersunk hole for filling and to reduce the risk of surface damage from hammer-blows. Decorators usually attend to the nail or screw holes and also fill, tape and refill the vertical joints within the indentation of the tapered edges, to achieve a finish.

13.9.7 Spacing of Studs

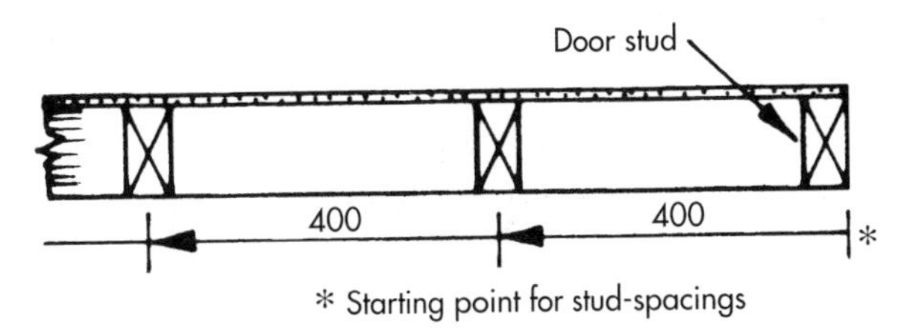

Figure 13.18(c) Spacing of studs

Figure 13.18(c): As illustrated, it must be noted that the spacing of studs should start from the *edge* of the door-stud, to the *centre* of the intermediate studs, to achieve correct centres and full coverage of the door-stud edge by the board material.

13.9.8 Fixing Noggings

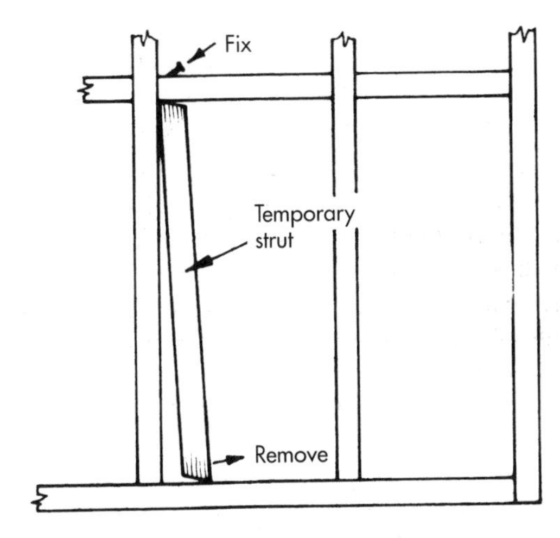

Figure 13.19 Alternative method of supporting noggings during fixing

Figure 13.19: As shown here, a temporary strut of 100 × 50 mm section, with bevelled ends to facilitate easy and quick removal, can be positioned to support a nogging during fixing with skew-nails, then lightly tapped at its base to remove it for the next fixing.

14

Geometry for Arch Shapes

14.1 INTRODUCTION

Brick or stone arches over windows or doorways, in a variety of geometrical shapes, can only now be seen mainly on older-type buildings. Present-day design favours straight lines for various reasons, including visual simplicity, cost, and structural requirements in relation to new materials and design. Curved arches in domestic buildings have been replaced mostly by various types of light-weight, galvanized, pressed-steel lintels as shown in Figure 14.1(a).

However, arches should not be regarded as old-fashioned or obsolete and will still be required to match existing work on property maintenance, conversions and extensions. Also, some architects nowadays are using geometrical shapes in modern design.

Arches for internal doorways, to be finished in plaster, are quite popular and can be formed traditionally with a structural brick-arch, although the modern practice is to use a lightweight, galvanized steel *archformer*, which is easily fitted and fixed within a standard pre-formed doorway, ready for plastering.

Brick or stone arches are built on temporary wooden structures called centres, dealt with in the next chapter.

14.2 BASIC DEFINITIONS

Figure 14.1(b):

- *Springing line*: a horizontal reference or datum line at the base of an arch (where the arch *springs* from).
- *Span*: the distance between the reveals (sides) of the opening.
- *Centre line*: a vertical setting-out line equal to half the span.
- *Rise*: a measurement on the centre line between springing and intrados.
- *Intrados* or *soffit*: the underside of the arch.
- *Extrados*: the topside of the arch.
- *Crown*: the highest point on the extrados.
- *Voussoirs* (pronounced *vooswars*): wedge-shaped units in the arch.
- *Key*: the central voussoir at the crown (i.e. the final insert that *locks* the arch structurally).

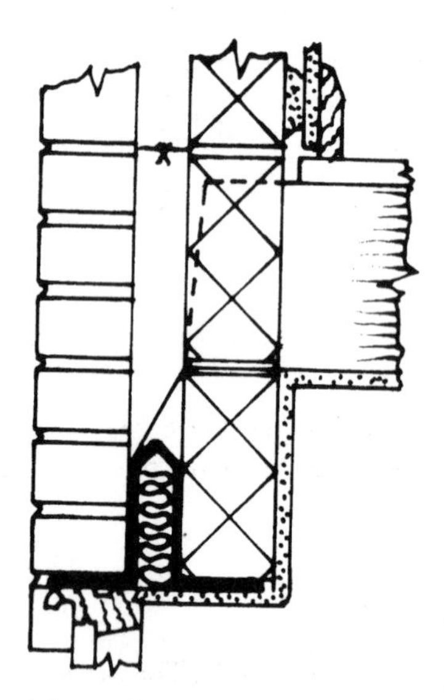

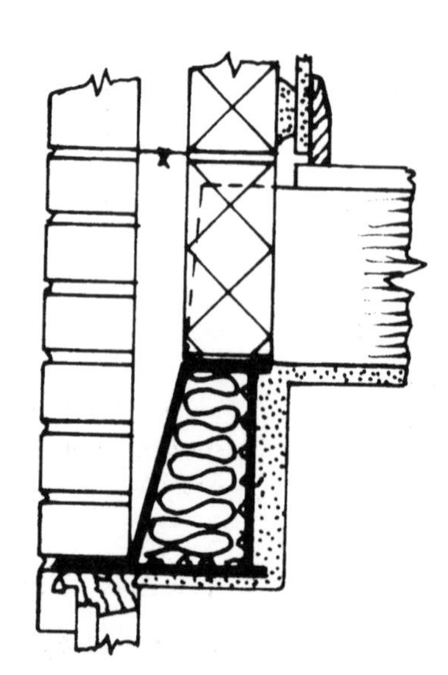

(a)

Figure 14.1(a) Modern steel lintels

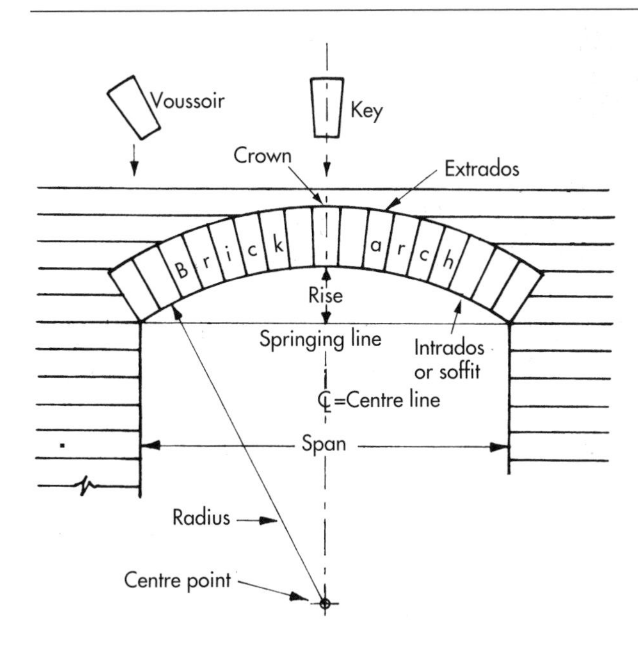

Figure 14.1(b) Basic definitions

- *Centre*: the pivoting or compass point of the radius.
- *Radius*: the geometrical distance of the centre point from the concave of a segment or circle

14.3 BASIC TECHNIQUES

Before proceeding, a few basic techniques in geometry must be understood.

14.3.1 Bisecting a Line

This means dividing a line, or distance between two points, equally into two parts by another line intersecting at right-angles. Figure 14.2(a) illustrates the method used. Line AB has been bisected. Using A as centre, set the compass to any distance greater than half AB. Strike arcs AC^1 and AD^1. Now using B as centre and the same compass setting, strike arcs BC^2 and BD^2. The

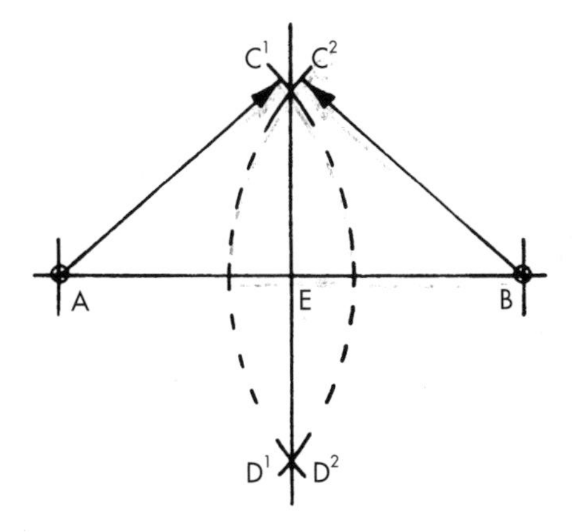

Figure 14.2(a) Bisecting a line

arcs shown as broken lines are only used to clarify the method of bisection and need not normally be shown. Draw a line through the intersecting arcs C^1C^2 to D^1D^2. This will cut AB at E into two equal parts. Angles C^1EA, BEC^2, AED^1 and D^2EB will also be right-angles (90°).

14.3.2 Bisecting an Angle

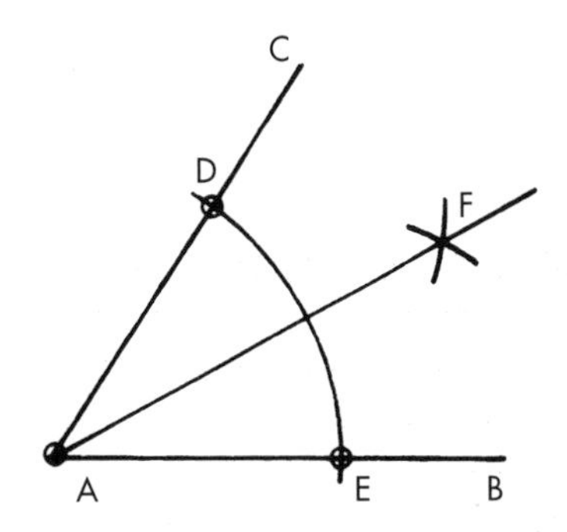

Figure 14.2(b) Bisecting an angle

This means cutting or dividing the angle equally into two angles. Figure 14.2(b) shows angle CAB. With A as centre and any radius less than AC, strike arc DE. With D and E as centres and a radius greater than half DE, strike intersecting arcs at F. Join AF to divide the angle CAB into equal parts, CAF and FAB.

14.3.3 Semi-circular Arch

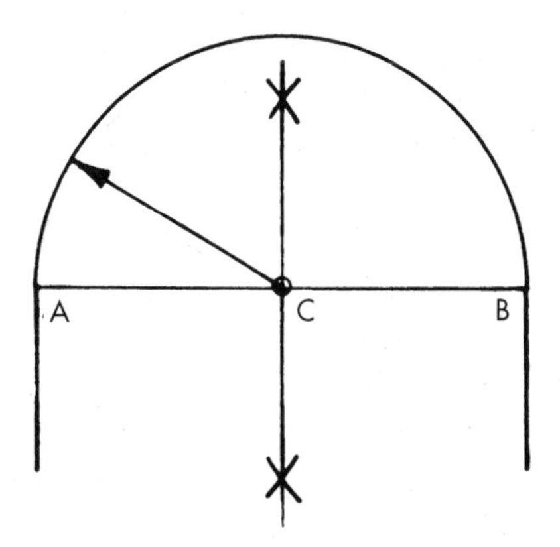

Figure 14.3 Semi-circular arch

Figure 14.3: Span AB is bisected to give C on the springing line. With C as centre, describe the semi-circle from A to B.

14.3.4 Segmental Arch

Figure 14.4: Span AB is bisected to give C. The rise at D on the centre line can be at any distance from C, but

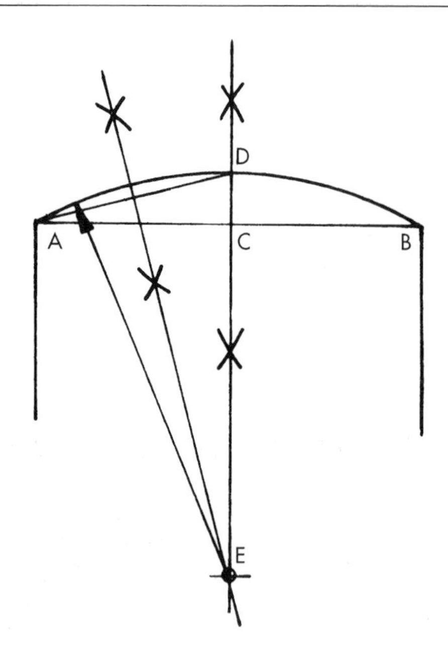

Figure 14.4 Segmental arch

less than half the span. Bisect the imaginary line AD to intersect with the centre line at E. With E as centre, describe the segment from A, through D to B.

14.3.5 Definition of Geometrical Shapes

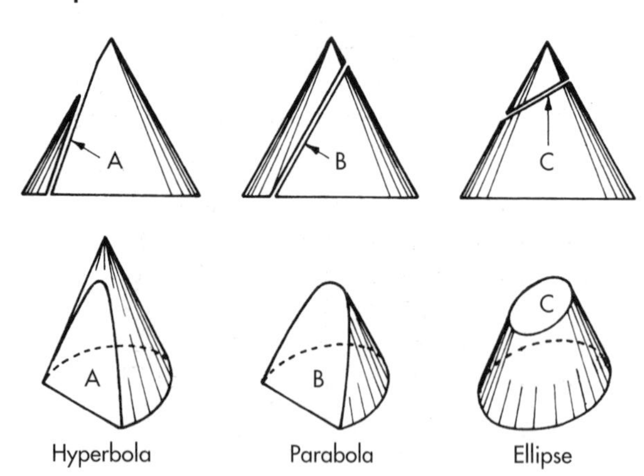

Figure 14.5(a) Definition of geometrical shapes

Illustrated in Figure 14.5(a) is an explanation for some of the other arch shapes to follow.

Hyperbola

This is the name given to the curve A, produced when a cone is cut by a plane (a flat, imaginary sheet-surface) making a larger angle with the base than the side angle of the cone (e.g. for a 60° cone use a 70–90° cut).

Parabola

This is the name given to the curve B, produced when a cone is cut by a plane parallel to its side (e.g. for a 60° cone use a 60° cut).

Ellipse

This is the name given to the shape C, produced when a cone or cylinder is cut by a plane making a smaller angle with the base than the side angle of the cone (or a cylinder). The exception is that when the cutting plane is parallel to the base, true circles will be produced.

Axes of the Ellipse

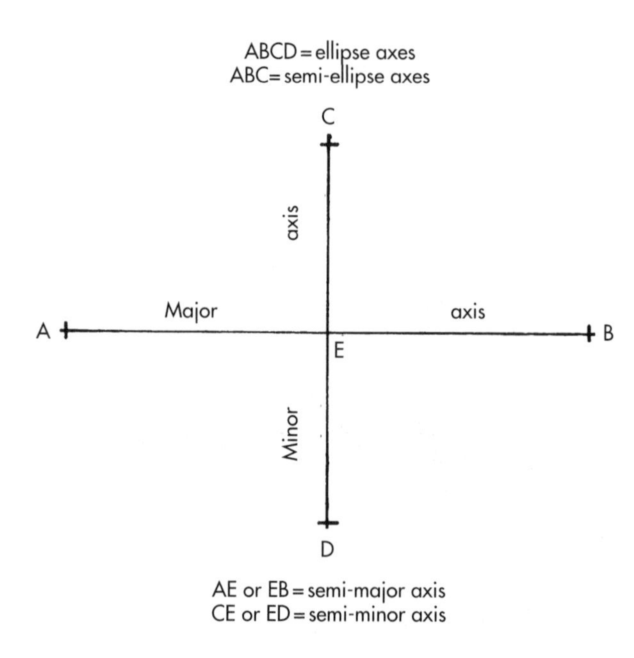

Figure 14.5(b) Axes of the ellipse

Figure 14.5(b): An imaginary line through the base and top of a cone or cylinder, that cuts exactly through the centre, is known as an axis. The shape around the axis (centre) is equal in any direction, but when cut by an angled plane – to form an ellipse – the shape enlarges in one direction, according to the angle of cut. For reference, the long and the short lines that intersect through the centre, are called the major axis and the minor axis.

The axes on each side of the central intersection, by virtue of being halved, are called semi-major and semi-minor axes. The semi-elliptical arch is so called because only half of the ellipse is used.

14.4 TRUE SEMI-ELLIPTICAL ARCHES

True semi-elliptical shapes are not normally used for brick arches, as the methods of setting out do not give

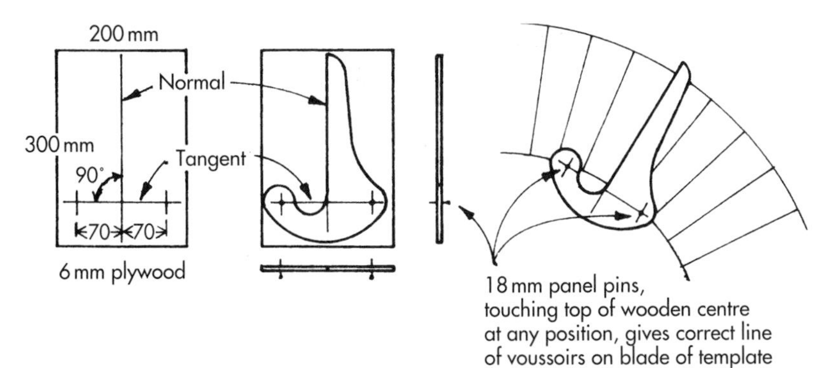

Figure 14.6(a) Tangent-template

the bricklayer the necessary centre points as a reference to the radiating geometrical-normals of the voussoir joints. However, the problem could be solved by using a simple, purpose-made tangent-template, as shown in Figure 14.6(a). Therefore, the *true* semi-elliptical arch methods shown here might only serve to build a complete knowledge of the subject being covered, leading on to the methods favoured by bricklayers.

14.4.1 Intersecting-lines Method

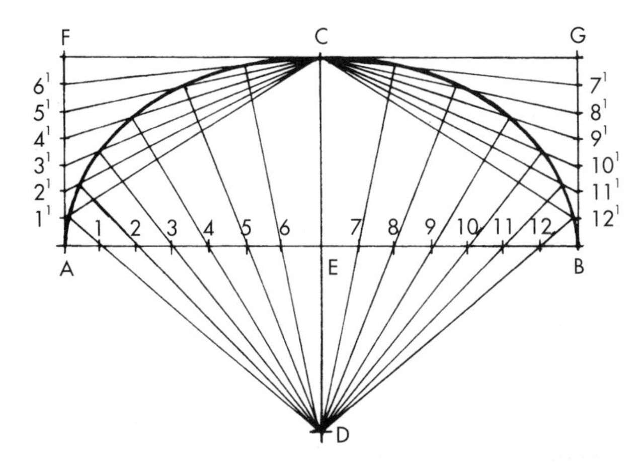

Figure 14.6(b) Intersecting-lines method

Figure 14.6(b): Span AB, given as the major axis, is bisected at E to produce CD, a lesser amount than AB, given as the minor axis. Vertical lines from AB and horizontal lines from C are drawn to form the rectangle AFGB. Lines AF, GB, AE and EB are divided by an equal, convenient number of parts. Radiating lines are drawn from C to 1^1, 2^1, 3^1, and so on to 12^1; and from D, through divisions 1 to 12 on the major axis, to intersect with their corresponding radial. These are radials 1 to 1^1, 2 to 2^1, 3 to 3^1, and so on. The intersections plot the path of the semi-ellipse to be drawn freehand or by other means, such as with the aid of a flexi-curve instrument.

14.4.2 Intersecting-arcs Method

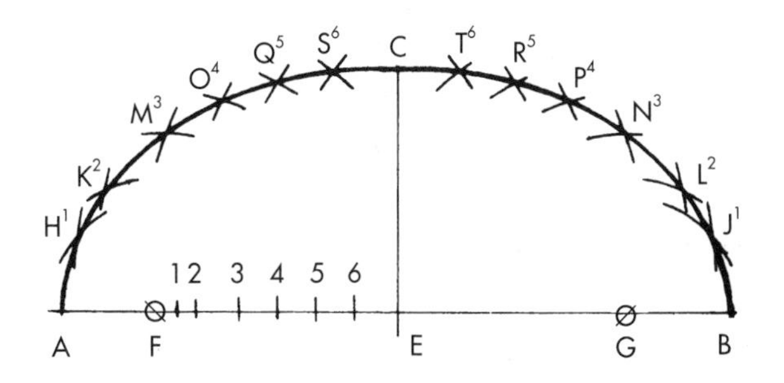

Figure 14.7 Intersecting-arcs method

Figure 14.7: Draw the major axis AB and the semi-minor axis CE as before. With compass set to AE or EB, and C as centre, strike arcs F and G on the major axis; these are known as the *focal points*. Mark a number of points anywhere on the semi-major axis between F and E; place the first point very close to F. Number these points 1, 2, 3, etc. Now with the compass set to A1, strike arcs H^1 from F, and J^1 from G. Reset compass to B1, strike arcs H^1 from G, and J^1 from F. Continue as follows:

compass A2, strike K^2 from F, L^2 from G
compass B2, strike K^2 from G, L^2 from F
compass A3, strike M^3 from F, N^3 from G
compass B3, strike M^3 from G, N^3 from F
compass A4, strike O^4 from F, P^4 from G
compass B4, strike O^4 from G, P^4 from F
compass A5, strike Q^5 from F, R^5 from G
compass B5, strike Q^5 from G, R^5 from F
compass A6, strike S^6 from F, T^6 from G
compass B6, strike S^6 from G, T^6 from F

These arcs plot the path of the semi-ellipse to be completed as before.

14.4.3 Concentric-circles Method

Figure 14.8: Draw the major axis AB and the semi-minor axis CE as before. Strike semi-circles radius EA and EC. Draw any number of radiating lines from E to cut both semi-circles. For convenience, the angles of the radials used here are 15°, 30°, 45°, 60° and 75°, each side

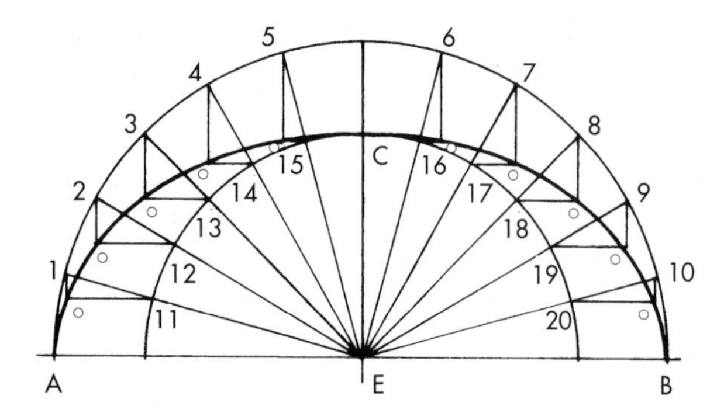

Figure 14.8 Concentric-circles method

of the centre line CE. Draw vertical lines inwards from points 1, 2, 3, etc. on the outer semi-circle, and horizontal lines outwards from points 11, 12, 13, etc. on the inner semi-circle. These intersect at points o, which plot the path of the semi-ellipse to be completed as before.

14.4.4 Short-trammel Method

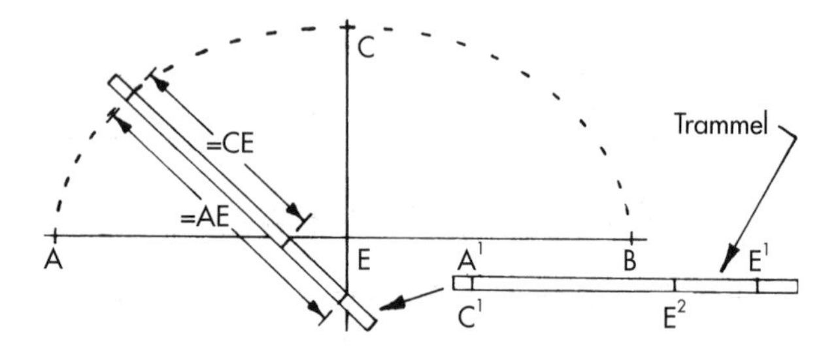

Figure 14.9 Short-trammel method

Figure 14.9: Draw the major and semi-minor axes as before. Select a thin lath or strip of hardboard, etc., as a trammel rod. Mark it as shown, with the semi-major axis A^1E^1 and the semi-minor axis C^1E^1. Rotate the trammel in a variety of positions similar to that shown, ensuring that marks E^1 and E^2 always touch the two axes, and mark off sufficient points at A^1/C^1 to plot the path of the semi-ellipse to be completed as before.

14.4.5 Long-trammel Method

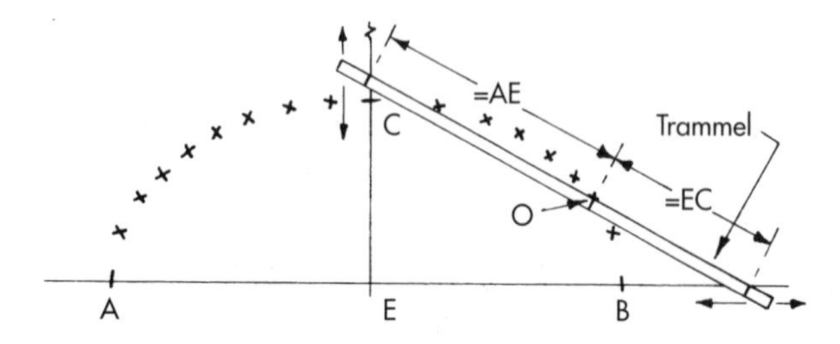

Figure 14.10 Long-trammel method

Figure 14.10: This is similar to the previous method, except that the semi-major and semi-minor axes form a continuous measurement on the trammel rod; the outer marks thereon move along the axes, while the inner mark, O, plots the path of the semi-ellipse. This method is better than the previous one when the difference in length between the two axes is only slight.

14.4.6 Pin-and-string Method

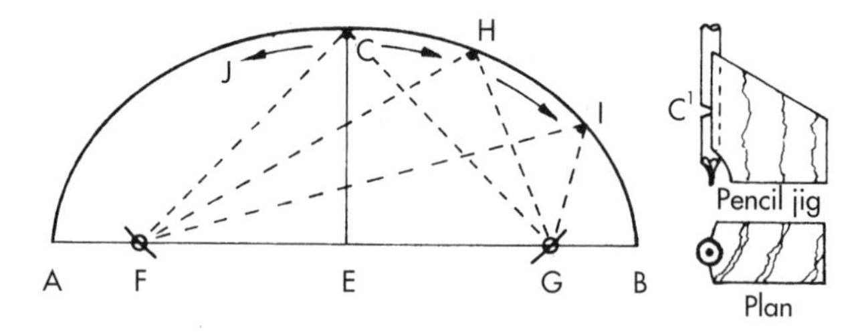

Figure 14.11 Pin-and-string method

Figure 14.11: This method uses focal points on the major axis. These are shown here as F and G, and either point equals AE or EB on the compass, struck from C to give F and G. This time, to describe the arch shape, drive nails into points F, C and G, pass a piece of string around the three nails and tie tightly. Make a pencil jig, if possible, and cut a notch in a pencil – as shown at C^1 – remove nail at C, replace with pencil and jig and rotate to left and right, as indicated at HIJ, to produce a true semi-ellipse.

14.5 APPROXIMATE SEMI-ELLIPTICAL ARCHES

These *approximate* semi-elliptical shapes, as previously mentioned, are preferred for brick or stone arches, as they eliminate the freehand flexi-curve, simplify the setting out and give the bricklayer definite centre points from which to strike lines for the radiating geometrical-normals of the voussoir joints.

14.5.1 Three-centred Method

Figure 14.12: Draw major axis (span) and semi-minor axis (rise) to the sizes required, as described before. Draw a diagonal line from A to C (the chosen or given rise). With centre E, describe semi-circle AB to give F. With centre C, strike an arc from F to give G. Bisect AG to give centres H and I. With centre E, transfer H to give J. Draw the line through IJ to give L. HIJ are the three centres. Draw segments AK from H, BL from J, and KL from I, to cut through the rise at C and complete the required shape.

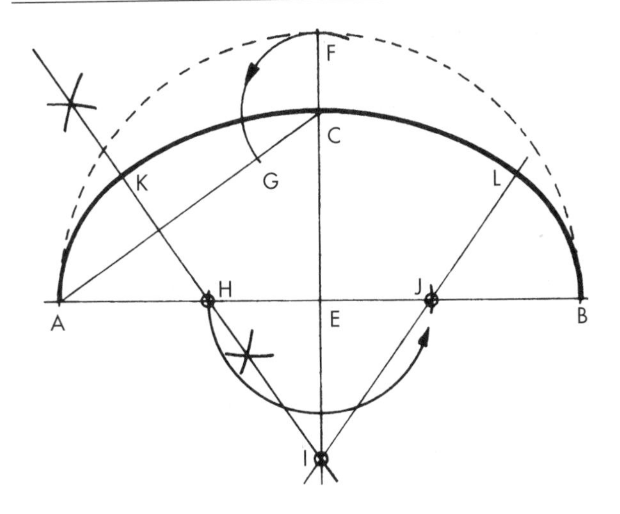

Figure 14.12 Three-centred method

14.5.2 Five-centred Method

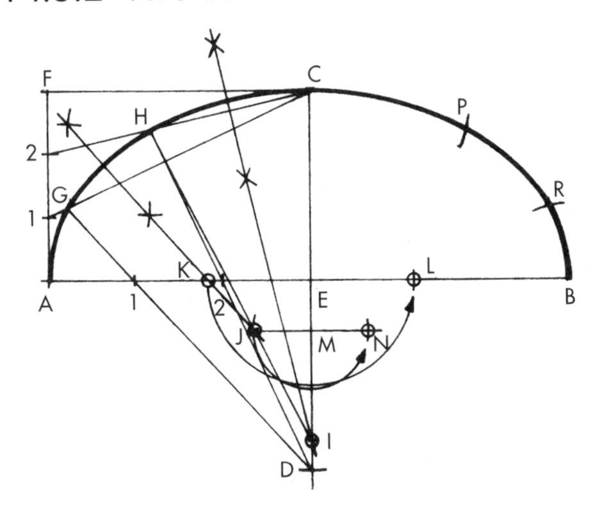

Figure 14.13 Five-centred method

Figure 14.13: Draw major and semi-minor axes as before. Draw lines AF and CF, equal to CE and AE, respectively. Divide AF by three, to give A12F. Draw radials C1 and C2. With centre E and radius EC, strike the arc at D on the centre line. Divide AE by three, to give A12E. Draw line D1 to strike C1 at G, and line D2 to strike C2 at H. Bisect HC and extend the bisecting line down to give I on the centre line. Draw a line from H to I. Now bisect GH and extend the bisecting line down to cut the springing line at K and line HI at J. IJK are the three centres to form half of the semi-ellipse. The other two centres are transferred as follows: with centre E, transfer K to give L on the springing line; draw a horizontal line from J to M and beyond; with centre M, strike an arc from J to give the centre N. To transfer the normals G and H, strike arc CP, equal to CH, and BR, equal to AG. To form the semi-ellipse, draw segments AG from centre K, GH from J, HCP from I, PR from N, and RB from L.

14.5.3 Depressed Semi-elliptical Arch

Figure 14.14: This arch uses a very small rise. The geometry is exactly the same as that used for the three-centred method explained in Figure 14.12.

14.6 GOTHIC ARCHES

14.6.1 Equilateral Gothic Arch

Figure 14.15: The radius of this arch, equal to the span, is struck from centres A and B to a point C. The line AD highlights a geometrical normal to the curve and a line at right-angles to this, as shown, is known as a tangent. Normals E, F, G, H, I, J, K, L, etc. are indicated by broken lines to form the voussoirs of the arch. Incidentally, points A, B and C of this arch, if joined by lines instead of curved arcs, form an equilateral triangle, where all three sides are equal in length and contain three angles of 60°.

14.6.2 Depressed Gothic Arch

Figure 14.16: This arch is sometimes referred to as an *obtuse* or *drop* Gothic arch. The centres for striking this

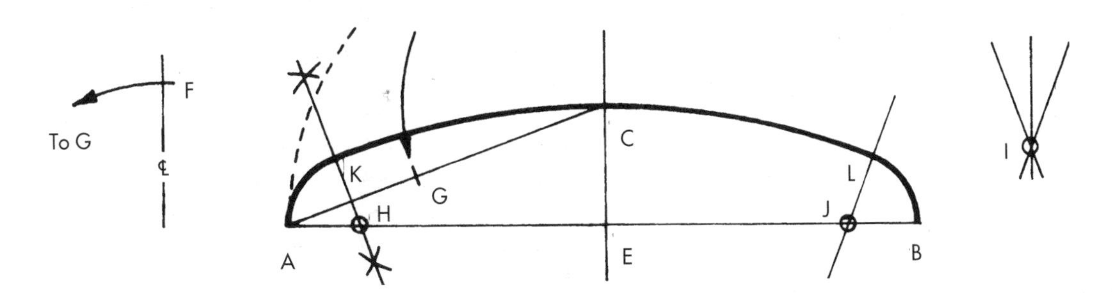

Figure 14.14 Depressed semi-elliptical arch

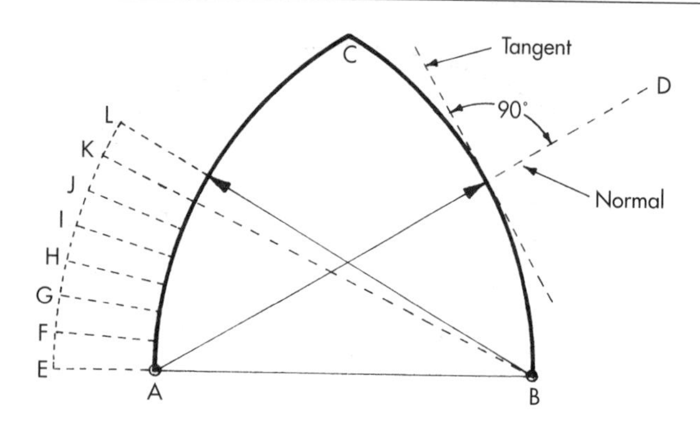

Figure 14.15 Equilateral Gothic arch

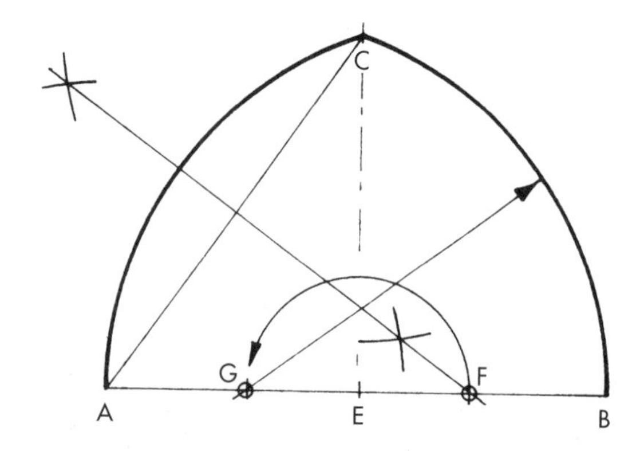

Figure 14.16 Depressed Gothic arch

arch come within the span, on the springing line. Bisect the span AB to give the centre line through E. With compass less than AB, strike the rise at C from A. Alternatively, mark the chosen or given rise at C from E. Draw line AC and bisect to give centre F on the springing line. With centre E, transfer F to give centre G. Strike segments AC from F and BC from G.

14.6.3 Lancet Gothic Arch

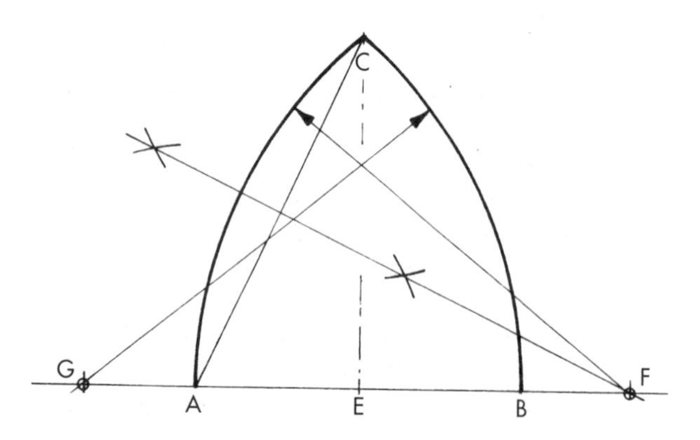

Figure 14.17 Lancet Gothic arch

Figure 14.17: The centres for this type of arch are outside the span, on an extended springing line. Bisect the span AB to give the centre line through E. With compass more than AB, strike the rise at C from A. Alternatively, mark the chosen or given rise at C from E. Draw line AC and bisect to give centre F on the extended springing line. With centre E, transfer F to give centre G. Strike segments AC from F and BC from G.

Note that line AC in the above is optional and need not actually be drawn.

14.7 TUDOR ARCHES

14.7.1 Tudor Arch – Variable Method

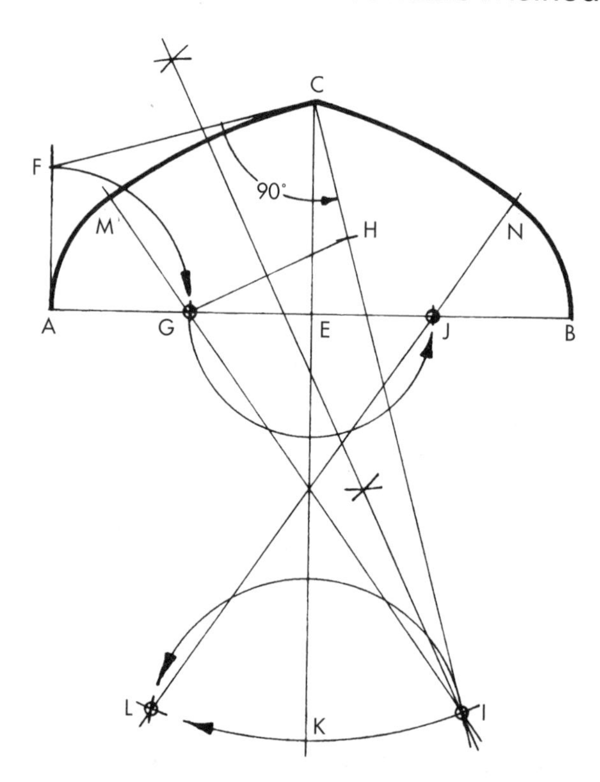

Figure 14.18 Tudor arch – variable method

Figure 14.18: This method is best and can be used to meet a variety of given or chosen rises. The geometry is usually mastered when practised a few times.

Draw the span AB and bisect it to give an extended centre line through E and down. Mark the rise at C. Draw vertical line AF, equal to two-thirds rise (CE). Join F to C. At right-angles to FC, draw a line down from C. With compass equal to AF, and A as centre, transfer F to give G. With the same compass setting, mark H from C on line CI. Draw line from G to H and bisect; extend bisecting line down until it intersects with line CI to give centre I. Draw line from I, extended through G on the springing line. With E as centre, transfer G to give J on the springing line. Again with E as centre, transfer I, through K, to strike arc at L. With K as centre, transfer I to give centre L. Draw line from

L to extend through J on the springing line. To complete, strike segments AM from G, MC from I, CN from L and NB from J.

14.7.2 Tudor Arch – Fixed Method

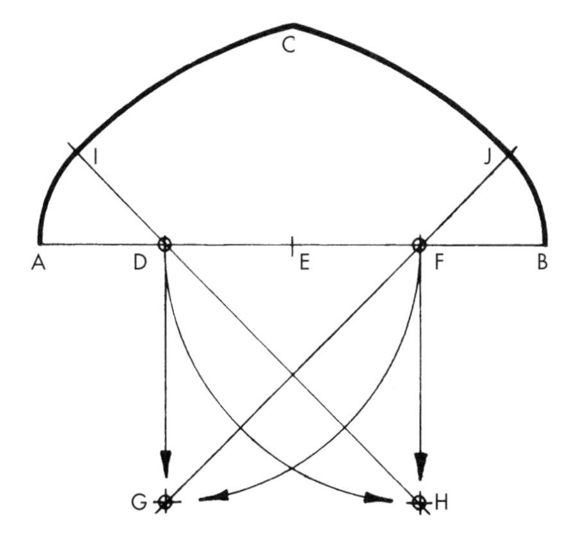

Figure 14.19 Tudor arch – fixed method

Figure 14.19: This method is simpler and can be used when the rise is not critical and the only information given or known is the span.

Draw span AB and divide by four to give DEF. Draw vertical lines down from D and F. With D as centre, transfer F to intersect the vertical line, giving G. With F as centre, transfer D to intersect the other vertical line, giving H. Draw diagonals from H and G, extending through D and F on the springing line. To complete, strike segments AI from D, IC from H, BJ from F and JC from G.

14.7.3 Depressed Tudor Arch

Figure 14.20: Draw span AB and divide by six to give DEFGH. Draw vertical lines down from E and G. With centre D, transfer H down to O, and with centre H, transfer D down to O. Draw diagonal normals through DO and HO, extending down to intersect the vertical lines at K and L, and extending up past the springing line to I and J. To complete the arch, strike segments AI from D, IC from L, BJ from H and JC from K.

Note that division of the span can be varied to achieve a different visual effect, as can the angles of the normals at D and H, drawn here at 60°. For example, 75° would make the arch more depressed.

14.7.4 Straight-top Tudor Arch

Figure 14.21: Draw span AB and divide by 9. Mark one-ninth of span from A to give D, and one-ninth

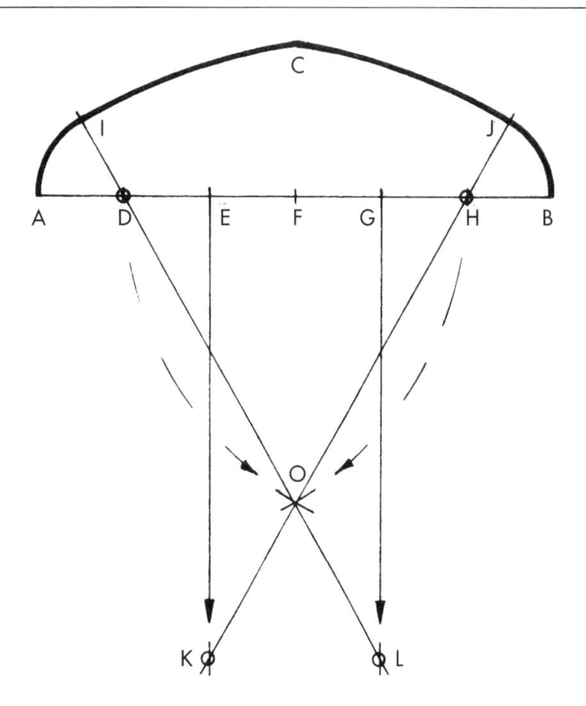

Figure 14.20 Depressed Tudor arch

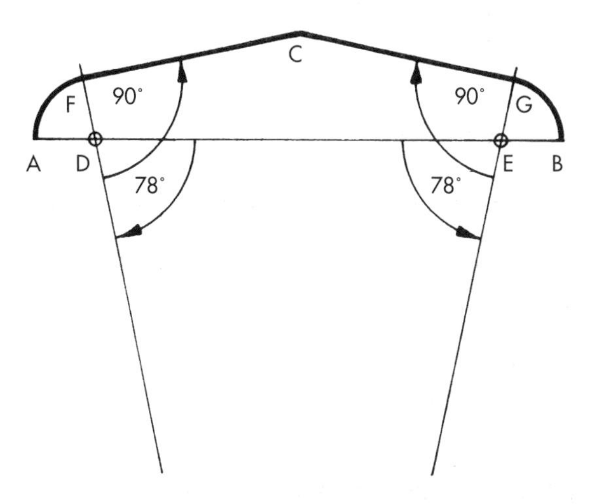

Figure 14.21 Straight-top Tudor arch

from B to give E. With a protractor, or a roofing-square containing a degree facility, set up diagonal normals passing through D and E at 78° to the horizontal. With centre D, strike arch curve AF, and with centre E, strike BG. From F and G, draw straight crown lines at 90° to the two normals, to intersect at key-position C.

Note that positions of the centres D and E can be varied to achieve a different visual effect, as can the angles of the normals at D and E, drawn here at 78°; however, the arch top or crown must always be tangential (at 90°) to the normals.

14.8 PARABOLIC ARCHES

14.8.1 Triangle Method

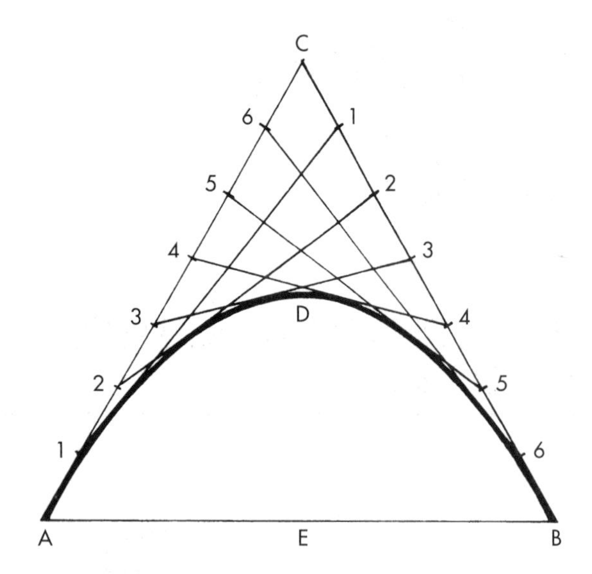

Figure 14.22 Parabolic arch – triangle method

Figure 14.22: Draw AB equal to the span, and vertical height EC equal to twice the rise. Join AC and CB to form a triangle. Divide each side of the triangle by a convenient number of equal divisions. This illustration uses seven divisions each side, numbered A, 1, 2, 3, 4, 5, 6, C, and C, 1, 2, 3, 4, 5, 6, B. Join 1 on AC to 1 on CB, 2 on AC to 2 on CB, and so on. These lines are tangential to the curve and give the outline shape of the parabola to be drawn freehand or by other means.

14.8.2 Intersecting-lines Method

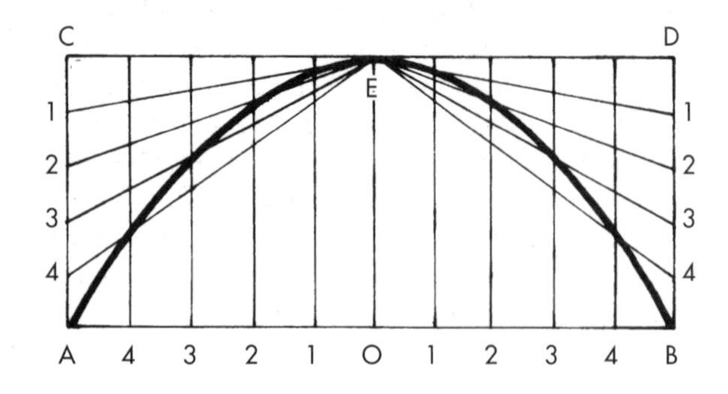

Figure 14.23 Parabolic arch – intersecting lines method

Figure 14.23: Draw a rectangle ABDC in which AB equals the span and AC equals the rise. Bisect AB to give vertical line EO. Divide OA into any number of equal parts, and OB, CA and DB, into the same number of equal parts as OA. Draw vertical lines up from horizontal divisions, and radiating lines from E to vertical divisions. The intersections thus formed, being base-vertical 1 (bv1) intersecting side-radial 1 (sr1), bv2 intersecting sr2, etc., on each side of the centre line, gives the outline shape of the parabola, to be completed as before.

14.9 HYPERBOLIC ARCH

14.9.1 Intersecting-lines Method

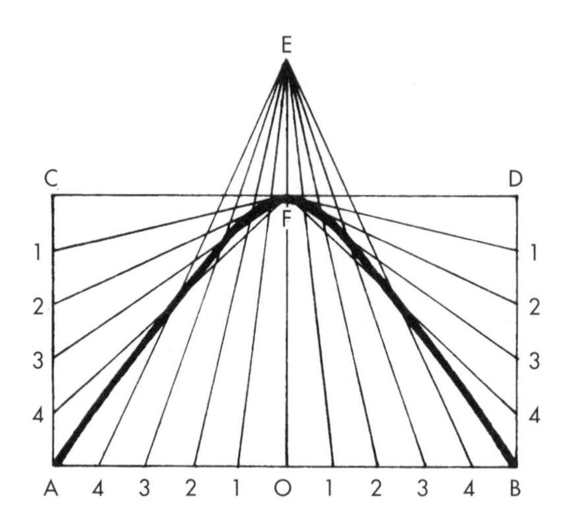

Figure 14.24 Hyperbolic arch

Figure 14.24: As before, draw a rectangle ABDC in which AB equals the span and AC equals the rise. Bisect AB to give vertical line EFO. Make apex E from F equal to half the rise (EF = FO/2). Divide OA, OB, CA and DB, as before and draw intersecting radials thus: base-radial 1 (br1) intersecting side-radial 1 (sr1), br2 intersecting sr2, etc., on each side of the centre line. The intersections give the outline shape of the hyperbola, to be completed as before.

Note that, as with true semi-elliptical shapes (Figures 14.6 to 14.11), parabolic and hyperbolic arches, if constructed of brick or stone, requiring centres and normals for alignment of joints, present the same problems. To overcome this, the tangent-template (see Figure 14.6(a)) could be used.

15

Making and Fixing Arch Centres

15.1 INTRODUCTION

The temporary wooden structures upon which brick arches are formed, are known as *centres*. They can be made in the joinery shop – taking advantage of a greater variety of machinery – or on site. The construction of the centre can be simple or complex, depending mainly on two factors: the span of the opening, and how many times the centre is to be used for other arch constructions.

15.1.1 Simple or Complex

For small spans up to about 1.2 m, the centre can be simple, of single-rib, twin-rib or four-rib construction. For spans exceeding 1.2 m, the centre becomes more complex, of multi-rib construction.

15.1.2 Practical Compass

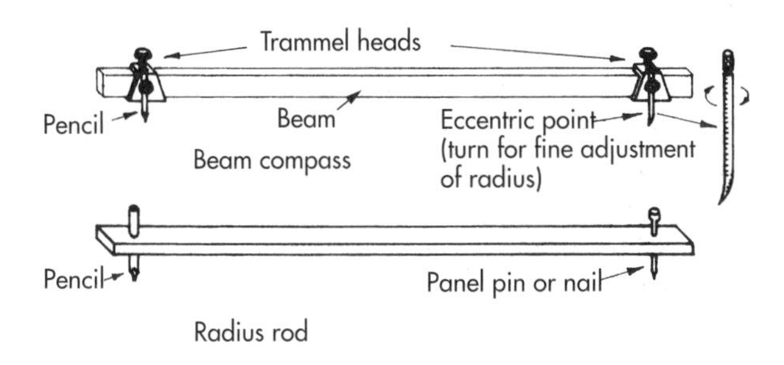

Figure 15.1 Radius rod and beam compass

Figure 15.1: A beam compass or a radius rod is required to set out the full-size shape of the centre, either directly onto the rib material (single and twin-rib) or onto a hardboard or similar setting-out board (four-rib and multi-rib centres), from which a template is made of the common rib shape. The beam compass consists of a pair of trammel heads and a length of timber, say of 38 × 19 mm section, known as a beam. To improvise, a radius rod can be easily made, consisting of a timber lath with a panel pin or nail through one end, the other end drilled to hold a pencil firmly.

15.2 SOLID TURNING PIECE (SINGLE-RIB)

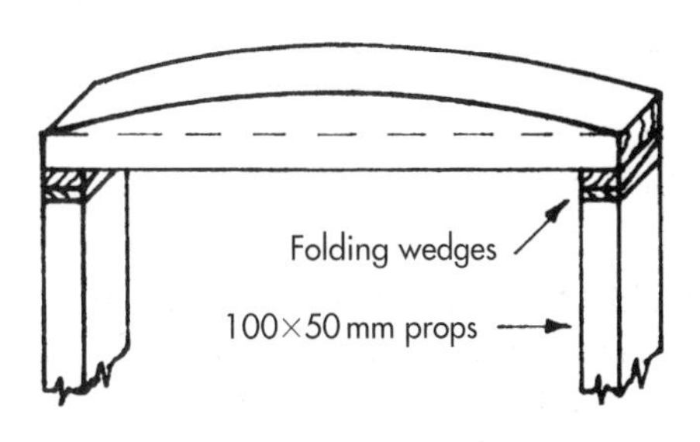

Figure 15.2(a) Turning piece

Figure 15.2(a): Solid turning pieces are used for segmental arches with small rises up to about 75 mm. If the rise is too slight (say 10 mm rise, 900 mm span), then a beam compass or radius rod would not be practical for drawing the curve and a triangular trammel frame or trammel rod should be used. As illustrated in Figure 15.2(b), to make the trammel rod or frame, mark the required span AB and rise CD on the rib material, place a board against CB and mark and cut line A^1C^1. As shown separately, make a sawcut at C^1 to take a pencil. To mark semi-segment CA, position pencil at C^1 and push the trammel against the protruding nails at A and B, moving to the left. Reverse the trammel and move to the right to mark semi-segment CB.

15.2.1 Cutting the Segment

Figure 15.2(c): Ideally, the curve for the solid turning-piece is best cut with a narrow band-saw machine, or, if on site, with a jigsaw. Alternatively, as shown in Figure 15.2(c), a series of tangential cuts can be made with a circular saw or handsaw, prior to shaping with a Surform and/or a traditional plane.

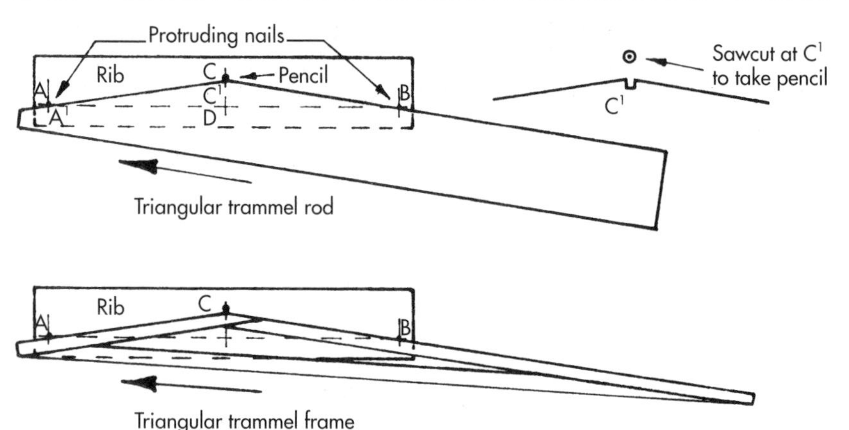

Figure 15.2(b) Triangular trammel rod and frame

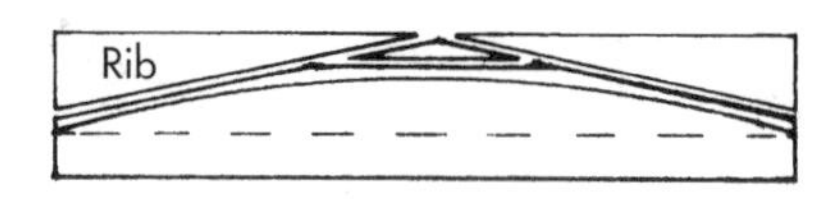

Figure 15.2(c) Tangential cuts

15.3 SINGLE-RIB CENTRES

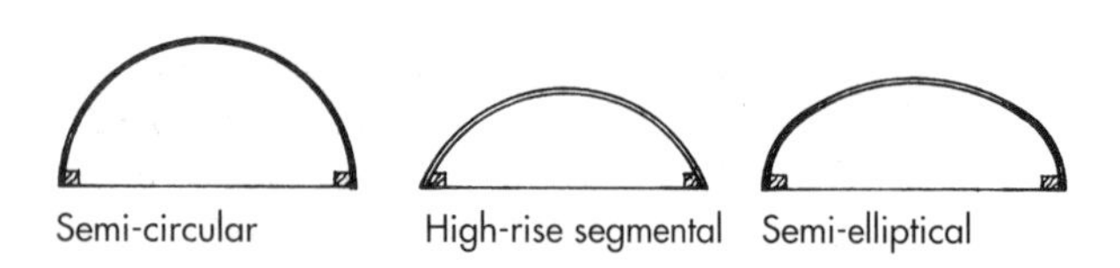

Figure 15.3(a) Single-rib centres

Figure 15.3(a): These centres follow an unconventional method of construction, but are surprisingly strong and effective. Their strength is dependent upon the stress of compression achieved in bending the hardboard or plywood skin over the curved rib. For this reason, they are more suitable for semi-circular, high-rise segmental and semi-elliptical centres – in that order of diminishing suitability – and less suitable for Gothic, Tudor and low-rise segmental centres.

15.3.1 Marking and Cutting the Rib

Figure 15.3(b): The plywood rib, preferably of 18 mm thickness, is set out along the springing line to the span, minus the skin thickness each side, and marked with a radius rod (or beam compass), as shown. The shape is best cut – as before – by band saw or portable jigsaw, but, if not available, the job can be done by using a sharp compass saw, finishing with a flat spokeshave, Surform and/or a smoothing plane.

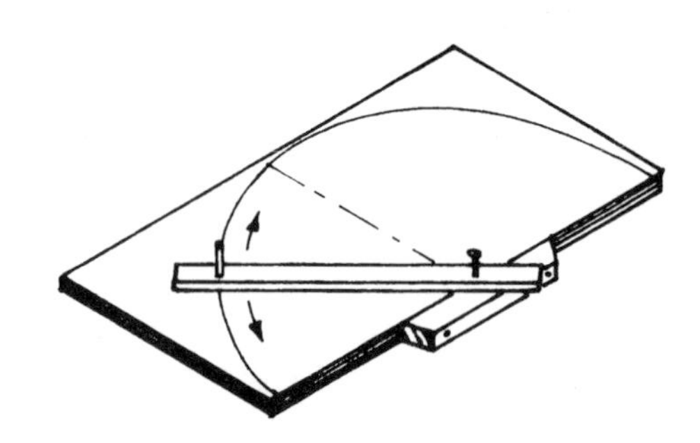

Figure 15.3(b) Setting out

15.3.2 Adding Springing Blocks

Figure 15.3(c) Adding springing blocks

Figures 15.3(c)–(e): Two 50 × 50 mm blocks, cut in length to the brick size of the arch minus, say, 18 mm inset-tolerance, are half-housed centrally to fit housings cut into the springing points of the rib on each side and fixed with 63 mm oval, lost-head nails or countersunk screws (Figures 15.3(c) and (d)). The skin, of 3.5–4 mm hardboard or plywood, long enough to cover the curved shape of the rib and as wide as the block-length, must have the rib thickness pencil-gauged through the centre (Figure 15.3(e)).

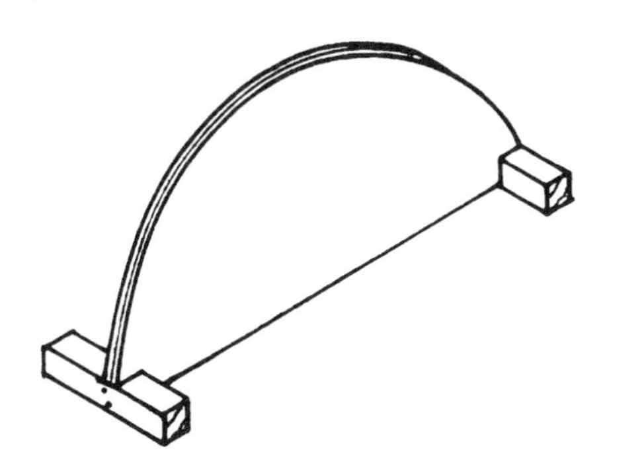

Figure 15.3(d) Springing blocks in place

15.3.3 Fixing the Skin

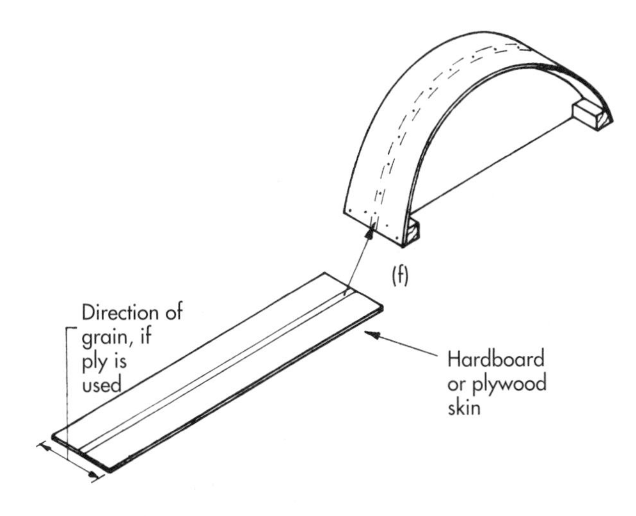

Figure 15.3(e) Skin. (f) Skin in place

Figure 15.3(f): When finally fixing the skin to the springing blocks and the rib, using 25 mm panel pins or 32 mm small-headed clout nails, it is important to ensure that the skin is taut and is centred on the rib, following the gauge lines, as indicated. Note that the length of skin can be calculated, or measured directly from the rib shape by encircling the finished curve with a tape rule.

15.4 TWIN-RIB CENTRES

Figure 15.4(a): These centres are superior to the single-rib type and are suitable for all arch shapes. The ribs may consist of 12–18 mm plywood – or other sheet material like chipboard or Sterling board, if the arch is not being reused too many times. Bearers, of 75 × 25 mm or 100 × 25 mm section, are fixed to the underside of the centre, at each end of the springing line. These act as spacers betwen the two ribs and as bearers supporting the centre on the props.

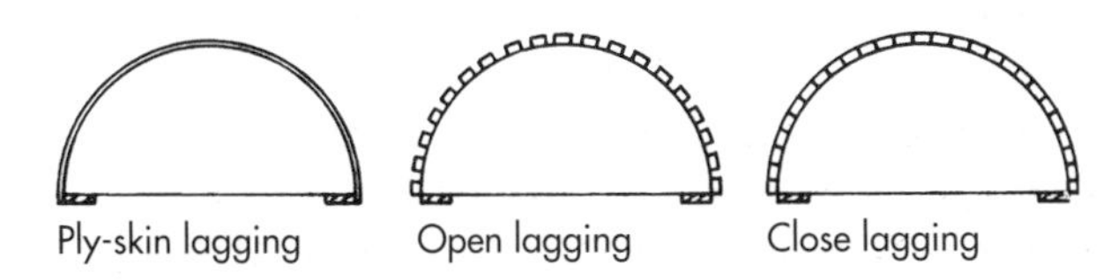

Figure 15.4(a) Twin-rib centres

15.4.1 Bracing-spacers

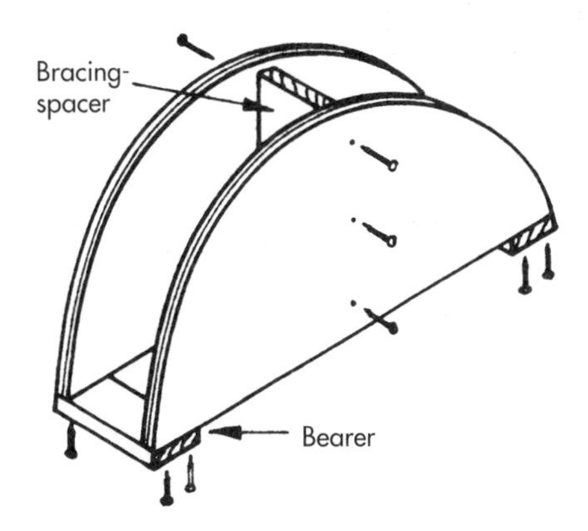

Figure 15.4(b) Bracing-spacer board

Figure 15.4(b): A bracing-spacer board, equal to the inside width of the centre, as indicated, can be inserted to keep the ribs initially parallel and square. The centre is then covered with a hardboard or plywood skin or, alternatively, with strips of timber known as *laggings*, illustrated in Figures 15.4(c) and 15.4(d). These can be placed close together, referred to as *close lagging*, or be spaced apart and referred to as *open lagging*, as indicated in Figures 15.4(d) and (e). Close lagging should be used

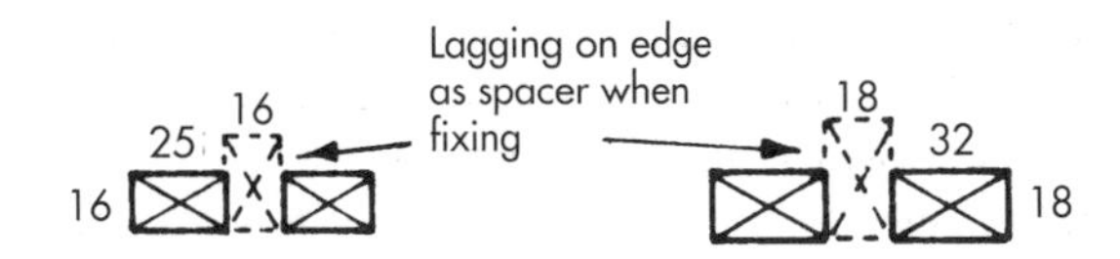

Figure 15.4(c) Lagging sizes for small spans

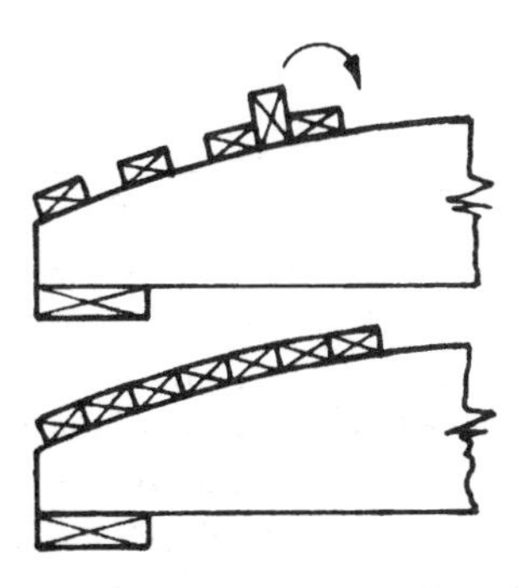

Figure 15.4(d) Open and close lagging

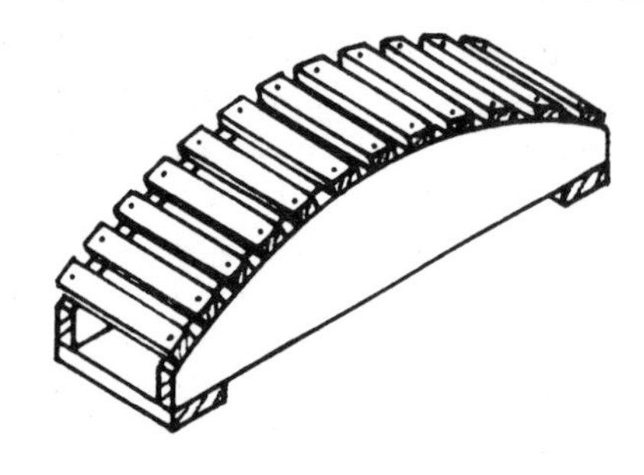

Figure 15.4(e) Centre ready for skin

for gauged arches with tapered bricks (voussoirs), and open lagging for common arches with ordinary bricks used as parallel voussoirs.

15.4.2 Making the Centre

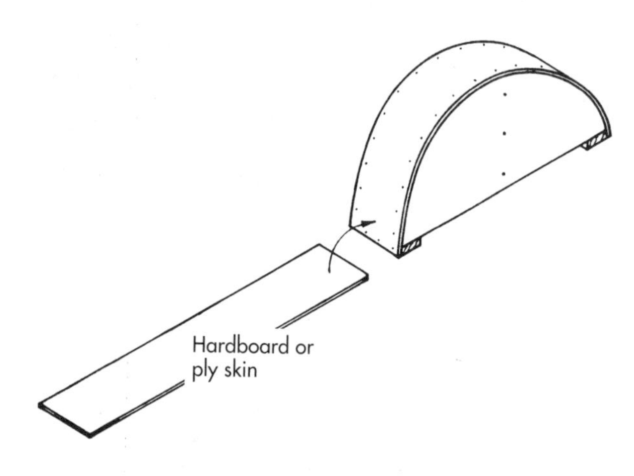

Figure 15.4(f) Applying the skin

Figure 15.4(f): To make the centre, set out and form one rib as previously described. Use the first rib as a template to mark out and form the second rib. Cut the two 75 × 25 mm bearers in length to the brick size of the arch, minus 18 mm inset-tolerance, and fix to the underside of the ribs at each end with minimum 50 mm round-head or lost-head wire nails. Cut and fix the bracing-spacer (Figure 15.4(b)). Cut the hardboard or plywood skin to width (equal to length of bearers) and fix to the ribs, as indicated in Figure 15.4(f), or cut the laggings and fix to the ribs with 38 or 50 mm lost-head oval nails.

15.4.3 Lagging Jig

Figure 15.4(g): A plan view of a simple jig for cutting the lengths of lagging quickly. Two 50 × 25 mm battens are fixed to a bench, stool top or scaffold board, etc., to form the jig. As indicated, a nail at one end positions the lagging strip without trapping sawdust being pushed

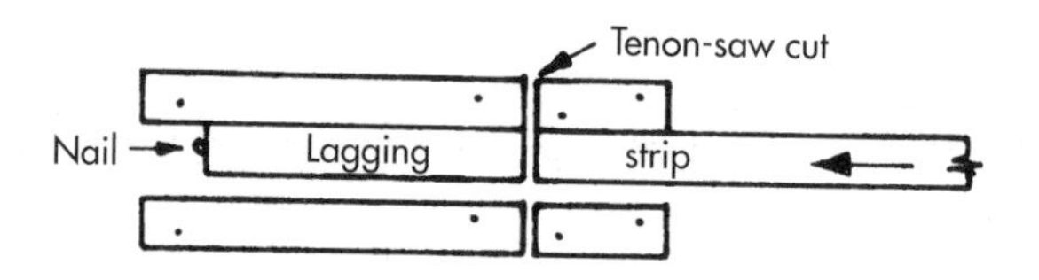

Figure 15.4(g) Lagging jig

down from the right; the saw at the other end cuts the required length repeatedly after feeding in the strip towards the nail.

15.5 FOUR-RIB CENTRES

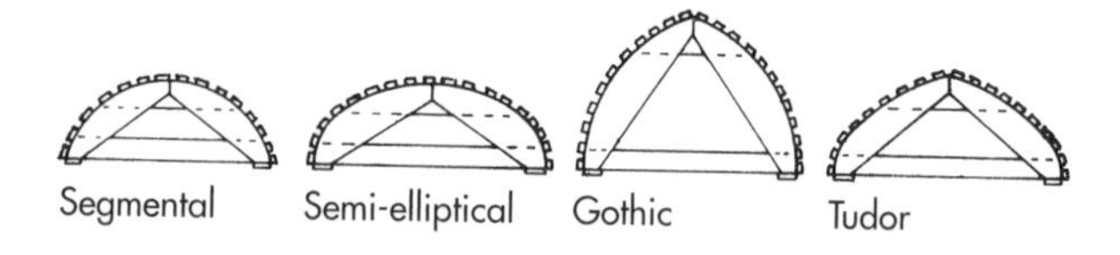

Figure 15.5(a) Four-rib centres

Figure 15.5(a): These centres follow traditional methods of construction and are suitable for all arch shapes except low rise, which require only one or two ribs for small spans, as shown previously in Figures 15.2(a) and 15.4(e). Because of the time involved in making four-rib centres, they are less favoured than twin-rib centres using sheet material. As seen in elevation and section A–A in Figure 15.5(b), four ribs are required, two tie beams, two collars, two bearers, two optional struts, an optional brace and laggings.

15.5.1 Construction Details

Figures 15.5(b) and (c): Each pair of ribs is connected to a top collar and a bottom tie beam by clench-nails. As

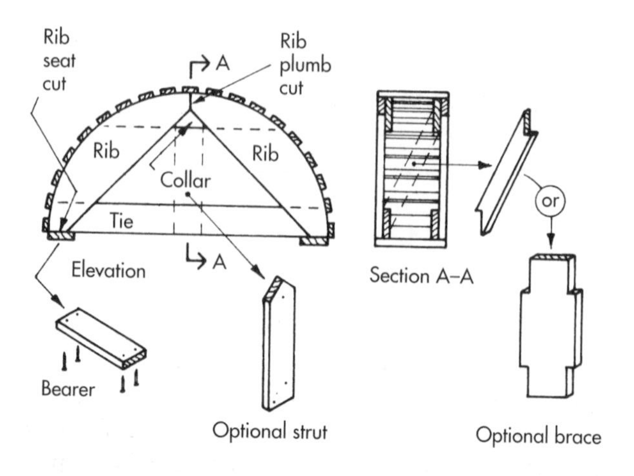

Figure 15.5(b) Semi-circular centre

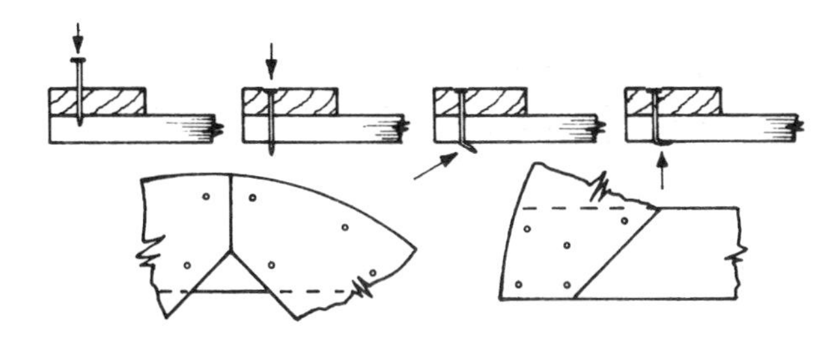

Figure 15.5(c) Staggered clench-nailing

indicated in Figure 15.5(c), when driven in, the nails purposely protrude by at least 6 mm and are then bent sideways and clenched over to secure the joint. For extra strength, optional struts can be fixed from the birdsmouth at the apex of the ribs, against the collar, to a central position on the face of the lower tie. The two bearers are fixed to the ends of the lower ties to line up the rib structures and act as bearers, supporting the centre when resting on the props and wedges.

15.5.2 Optional Skins

Hardboard or plywood skins can be used to cover the centre, in single or double thicknesses – or traditional laggings can be used, efficiently cut to length as previously described and fixed to the curved ribs, as before with 38 or 50 mm lost-head oval nails.

Note that, if considered necessary, an optional brace, indicated in section A–A of Figure 15.5(b), can be used to achieve squareness in width and greater rigidity between the two rib structures.

15.5.3 Setting Out

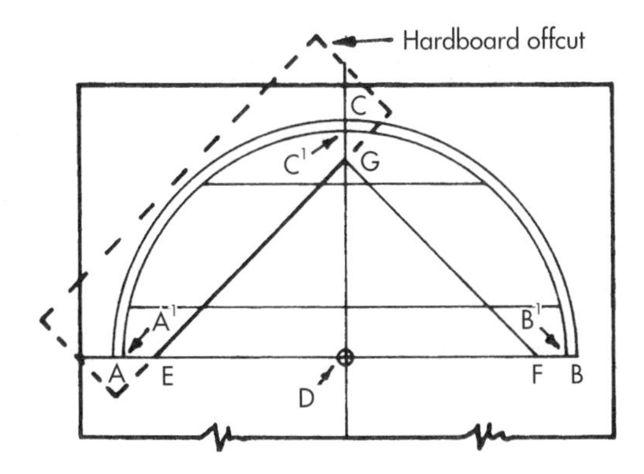

Figure 15.5(d) Setting-out board

Figure 15.5(d): To produce the rib shape for this type of arch and also to ensure that the finished centre is true to shape, a setting-out board is required. The figure indicates the setting out for a semi-circular centre. The first line to be drawn is line AB on the springing line, representing the span minus 2–3 mm (this reduction is a practical tolerance to allow for brick reveals being slightly out of square or irregular). Next, bisect line AB to produce the centre line CD. Then set the beam compass or radius rod to AD and describe the semi-circle ACB. Reset the compass or radius rod to AD minus the lagging thickness and describe the second semi-circle, producing A^1, C^1, and B^1, representing the rib curve. Now measure down at least 60 mm from C^1 to G, as the depth of the rib's plumb cut and measure inwards at least 60 mm from A^1 to E, and B^1 to F, being the length of the rib's seat cuts. Draw tangential lines EG and GF, representing the underside of the ribs. Draw two lines parallel to the springing, representing the top collar and bottom tie beam; these timbers should be about 100–150 mm wide × 25 mm thickness.

15.5.4 Marking and Cutting the Ribs

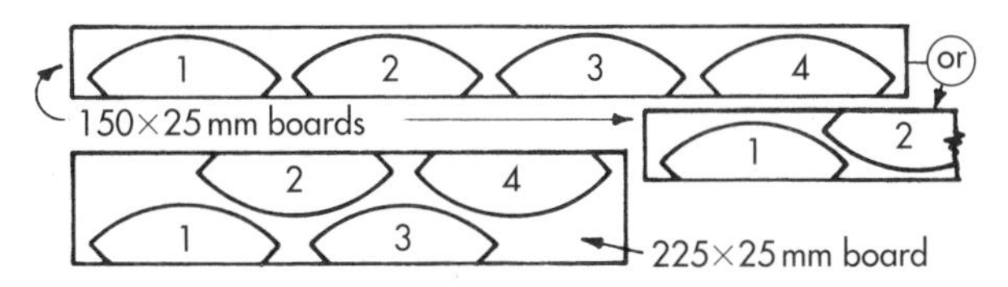

Figure 15.5(e) Method of marking out ribs

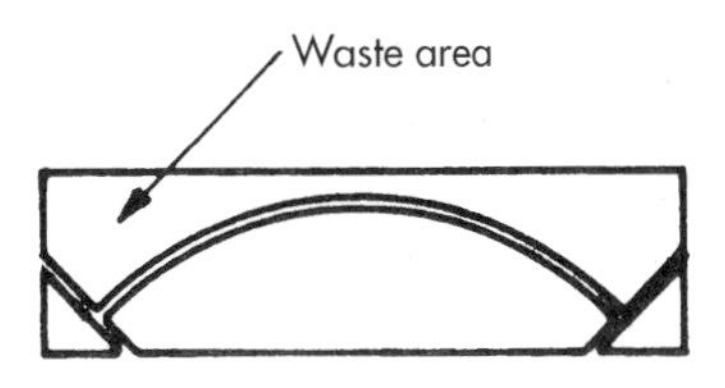

Figure 15.5(f) Producing a hardboard template

Figures 15.5(e) and (f): The rib segments can be cut from a 150 × 25 mm or 225 × 25 mm board, as illustrated. A hardboard template, shown in Figure 15.5(f), facilitates the marking out of the ribs onto the board economically. To make the template, lay a piece of hardboard on the setting-out board, as shown by the dotted outline in Figure 15.5(d), strike the compass or rod from centre D to describe the quadrant A^1C^1, mark the plumb cut CG and the seat cut AE, then remove and cut to shape.

15.5.5 Marking the Collar and Tie

Figure 15.5(g): The collar and tie are laid in position on the setting-out board and marked with the compass or radius rod setting of A^1D, struck from centre D^1, squared up from D. As shown, a small temporary wooden block can be fixed to the tie, to retain the point of the radius rod or compass.

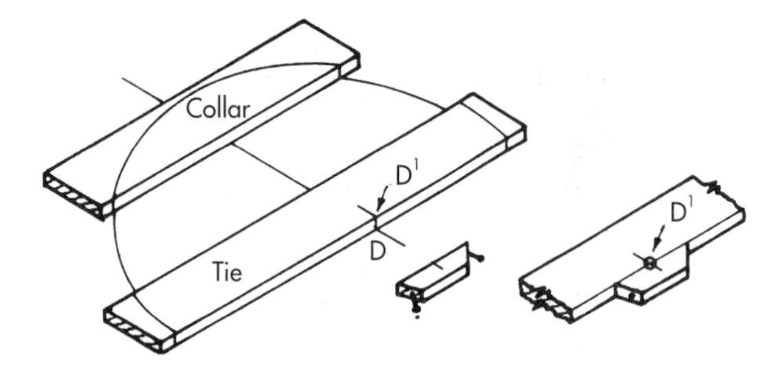

Figure 15.5(g) Marking out collar and tie

15.6 MULTI-RIB CENTRES

These centres also follow traditional methods of construction, simply because their multi-rib design allows a variety of configurations suitable for very wide arch-spans. They are mainly used for high-rise segmental, semi-circular and semi-elliptical arches. According to the span, they may have to be set out directly on the floor, or onto two or more setting-out boards placed side by side. The principles of setting out and producing rib templates are similar to those described for four-rib centres

15.6.1 Semi-hexagonal Configuration

Figure 15.6(a): An elevation and sectional view of a semi-circular centre suitable for spans up to about 4 m are illustrated. The joints of each rib radiate from the centre point, comprising six sector-shapes with angles of 30° (6 × 30° = 180°). Given joint-lengths of about 60mm, it follows that the underside of the ribs is tangential to the curve and forms two semi-hexagonal shapes. This is seen more clearly in Figure 15.6(b). For extra strength, struts should be added as shown.

15.6.2 Semi-octagonal Configuration

Figure 15.7(a): An elevation and sectional view of a semi-elliptical centre are shown (drawn by a three-centre method), suitable – as before – for spans up to about 4 m. The joint-lines radiate from the centre point, comprising eight sector shapes with angles of $22\frac{1}{2}$° (8 × $22\frac{1}{2}$° = 180°). (Alternatively, the joints and struts can radiate from the three centres of the semi-ellipse.) Measuring equal amounts of about 60 mm along each joint-line, to determine the line of the ribs, the underside of the ribs forms two distorted semi-octagonal shapes; true octagonal shapes occur with semi-circular centres, as seen more clearly in Figure 15.7(b).

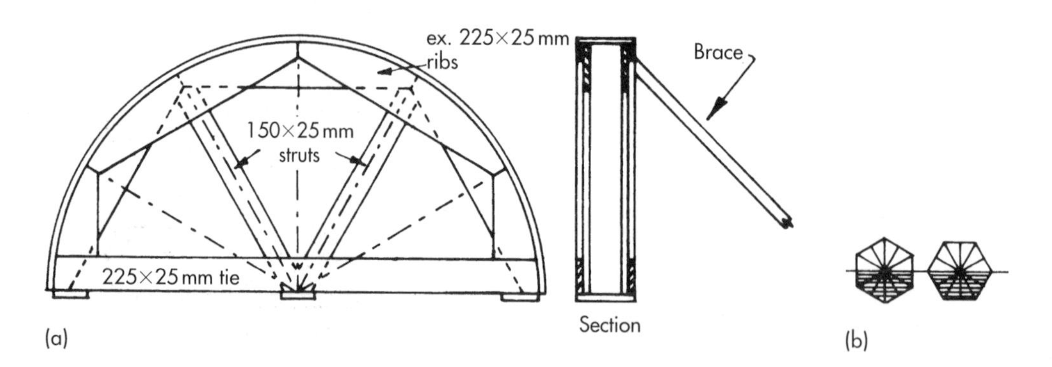

Figure 15.6(a) Multi-rib semi-circular centre. (b) Semi-hexagons

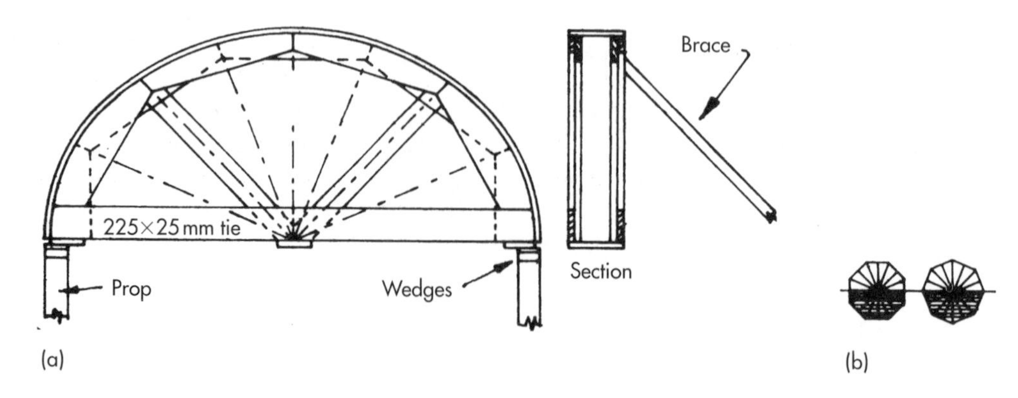

Figure 15.7(a) Multi-rib semi-elliptical centre. (b) Semi-octagons

15.6.3 Props and Folding Wedges

Props are required to give support to the arch centre and its temporary load. The timber sizes and arrangements of these props will vary according to the size of the centre and arch to be supported. Folding wedges, under each end of the arch-centre and on top of the props, are required for three reasons. First, they can be adjusted to help set the centre's springing line and height to the correct level. Second, soon after the arch has been 'locked' by the insertion of the final key brick or voussoir – and without being left overnight – the wedges are very gently eased to drop the centre by about 3–4 mm, allowing the arch to settle without cracking. Third, when the centre is finally being removed, further easing and removal of the wedges will make the job easier and less hazardous.

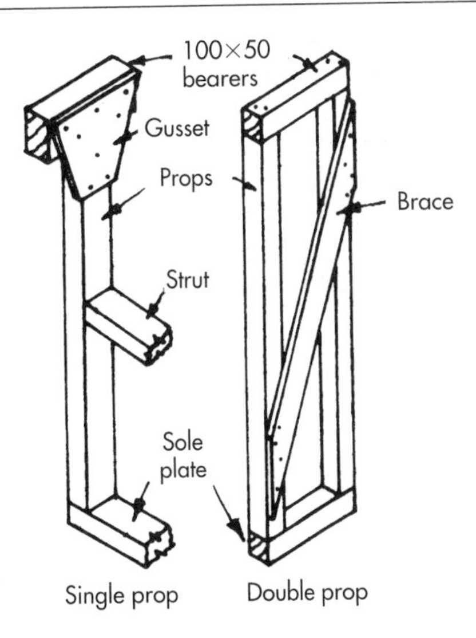

Figure 15.8(b) Props for large centres

15.6.4 Prop Arrangements

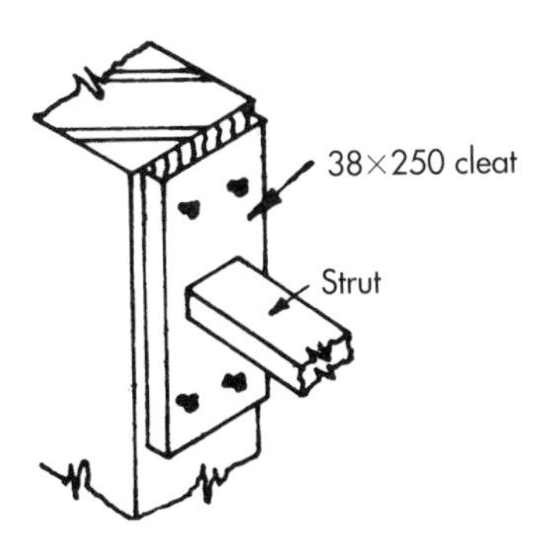

Figure 15.8(a) Prop for small centres

Figure 15.8(a): An unconventional but effective method of temporary support for small centres is shown. Short-lengths of timber cleats are fixed on each reveal of the opening, nailed with heads protruding and strutted as shown. The strut is vital to this arrangement. Figure 15.8(b) shows a slightly modified traditional method, using 100 × 50 mm single props each side, with plywood gussets and 100 × 50 mm bearers – and another method is shown using double props each side. Both of these methods are for large spans and each should have some form of strutting.

15.6.5 Wedge Shapes and Lagging Sizes

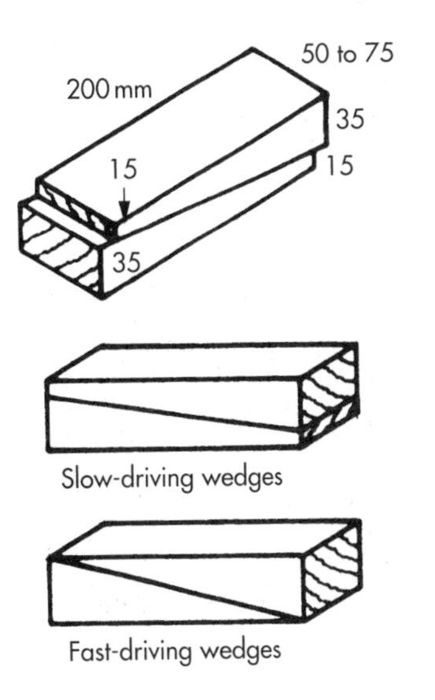

Figure 15.8(c) Folding wedges

Figure 15.8(c): Folding wedges, as illustrated, should be 'slow-driving' (be cut with a shallow angle), as these are beter for fine adjustments, non-slip bearing, and easing (slackening) prior to striking (removing) the centre. Lagging sizes for large centres are usually 25 × 38 mm or 25 × 50 mm.

16

Fixing Architraves, Skirting, Dado and Picture Rails

16.1 ARCHITRAVES

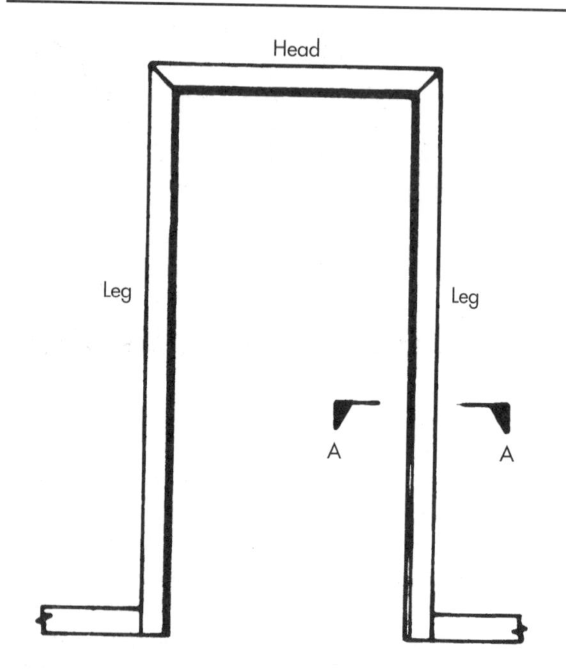

Figure 16.1(a) Elevation of a set of architraves

Figure 16.1(a): This is the name still used to describe the cover moulds of various designs and sizes that are fixed around the edges of door linings and door frames, etc., primarily to cover the joint between the lining or frame within the opening and the plaster (or plasterboard) surface of the wall. Even though present-day architraves are usually plain in design, architraves also add a visual finish to the opening. They are referred to as being in sets. A set of architraves for a door opening consists of two uprights, called jambs or legs and a horizontal piece called a head. A door lining or frame normally requires two sets of architraves, one on each side of the wall.

16.1.1 Number of Sets Required

On a second-fixing operation, the architraves need to be fixed before the skirting, so that the skirting on each side of a doorway can be butted up against the back of the architrave leg. To do the job efficiently, count the number of doorways in the dwelling and cut all of the sets required in one operation. For example, if there are five doorways, then ten sets of architraves are required, which is 10 heads and 20 legs. These are initially cut up squarely with an allowance in length for mitring. The length of the legs will be the door-opening height plus the width of the architrave-head plus about 30 mm allowance. The length of the heads will be the door-opening width plus the width of the architrave-leg each side plus about 40 mm allowance. Some of this allowance is required for the margins.

16.1.2 Margins

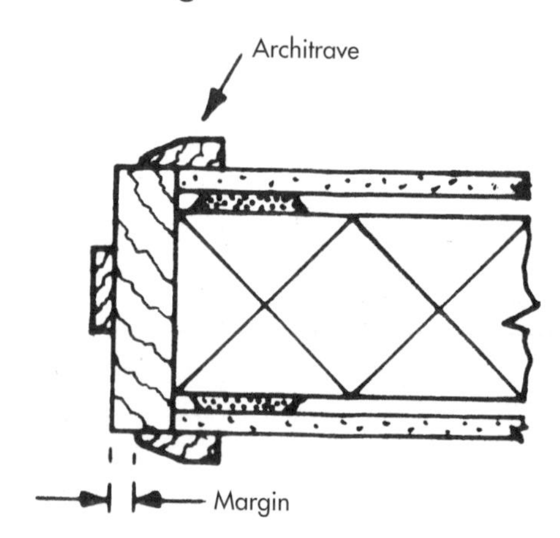

Figure 16.1(b) Section A–A

Figure 16.1(b): The architraves are always set back from the edge by a small amount, referred to as a margin. The margin is usually either 6 mm or 9 mm and whatever amount is decided upon or required, it must be consistently maintained around the opening. Inconsistent margins, often varying in size between the legs and the head and tapering from, say, 6 mm to 3 mm, are too often seen and reflect a very low standard of workmanship, spoiling the appearance of the finished job. The best way to establish a consistent margin, is to mark the required amount around the lining or frame's edge with a sharp pencil. It need not be a continuous line; broken lines of about 50–150 mm length, marked at about 450 mm intervals will be enough to achieve a good margin, even if the architrave leg is 'sprung' out of shape.

16.1.3 Margin-template

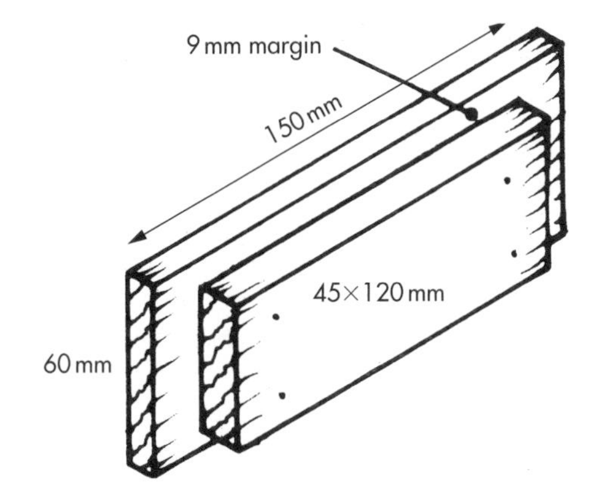

Figure 16.2(a) Shouldered margin-template

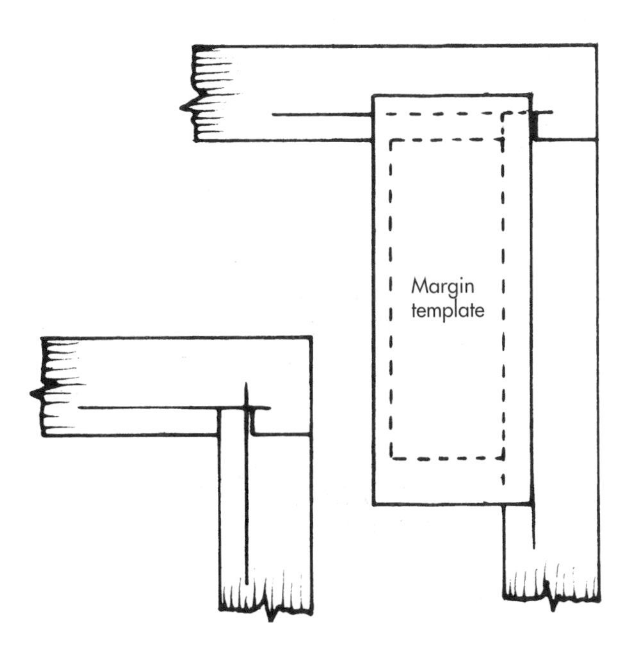

Figure 16.2(b) Use of the margin-template

Figures 16.2(a) and (b): Measuring and marking these margin lines would be tedious and time-consuming, so they need to be gauged and this is best done with a *shouldered margin-template*. This can be made quite easily by nailing two pieces of timber together, as illustrated, with a 6 mm margin on one side and an alternative 9 mm margin on the other. The ends are also shouldered to allow the template to over-run when marking against each top corner. These over-running marks (Figure 16.2(b)) create an overlapping intersection for marking the leg and head mitres accurately. Holding the template with one hand, the pencil in the other, the template is held horizontally against the head and the corner and marked, then slid to the mid-area and marked, then to the other corner and marked. On each side jamb, the template is held vertically against the head and marked (to create the intersection), then slid down to mark the mid-area, the bottom and intermediate positions between these three, making five marks on each side and three on the head. This method of marking should take no more than one minute per set of architraves.

16.1.4 Mitring and Fixing Technique

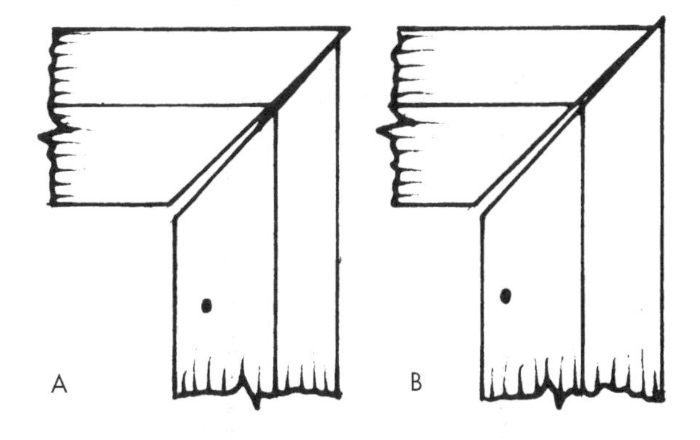

Figure 16.3 Ill-fitting mitres

Figure 16.3: Next, two sets of architraves can be placed near each doorway, ready for mitring and fixing. Starting with the first set, stand each leg in position (one at a time) and mark the inner point of the mitre from the margin-intersection point, onto the inner edge of each leg. These mitres are then cut carefully on the waste side of the mark and, without the need for marking, the left-hand mitre of the head is also cut. Next, the left-hand leg is fixed carefully to the margin marks, at top, bottom and centre positions, with three nails only, their heads left protruding. The left-hand mitred head is then tried in position. If the mitre is a good fit, then the head is held firmly while the right-hand mitre is marked at the intersection and then cut carefully. The right-hand leg is then fixed, again with three nails only, their heads left protruding. Next, the head is placed in position and the mitres are checked. If the appearance is as at A, then the head-mitres need easing with a plane, if as at B, then the leg-mitres need easing, which means releasing the provisionally nailed leg(s).

16.1.5 Final Fixing

If the mitres are a good fit, then the head can be fixed with three or four nails, 50–60 mm from the mitre-point at each end and one or two between. Also, the nailing of the legs can be completed with two more nails between the spacings of the nails already used, making seven nails in all on each leg, about 50–60 mm up the leg and the same distance from the mitre-point, then spaced approximately 300 mm apart. The nails used throughout are 38 mm oval nails, preferably the lost-head type, which are easier for punching in. The mitres are sometimes nailed through the top or side edge to achieve flushness on the face side and to close and hold the mitre. These mitre-fixings and the face-fixings nearest to the mitres, being six nails in all, are most likely to cause splitting and it is advisable to blunt their points to reduce this risk. This is done by holding the nail between forefinger and thumb, standing its head on a solid metal object or a concrete/brick surface and tapping the point with a hammer.

16.1.6 Mitre Saw or Mitre Box

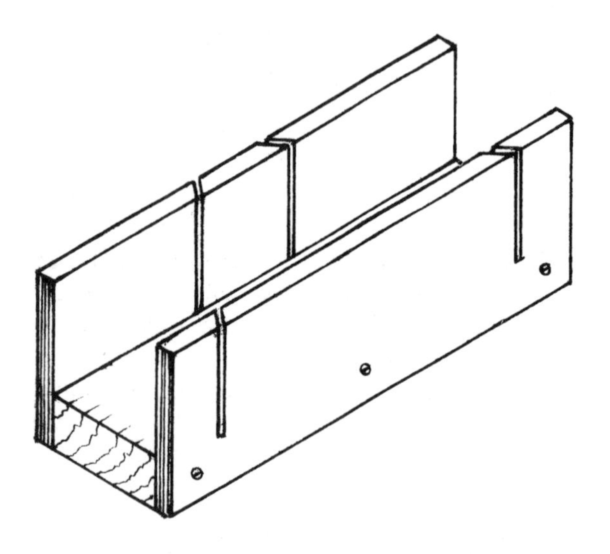

Figure 16.4(a) Mitre box

Figure 16.4(a): Traditionally, mitres used in second-fixing operations were always cut in a purpose-made mitre box, but nowadays, as mentioned in the chapter on tools, mitre saws are quite commonly used and may be power-driven or hand-operated. However, mitre boxes are still used and if one is to be made, it should be at least 600 mm long, have a solid timber base, as illustrated, of (ideally) 45 mm thickness and have plywood, MDF or Sterling board sides of 18 mm thickness. The width of the base and sides must be parallel and the base must be wide enough to take a piece of architrave lying flat with at least 12 mm tolerance. The sides must be wide enough to allow for attachment to the base with sufficient upstand to accommodate the skirting-height plus at least 10 mm tolerance. The sides are nailed to the base with 50 mm nails or screwed with 38 mm ($1\frac{1}{2}$ in) countersunk screws. The 45° mitres must be marked and cut carefully with a panel or other fine-toothed saw. Once the cut is about 10 mm deep, to avoid 'cutting blind', only let the saw run in the opposite cut while increasing the depth of the nearest cut, then turn the box around and repeat this operation as many times as necessary to control the plumbness of the cut.

16.1.7 Mitre Block

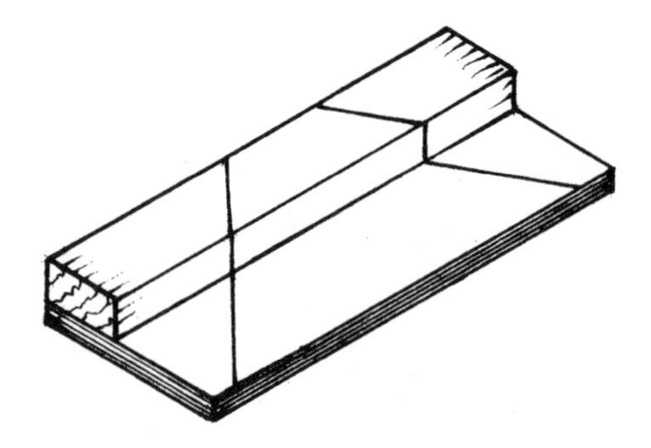

Figure 16.4(b) Mitre block

Figure 16.4(b): This is shown here for comparison with the mitre box and although architraves can be cut on it, the mitre block is more suitable for mitring smaller sections like glazing beads and quadrant moulding, cut with a tenon saw. When making one of these, more precise mitring will be achieved if, as illustrated, the solid top block is kept wide enough to give better control of the back-saw within the extended mitre cut. A piece of ex. 75 × 50 mm timber, about 450 mm long can be used, lying flat on a 150 mm wide base of 18 mm ply, MDF or Sterling board. The mitres are marked before screwing or nailing up through the base, so that the fixings can be placed strategically to avoid clashing with the 45° mitred saw cuts.

16.1.8 Splicing

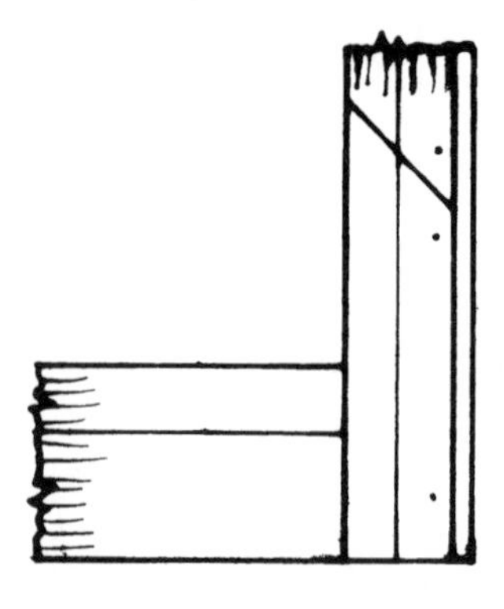

Figure 16.5 Leg-splicing

Figure 16.5: Ideally, each architrave member should be in one piece, but sometimes there is a need to use up the offcuts. The joining of two pieces is known as *splicing*, which is done at 45° across the face. Splicing should never be done on a head piece and only be done sparingly on the legs, as low as possible – well out of eye level – and the splice, as illustrated, should be cut to face downwards in the doorway. This tends to make it less obvious.

16.1.9 Scribing

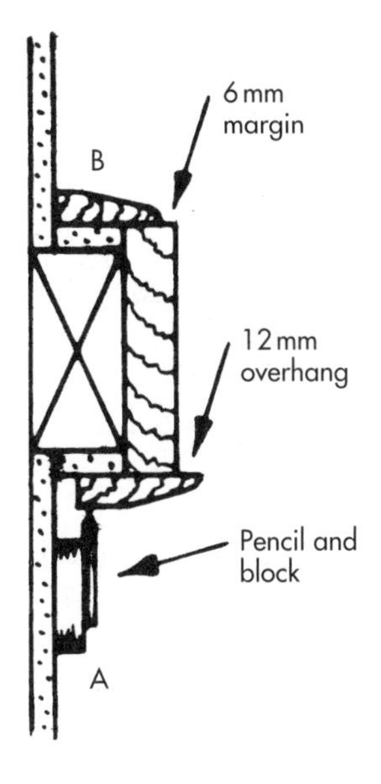

Figure 16.6(a) Horizontal section showing scribing technique

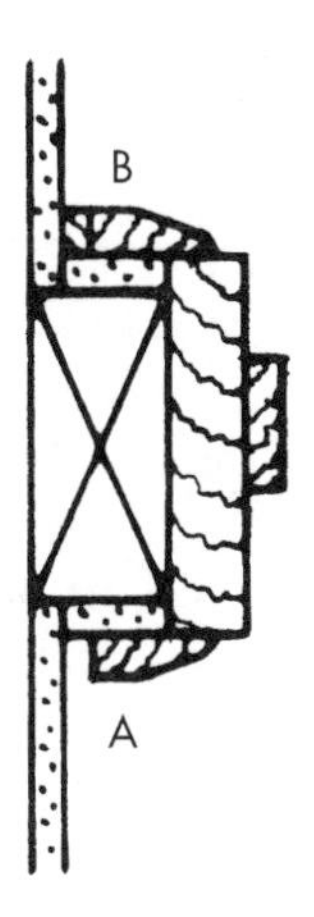

Figure 16.6(b) Horizontal section showing A gap and B the infill

Figures 16.6(a) and (b): Where doorways are close to an adjacent wall, very often the architrave leg on that side touches the wall surface (B in Figure 16.6(a)) or, as illustrated as A in Figure 16.6(b), leaves an undesirable gap. In the first instance, the carpenter usually has to 'scribe' the architrave to the wall surface to achieve a good fit thereto and, at the same time, achieve the required margin. The scribing technique for this is indicated at A, in Figure 16.6(a) and is described as follows:

1. With a minimum of temporary fixings, say two or three, fix the unmitred leg to the lining's edge and establish a constant overhang of any amount, but say 12 mm from the lining's face.
2. Add the margin required, say 6 mm, to the overhang-amount, making 6 + 12 = 18 mm. This is the amount to be scribed from the wall-surface to mark the architrave-cut. Most text books suggest that this can be done with a pair of dividers, which is theoretically possible, but not very precise in practice. A good practical way is to cut a small wooden block, equal in thickness to the scribe-amount (18 mm), but minus 3 mm for half the pencil's diameter, so that the pencil held on the 15 mm thick block measures 18 mm to the pencil point.
3. Holding the scribing block's edges between the second index finger and the thumb, place the forefinger on the projecting pencil and mark/scribe the architrave leg from top to bottom.
4. Release the leg from its fixings and very carefully rip down on the waste side of the line, achieving a slight undercut with the saw.
5. Try in position, then ease with a plane or Surform file, if necessary. Mark and cut the mitre.
6. Fix in position. In the case of any small, undesirable gaps between the leg and the wall's surface (as at A in Figure 16.6(b)), the standard of work would be improved if a small section of timber was glued to the back edge and the increased width of architrave was scribed as before, finishing up like B in Figure 16.6(b).

16.1.10 Double Architraves

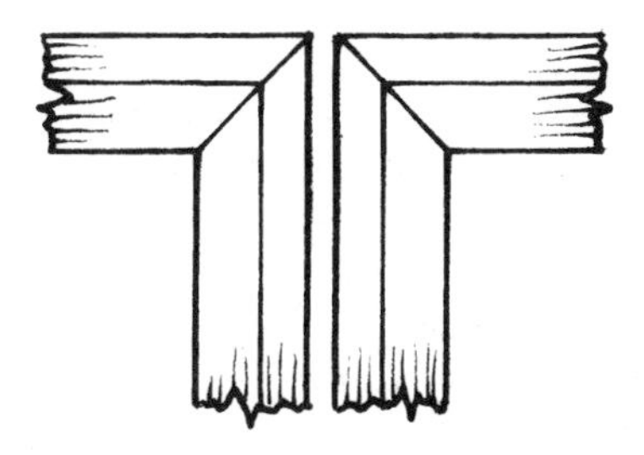

Figure 16.7(a) Close doorways

Figures 16.7(a) and (b): In the case of two doorways close together, separated by a partition wall, the architraves often come very close to each other on the side of the double doorway and again, small, undesirable gaps can result, indicated in Figure 16.7(a). A small section of timber can be glued in the gap, or as in Figure 16.7(b), a double architrave can be produced or built-up and glued together on site. This can be scribed or partly

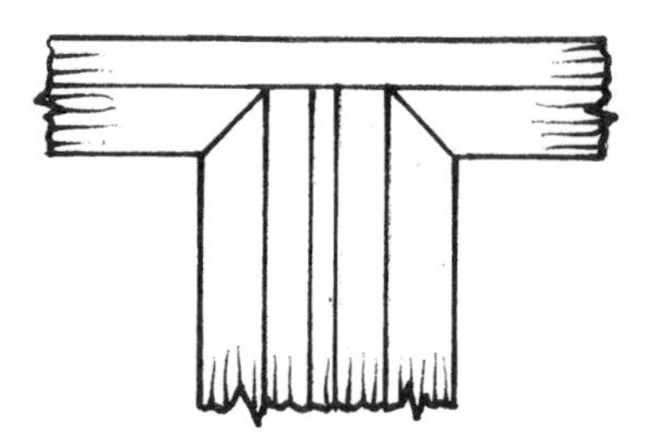

Figure 16.7(b) Double architrave

mitred into a double door-head placed across both doorways.

16.1.11 Storey-frame Architraves

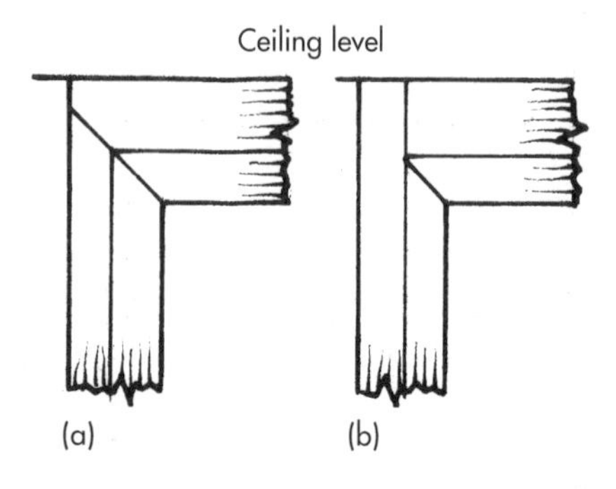

Figure 16.8(a) End-grain exposed. (b) End-grain concealed

Figure 16.8: Architrave legs that run up from floor to ceiling are required in the case of storey frames or linings. These are doorways that incorporate a glazed 'light' above the door transom. Very often, the architrave-head is required to be deeper to fit the ceiling without violating the head-margin. This means that the mitred legs will show unacceptable end-grain (Figure 16.8(a)), although the wider head can be partly mitred into the legs (Figure 16.8(b)).

16.1.12 Plinth Blocks

Figure 16.9(a): The need for these may only now be found on refurbishment works. Plinth blocks were used when deep, built-up skirtings, often involving one or two stepped face-boards, could not be mastered at the doorway by the relatively thinner architrave. Plinth blocks, therefore, are to accommodate the side abutment of any skirting member which would otherwise be unsightly in sticking out past the face of the architrave leg. The skirting was housed about 6 mm into the plinth-block side to offset any shrinkage across the block, and the base of the architrave leg was half-lap jointed and screwed into the back of the block, to make the block an integral part of the leg.

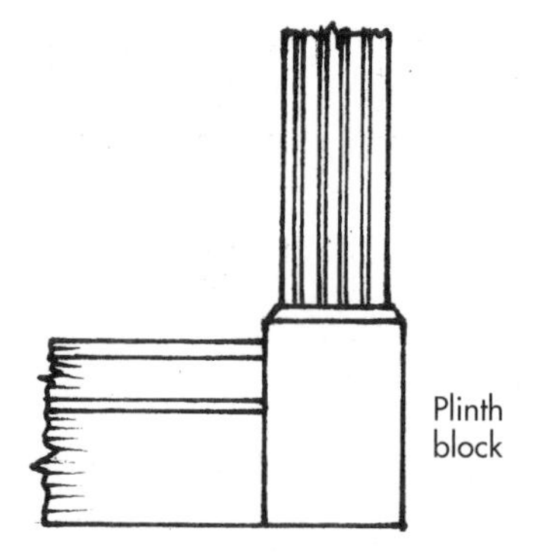

Figure 16.9(a) Use of plinth blocks

16.1.13 Cornice Blocks

Figure 16.9(b) Use of cornice blocks

Figure 16.9(b): As with plinth blocks, cornices are not found in modern buildings. They were at one time used as an alternative to the mitred corners of a doorway, especially on higher class work where hardwood linings, architraves and skirting, etc. were being used. The cornice block, usually slightly thicker than the architrave, was often carved or routered out in a decorative way. The leg and head architraves were simply butted up squarely to it. The design of architraves used on jobs involving cornices were often, as illustrated in Figure 16.10, either fluted, or reeded.

16.1.14 Architrave Shapes and Sizes

Figure 16.10: Architrave sizes vary, but those quite commonly used are ex. 50 × 19 mm, ex. 63 × 19 mm and ex. 75 × 19 mm. On large jobs, architrave is usually ordered by *the metre run*, so it will arrive in random lengths, hence the occasional need for splicing. On small jobs, it may be obtained in specified lengths with head-lengths added together. A variety of modern and traditional shapes still in use are illustrated.

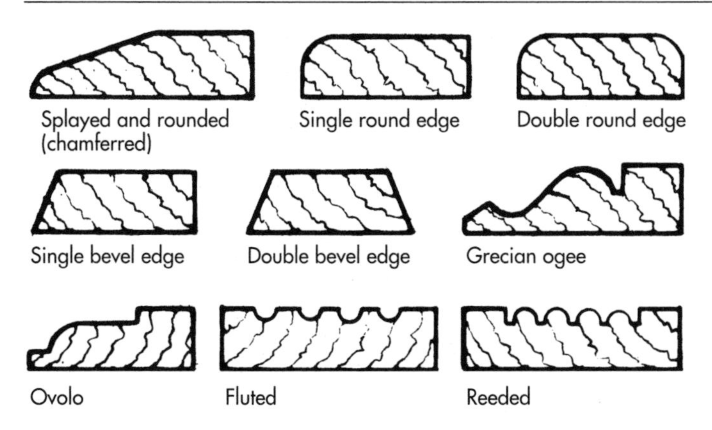

Figure 16.10 Architrave designs

16.2 SKIRTING

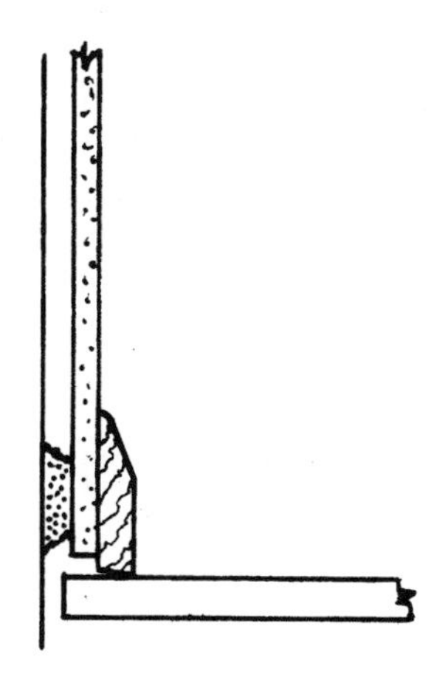

Figure 16.11 Skirting

Figure 16.11: Skirting is a protective board fixed at the base of plastered or plasterboard (dry-lined) walls, which also covers the joint between the wall surface and the floor. The depth of the skirting boards, for visual balance, is usually at least 25 mm more than the width of the architraves used; and the shaped or moulded face-edge usually matches or is compatible with the architrave.

16.2.1 Scribing to the Floor

If the skirting boards are reasonably straight-edged in length and the floors are not uneven, it is theoretically possible to fix the skirting without having to fit it to the floor. Otherwise, it should be scribed. This is done by positioning a length of skirting, then laying a pencil on the floor touching its face and running a line along it. Next, the bottom edge is under-shot (meaning more planed off the back edge than the front) to the line with a smoothing plane or Surform, used diagonally across the edge, rather than along it, while the skirting is laying face-up on a saw stool and the planing action is aimed at the floor.

16.2.2 External and Internal Angles

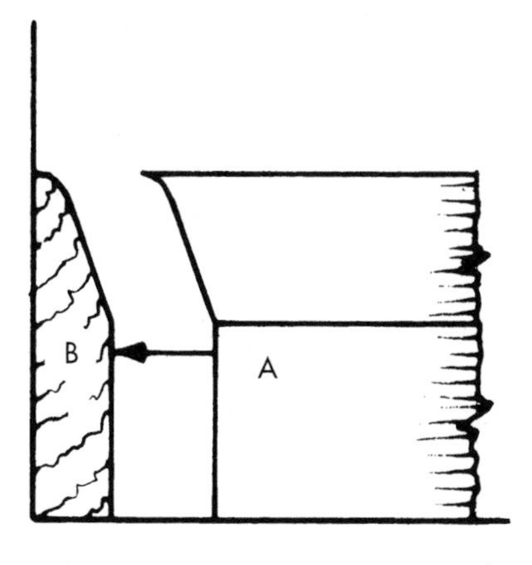

Figure 16.12 Internal scribe

Figure 16.12: External angles are mitred, either with the mitre saws mentioned previously, or with the purpose-made mitre box, but internal angles should always be scribed. In effect, this means that the moulded profile of the skirting is cut from the end of one piece, A, to fit the moulded face of another piece, B, already fixed squarely into the internal corner of the adjacent wall.

16.2.3 Scribing Internal Angles

Figure 16.13: The professional technique for scribing an internal angle in a 90° corner is first to cut the end of the moulded board to be scribed with a 45° mitre cut A. This actually produces a sawn outline or profile of the skirting's moulded shape, regardless of whether it is a simple, i.e. modern or complex traditional design. Next, the scribe is produced by cutting in the waste area of the mitred profile with a coping saw, keeping very close to the outline with a slightly undercutting angle, as indicated in B.

16.2.4 Sequence of Fixing

Figure 16.14: Even though well-fitted skirting scribes may have been achieved, scribes should always face away

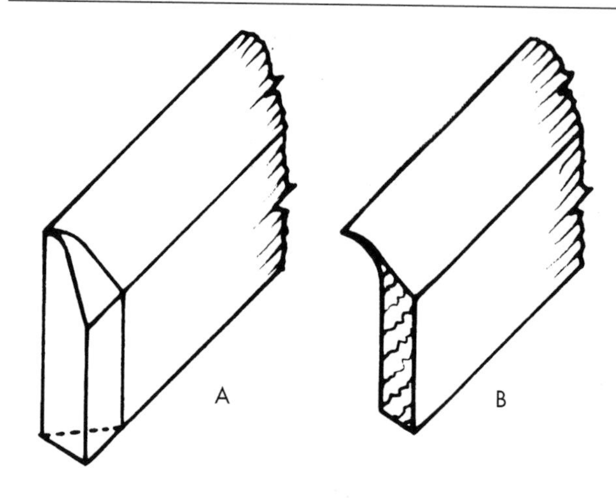

Figure 16.13 Scribing technique

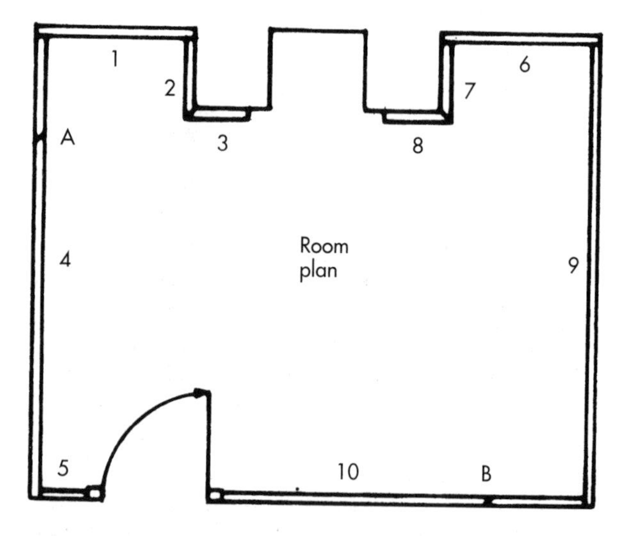

Figure 16.14 Sequence of fixing

from the doorway, so that any slight gaps are not looked directly into. This will always be possible, providing a certain sequence of fixing is followed, from 1 to 10 as illustrated in Figure 16.14. In the illustration, it can be seen that skirting-pieces 1 and 6 are square-ended at each end and pieces 4, 5, 9 and 10 have only one scribed end – as do pieces 2 and 7, which are also mitred. The message here is that you should always avoid having two scribed ends on one piece of skirting board – because it is much more time-consuming in demanding a higher degree of precision and skill for it to be successful. Avoiding double-ended scribes, therefore, is another good reason for adopting a sequence of fixing.

16.2.5 Splicing

As with architraves, lengthening-joints (splicing) should be minimal or non-existent. However, if they are unavoidable – perhaps because of the long length of a particular wall, or because there is an excessive amount of offcuts to use up – the splice, which is made with 45° cuts across the top edges, should be treated like the corner-scribes and always face away from the direct approach viewed from a doorway, as indicated in Figure 16.14 at A and B.

16.2.6 Mechanical Fixings

Each piece of skirting is fixed as it is fitted, except when a mitred corner is involved, when it is wise to fit the two mitred pieces before fixing one of them. This way, any adjustments to either mitre can be made more easily. Mechanical fixings, such as nails or screws, should be made at approximately 600–800 mm centres. Nail fixings must be punched in and screws should be countersunk or – especially in the case of hardwood skirtings – counterbored and pelleted. Skirting fitted to timber stud partitions can be fixed with 63 mm lost-head type oval nails – and 63 or 75 mm cut clasp nails are still good fixings for skirting being fixed to walls built with receptive aerated building blocks. Walls built with dense concrete blocks or brickwork can be drilled for screw fixings. This is made easier nowadays by drilling directly through the *in situ* skirting with a masonry drill to take Fischer-type nylon sleeved Frame-fix screws.

16.2.7 Adhesive Fixings

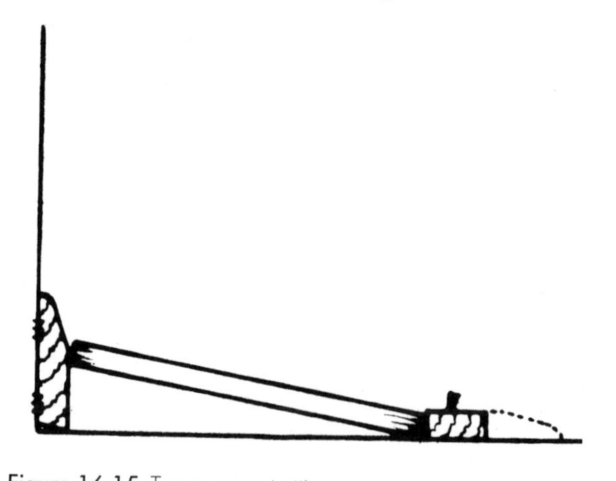

Figure 16.15 Temporary strutting

Figure 16.15: Present-day practices include fixing skirtings with so-called *gap-filling adhesives* such as Laybond's Gripfill or Connect. Gripfill is a solvent-borne, filled rubber resin and Connect is a solvent-free gap-filling adhesive specially formulated to provide good initial 'grab' or suction. Although either of these two adhesives may be used, the Laybond Connect is recommended when bonding to vertical surfaces. These adhesives are in 310/350 ml cartridge tubes which fit into silicone/mastic guns. Working speedily, one or two 6 mm diameter continuous beads should be applied to the back of the skirting board, 25 mm in from the edges, then it should be quickly placed and pressed (slid slightly, if possible) into position. A few temporary nails or pins may be needed if the skirting appears to come away from the wall in places. These should be left in overnight. Alternatively, as illustrated, small struts (maybe offcuts of skirting) can be wedged against the skirting, taking a foothold from short battens or offcuts fixed temporarily to the floor.

16.2.8 Skirting Shapes and Sizes

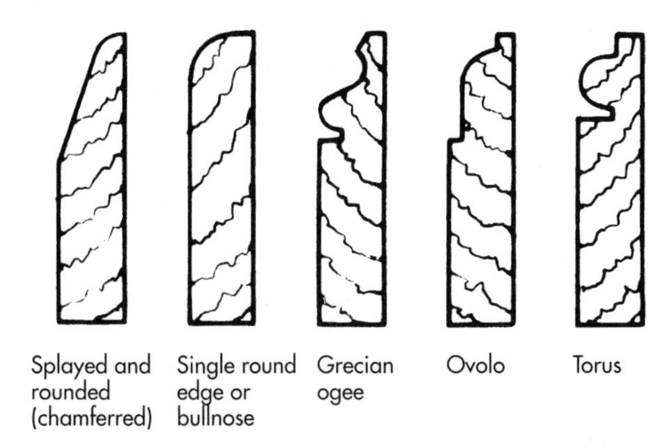

Figure 16.16 Skirting designs

Figure 16.16: Like architraves, skirting sizes vary, but those most commonly used are ex. 75 × 19 mm, ex. 100 × 19 mm and ex. 150 × 19 mm. If required, the thickness of these skirting boards can be obtained increased to ex. 25 mm, but this would mean increasing the architrave-thickness. Again, like architraves, it is usually ordered *by the metre run*, so it will arrive in random lengths. A variety of modern and traditional shapes still in use are illustrated. Also of course, on certain jobs where simplicity is sought and moulded shapes are being avoided, plain square-edged boards may be used.

16.3 DADO RAILS AND PICTURE RAILS

Figures 16.17(a) and (b): In recent years there has been a degree of revival of these traditional items, especially on home-improvement schemes. Basically, the same rules and techniques regarding mitring, scribing, splicing, fixing, etc., apply to dado and picture rails as already covered regarding the fitting and fixing of skirting. Dado rails of usually ex. 75 × 38 mm section, of whatever moulded design, are fitted and fixed around the walls of a room at about 900 mm up from the floor. Whatever height is used, the dado rails should be kept parallel to the floor, regardless of exact levels. Picture rails of about ex. 50 × 25 mm section, of whatever moulded design, but with the essential grooved or rebated top edge (as in Figure 16.17(b)) to hold the picture hooks, are fitted and fixed around the walls of a room, usually at the level of the top of the architrave-head of the room's doorway. Whatever height is used, the picture rails should be kept parallel to the ceiling, again regardless of levels.

Figure 16.17(a) Dado rail

Figure 16.17(b) Picture rail

17

Fitting and Hanging Doors

17.1 INTRODUCTION

Door-hanging is a vital part of second-fixing carpentry and requires a good standard of workmanship and speed. One, to one and a half hours is the established hanging-time for a lightweight internal door and two and a quarter hours for a heavy external type or fire-resisting door. When hanging a door on a solid, rebated frame, this time includes fitting the door properly into the rebates – and on a door lining, it includes adjusting and fixing the planted door stops after the door has been hung. When more than one door is to be hung in the same locality, time will be saved by treating the fitting of locks and door furniture as a separate, secondary operation.

17.2 FITTING PROCEDURE

17.2.1 Removal of Horns

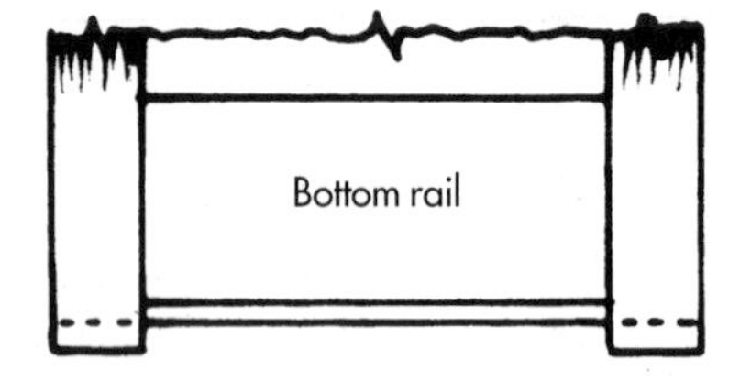

Figure 17.1 Increasing door-height

Figures 17.1 and 17.2: Traditionally, all doors arrived on site with horns left on to give protection to the corners of the stiles. This also made it possible to increase the height of a door, if required, by gluing an additional piece of timber to the bottom rail, between the horns (Figure 17.1). Without the horns, the additional timber would have to cross the opposing grain of the stiles, which is bad trade practice, not allowing for natural shrinkage (Figure 17.2).

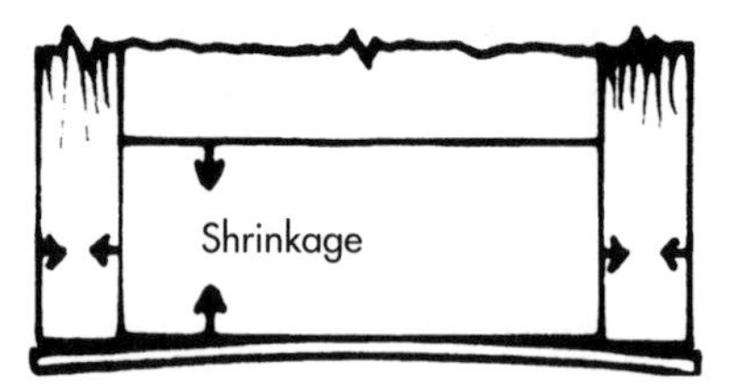

Figure 17.2 Bad practice

17.2.2 Checking the Opening Size

Nowadays, with a few exceptions, horns are usually non-existent or very minimal in size. Although openings with non-standard height may only be met occasionally – usually on older-type property with odd-sized doors – it will be sensible on certain jobs other than new works, not to cut the bottom horns (if they exist) until the height of the door opening is checked. This can be done with a pinch rod or tape rule, although, if no major discrepancy is suspected, it will be quicker to hold the door against the opening and mark the top and side edges onto the frame, thereby gaining visual evidence of the amount to be planed off.

17.2.3 Checking the Hanging Side

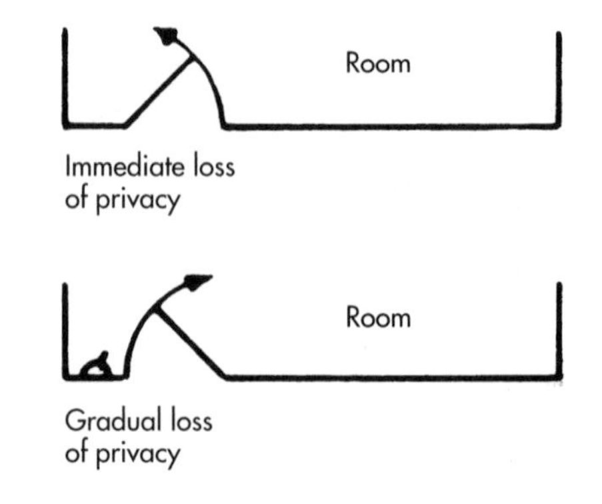

Figure 17.3 Hanging side

Figure 17.3: On which side the door hangs and whether it should open in or out, must be known. This information can be found on the *plan* views of the contract drawings, although sometimes a member of the site management team takes the responsibility of checking this out and marking the hinged side of the lining or frame with the letter H, at about one metre from the floor. Another guide, if needed, is that the lock edge of the door should be on the side nearest to the light switch at the side of the opening. Of course, on new contracts, this assumes that the electrician has put the switch in the correct position. On small contracts, where the builder or carpenter himself might decide on which side to hang the door, it may help to know that doors should give maximum privacy to a room while being opened (Figure 17.3). Another point to bear in mind on this subject, is that normally, external doors should open inwards, to avoid damage if caught by the wind in a partly open position.

17.2.4 Tools Required

The following tools are required to cover a variety of door-hanging jobs: pencil, tape rule, combination square, panel-type saw, size $5\frac{1}{2}$ jack plane and/or a $4\frac{1}{2}$ smoothing plane, optional power planer, one or two marking gauges, 32 mm bevel-edged chisel, claw hammer, optional mallet (for chisels with boxwood handles), large-sized bradawl with square tapered point, medium to large-sized spiral pump screwdriver, optional cordless screwdriver, bullnose rebate plane, or a shoulder plane with a removable front section (for easing rebates on solid timber frames), nail punch (for planted door stops), and a sharp marking knife (for scoring across the plies or veneers of flush doors, if they are to be reduced in height).

17.2.5 Equipment Required

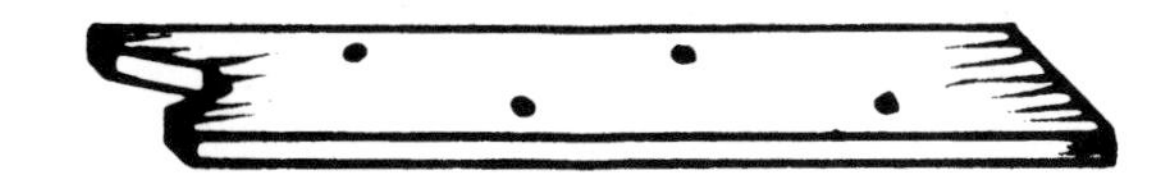

Figure 17.4 Extension stool-top

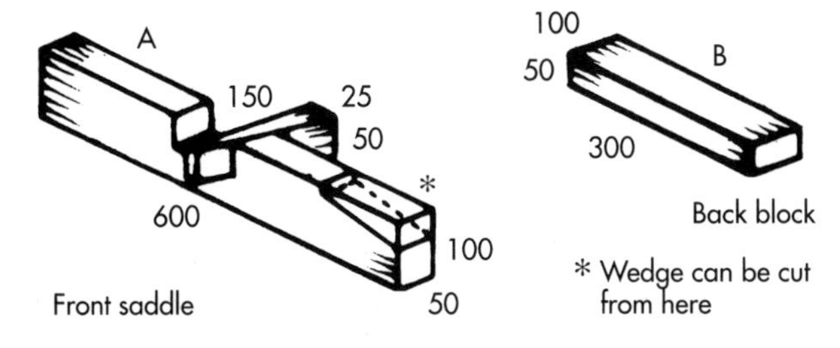

Figure 17.5 Saddle and block

Figures 17.4–17.6: The equipment required for door-hanging can be as simple as one saw stool and an ex. 100 × 25 mm board of about 1 m length with a V cut in one end. The board, shown in Figure 17.4, can be

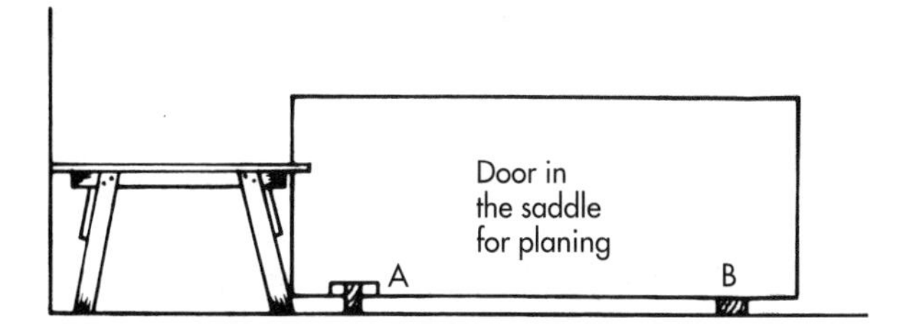

Figure 17.6 Door held firmly against saw stool

screwed onto the top of the stool as and when required and helps to hold the door on its edge during the planing or *shooting-in* operation indicated in Figure 17.6. Additionally, a device known as a *saddle and block*, detailed in Figure 17.5, and easily made from two pieces of 100 × 50 mm timber, provides a very simple but effective way of holding the door firmly while shooting-in or cutting out the housings for the hinges, as indicated in Figure 17.6. A second saw stool is occasionally required to act as a trestle should it be necessary to lay the door across both stools to cut the horns or the bottom of the door.

17.2.6 Skilful Planing Requirement

When shooting-in the door, it is good practice to remove an equal amount from each side. This is done by judgement, rather than measurement. If the amount to be removed is in excess of a few millimetres, a power planer, if available, would save a lot of effort. However, to eliminate the unsightly rotary cutter marks that can be left on the door-edges, especially if the planer is pushed along too speedily, the edges should be finished off with a jack or smoothing plane. Although, unskilful hand planing – by not putting the correct pressure and momentum to the plane, or by not lifting off correctly (by raising the heel) – produces ridges and chatter marks which can be as unsightly as the pronounced rotary-cutter marks.

17.2.7 Closing Edge

Figure 17.7(a): The closing or lock edge of a door requires a slight angle of about 87–88° to clear the frame or lining's edge effectively. This should be achieved while shooting-in and not added afterwards. Again, this is done by judgement, rather than measurement. Although, until experience is gained, a sliding bevel could be set up and used for testing the edge while planing. The clearance angle in relation to the jamb's edge is indicated in Figure 17.7(a).

17.2.8 Clearance Joints

Figure 17.7(b): A consistent and unwavering 2–3 mm joint (gap) should be achieved around the door and

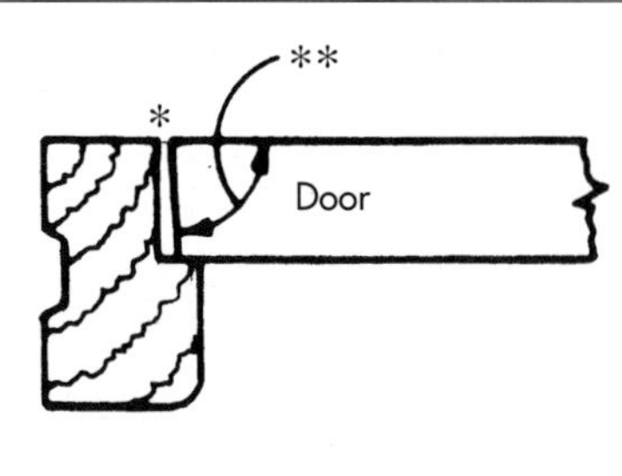

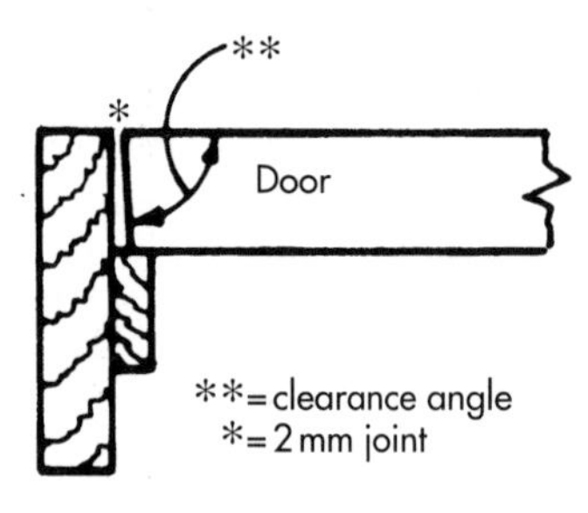

Figure 17.7(a) Horizontal sections showing closing edge of door

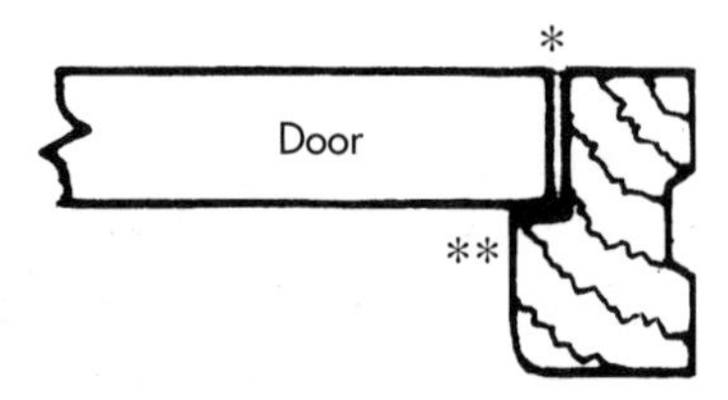

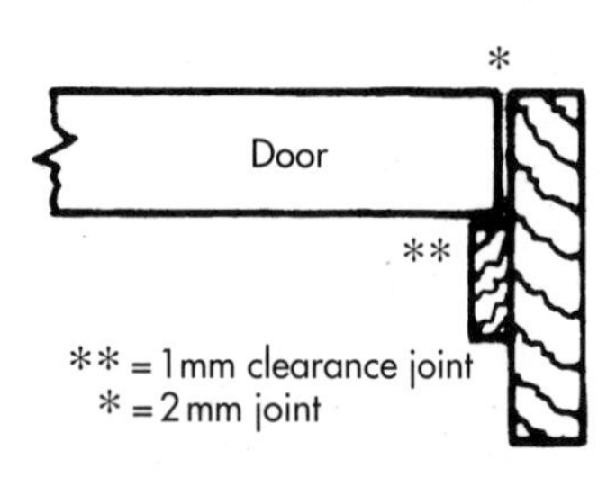

Figure 17.7(b) Horizontal sections showing hanging edge of door

frame or lining (side-edges and top) and the joint at the bottom should be a minimum 3 mm and a maximum 6 mm, unless extra allowance is required for floor covering, such as carpets. A two-pence coin is sometimes used as a feeler gauge for testing the top and side joints.

17.2.9 Arrises

After shooting-in the door and *before* screwing on the butt hinges, it is important to plane off the 'arrises' (sharp corner edges) from the top, bottom and side edges on both sides of the door. The appearance of this should be like a miniature chamfer, measuring no more than 1–1.5 mm across the 45° chamfer. Usually, one or two strokes with a smoothing plane will accomplish this.

17.2.10 Planted Door Stops and Sunken Rebates

Figure 17.7(b): The door stop or the sunken-rebate shoulder must be 1 mm clear of the door on the hanging side, to avoid a fault known as *binding*, which happens if these edges touch. The size of planted door stops varies in section between ex. 50 × 12 mm and ex. 38 × 12 mm, or ex. 50 × 16 mm and ex. 38 × 16 mm and requires fixing every 225 mm approximately, with 38 mm lost-head oval nails, staggered and punched under the surface by at least 1 mm. Solid frames with sunken rebates can present more problems than linings with planted stops, because the door must fit well into the rebates and if it – or the frame – is twisted, this will involve easing the shoulders of the rebates with a shoulder or bullnose plane.

17.2.11 Various Points to Note

1. When hanging hardwood doors, especially hardwood *flush* doors, or any door which may be easily surface-damaged, the door should be protected from being scratched or bruised by covering any door-bearing stools with dust sheets or soft fibreboard. For the same reasons, the door-side of the housing in the saddle block and the door-side of the wedge should be covered with masking tape.
2. When removing horns, or cross-cutting plywood or veneer-faced doors to reduce the height, *spelching out* of the fibres can be eliminated if the amount to be cut off is heavily scored with a sharp marking knife or chisel on both sides, then, after being cut very close to the line, is finished off by planing down to the knifed edge.
3. If more than 6 mm has to be removed from the bottom of a door, it is advisable to rip this off by saw. Any lesser amount should be planed, but first consider removing the cross-grain of the stiles, if practicable, with a fine saw, to make easier work of the planing, as indicated in Figure 17.8.

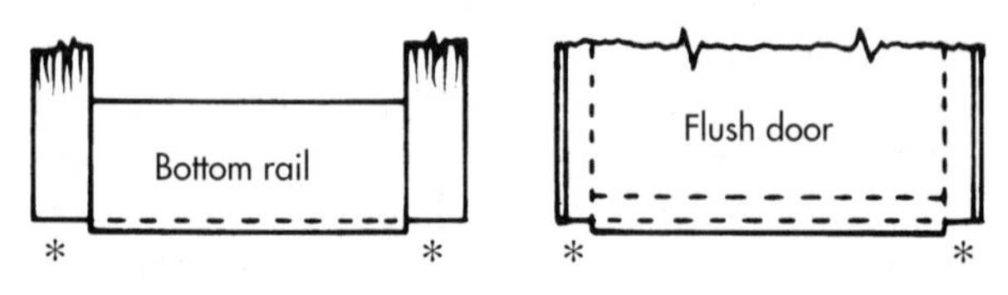

Figure 17.8 Reducing door height

17.2.12 Doors on Rising Butts

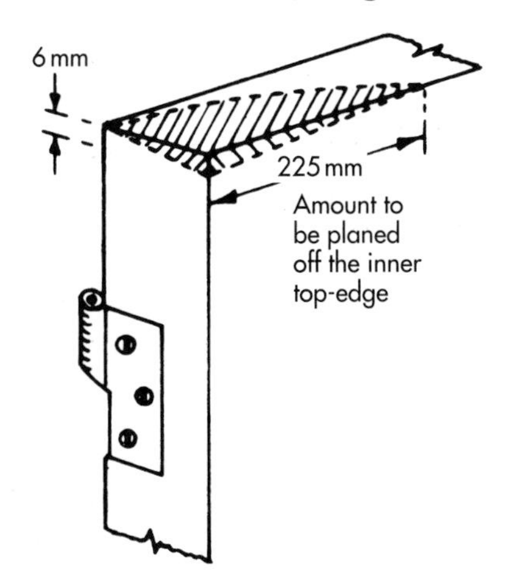

Figure 17.9 Rising butts

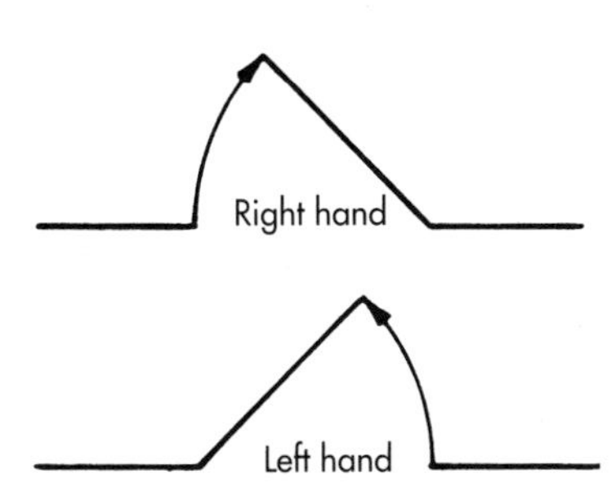

Figure 17.10 Handed hinges

Figures 17.9 and 17.10: Doors being hung on rising butt hinges require the inner top edge of the door to be shot off on the splay. The amount of splay to be planed off is shown in Figure 17.9. Rising butt hinges are either left-handed or right-handed and a satisfactory way to identify which hand is needed is to name the hand facing the door as it opens away from you, as shown in Figure 17.10.

17.2.13 Hinge Positions

Figure 17.11: There are no hard and fast rules about hinge positions, but they should always be less at the top than the bottom and must be clear of the end grain of any mortices (or their wedges) by at least 12 mm. Positions of 150 mm down to the top of the butt and 225 mm up to the bottom of the butt are the usual settings – but this may be governed by other doors already hung in the vicinity by others who used settings of 175 mm down and 250 mm up; in this case, their settings are followed. If three butts ($1\frac{1}{2}$ pairs) are specified or decided upon for a heavy door, the additional butt must be equidistant between top and bottom butt-positions, as in Figure 17.11.

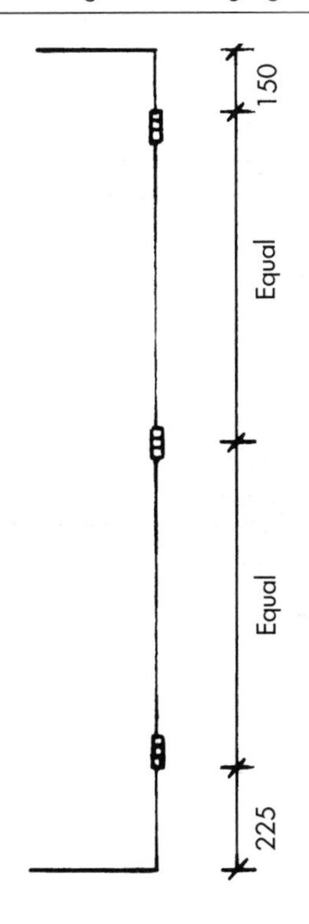

Figure 17.11 Hinge positions

17.2.14 Hinges and Screws

Butt hinges range in type and in size from 25 mm to 150 mm (75 mm and 100 mm being most commonly used), and come in various kinds of metal. Screws should always match and may be recessed for Supadriv/Pozidriv, Phillips or slotted screwdrivers. Mild or bright steel butts must be painted in with the door, but brass or other non-ferreous metal butts, used on exterior or hardwood doors, should not be painted. The screw gauge must suit the countersinking in the butts, but the screw length is usually from 32 mm ($1\frac{1}{4}$ in) to 38 mm ($1\frac{1}{2}$ in). On lightweight doors, 25 mm (1 in) screws are often used. A reasonably accurate way of determining the gauge of a screw is simply to measure across the head of the screw in millimetres, i.e., 6 mm diameter = gauge 6, 8 mm diameter = gauge 8, 10 mm diameter = gauge 10 and so on. Another way is to measure the diameter of the shank in millimetres and double it to give the gauge, i.e. 3 mm diameter shank × 2 = gauge 6 screw.

17.2.15 Knuckles In or Out

Figure 17.12: This illustrates different settings for butt hinges in relation to the knuckle part of the hinge, as follows.

(a) The butts are housed with a full knuckle projection. This is now standard practice to save time.
(b) The butts are housed-in to lessen the knuckle projection for aesthetic reasons, but, to enable the door to open beyond 90°, the centre of the knuckle must not be further in than the surface of the door.

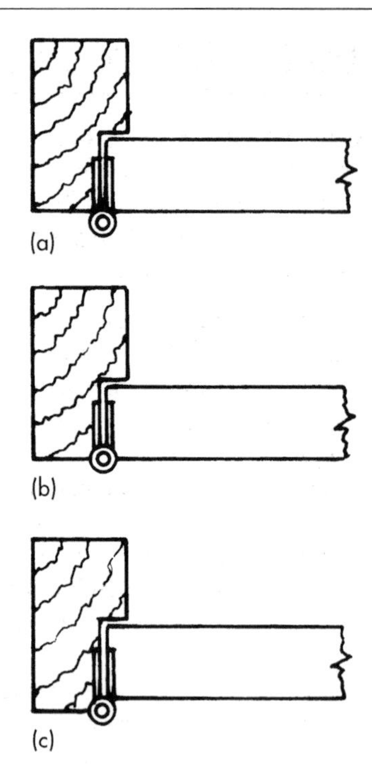

Figure 17.12 Knuckle projections

Although chiselled chamfers were used to accommodate the recessed knuckle, excessive time was added to the door-hanging operation and, therefore, this practice has not survived.

(c) In this example, only the knuckle on the frame side is partly housed-in to carry the door further into the shoulder of the jamb, when a rebate is deeper than the door thickness (which is met occasionally). If ignored, an excessive clearance-joint between the shoulder and the door would look unsightly.

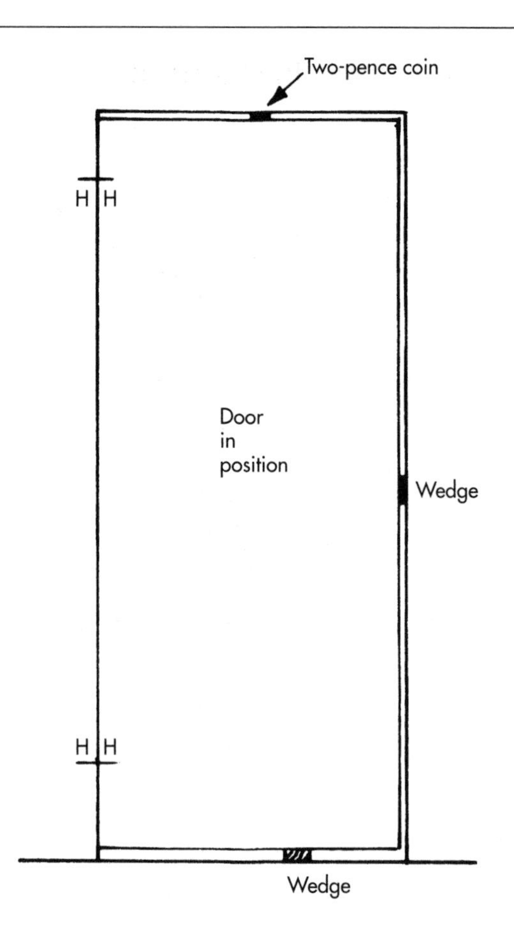

Figure 17.13 Marking hinge-positions

17.2.16 Marking the Hinge Positions

Figure 17.13: When the door has been shot-in and the clearance-joints have been achieved, set up the door in the opening as follows: insert an off-centre wedge, chisel or bolster under the door and lightly tighten up to a two-pence coin placed in the top, as illustrated, and then insert a small wedge at mid-height in the joint of the closing stile (the doubled-joint obtained this way can be tested, if necessary, with two two-pence coins held together and tried in several places (or slid along) above and below the wedge). Mark the hinge positions now, squarely across the frame or lining and the door. Alternatively, the butts could be already housed-in and screwed to the door before it is positioned in the opening and the unfolded leaves of the butts could then be marked onto the frame or lining.

17.2.17 Marking the Housings

Figure 17.14(a): The door is now removed from the opening and set up in the saddle and block ready to be hinged. Place an open butt on the edge of the door, with the protruding knuckle inverted to rest against the edge, locate carefully to the correct side of the face mark and scribe around the leaf with a sharp pencil. Repeat this on each of the four butt-positions. Next, adjust the leaves of the butt until parallel (as illustrated), measure the overall thickness, deduct 2 mm for the joint, then divide by two to give the depth of housing required and set this on the marking gauge. Mark the butt-positions with this on the face of the door and the edge of the frame or lining. Then reset the gauge to the leaf-width of the butt and mark this on the edge of the door and the face of the frame or lining.

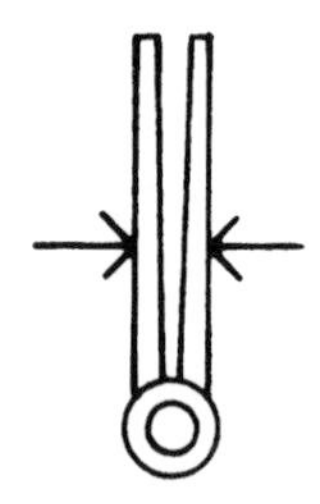

Figure 17.14(a) Determining housing-depth

17.2.18 Cutting the Housings

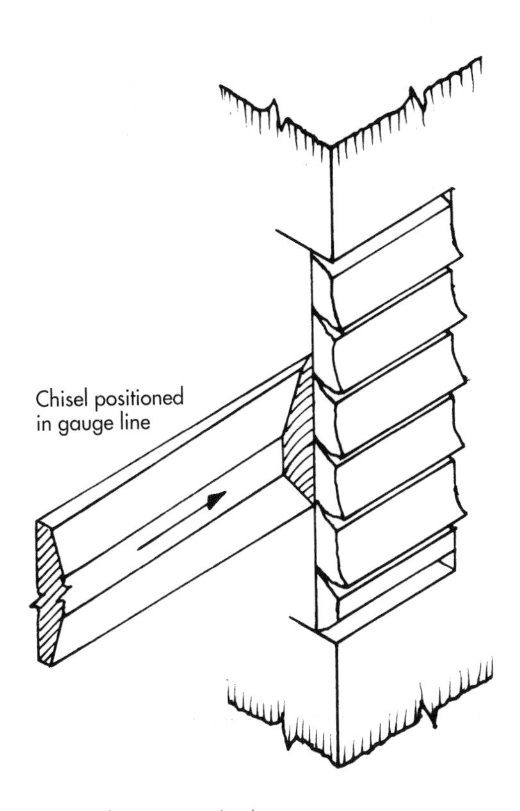

Figure 17.14(b) Cutting the housings

Figure 17.14(b): Now cut the housings, using the chisel-chopping method, prior to chisel-paring. When leaning the chisel (at about 45–50°) for the cross-grain chopping action, the chisel's bevelled-edge should be on the side of the acute angle formed between the chisel and the timber. Note that if the housings are cut too deeply, the hinge-side joint will be lost and binding may occur between the door-edge and the face of the frame or lining. Also, if the wrong gauge screws are used and the countersunk heads protrude, binding may occur between the opposite screw heads.

17.3 HANGING PROCEDURE

Finally, speed and skill in door-hanging is gained from the experience of repetition and it will assist if you adopt the set hanging procedure described below. (Note that the references are to a door-lining, but the procedure would also suit a door-frame.)

1. Check the opening size and hanging edge. Mark H on the lining face, mark T (for 'top') on the face of door (this is to avoid mistakenly changing the face-side during the shooting-in procedure) and remove horns, if necessary.
2. Offer the door up in position, judge the initial amount to remove from the edges and shoot-in.
3. Offer up again – if it now fits into the opening, concentrate on marking and planing the left-hand stile only, until a good fit is achieved against the lining-leg.
4. While keeping the door pressed or wedged against the left, lever or wedge up the door and check the fit against the underside of the lining-head. Remove the door and plane the top, if necessary, until another good fit is achieved.
5. Now concentrate on judging, marking and shooting-in the right-hand stile only, until (with the door pressed hard to the left) you are able to test for a double-joint with the aid of two two-pence coins held together.
6. If satisfied, re-establish the side and bottom wedges (Figure 17.13), after inserting a two-pence coin in the top joint. Now mark the hinge positions on the door *and* lining-edge.
7. Dismantle and remove the arrises from all door edges *before* fixing the butts to the door.
8. Mark and cut out the housings for the butts, try the butts in lining-housings and make a pilot hole for one screw in each butt.
9. Screw butts to door, hang door and try closing and check the all-round fit.
10. Remove and adjust, if necessary, or finish screwing to the lining.
11. Fix top door stop (1 mm clear on hinge-side only), then closing edge door stop, then hanging edge door stop. Finally, punch in all nails.

18

Fitting Locks, Latches and Door Furniture

18.1 LOCKS AND LATCHES

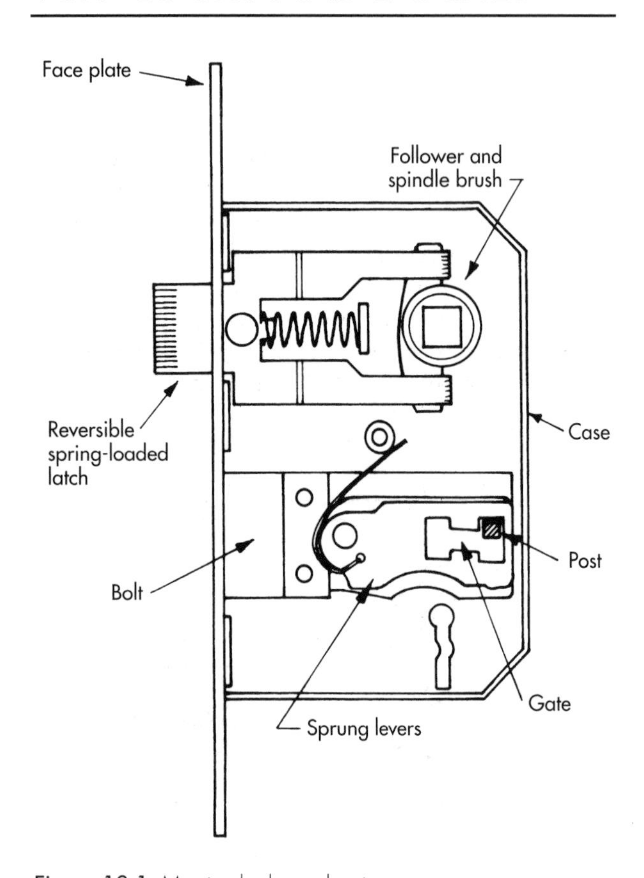

Figure 18.1 Mortice-lock mechanism

Figure 18.1: The types of lock used on dwelling houses are few and vary between locks, latches and combinations of these two. They may be mortice locks morticed into the door-edge, or various types of rim lock fixed on the inside face-edge of the door. The actual latch part of a lock is usually spring-loaded, quadrant shaped, or has a round-edged roller bolt, which holds the door closed (latches it) without locking it – unless it is a type of cylinder night latch,, which requires a latch-key to open it on the cylinder-side of the door. The locking mechanism of a mortice lock is usually an oblong-shaped bolt which is shot in or out by inserting and turning a key.

The concealed part of the bolt, as illustrated, has a small metal post protruding from it, which must be moved through an open *gate* cut in the middle area of a specified number of sprung levers. This happens when the key lifts the levers, gains access to the edge of the bolt and moves it. The more levers a lock has, the greater the security. The quadrant-shaped latch is usually reversible to enable the hand of the lock to be changed from left to right, or vice versa.

18.2 MORTICE LOCKS

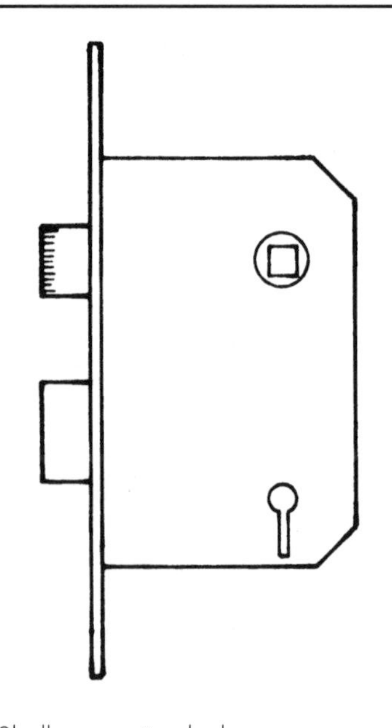

Figure 18.2(a) Shallow mortice lock

Figure 18.2: These locks vary in length (depth) between 64 mm and 150 mm. The deeper locks (Figure 18.2(b)) are of traditional size to receive door-knob furniture which consists of a spindle, two knobs (sometimes one of these is already attached to a metal spindle, the other being removable via a small grub screw), two rose plates, and two escutcheon plates (Figure 18.2(e)). The shallow-depth lock (Figure 18.2(a)) has its keyhole and spindle hole vertically in line to receive a set of lock lever-

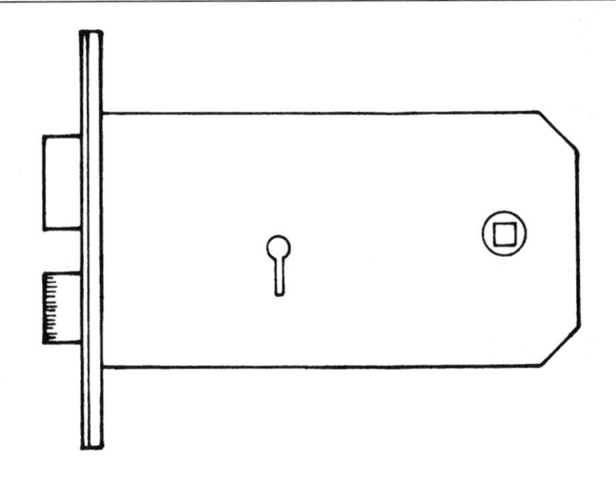

Figure 18.2(b) Deep mortice lock

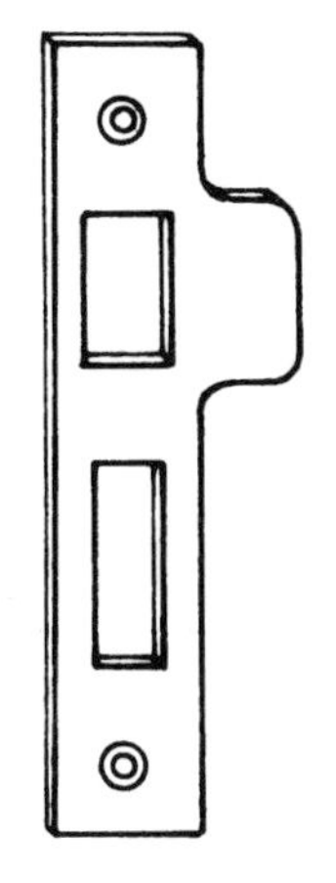

Figure 18.2(c) Striking plate (brass or chromed steel)

Figure 18.2(d) Lock lever-furniture

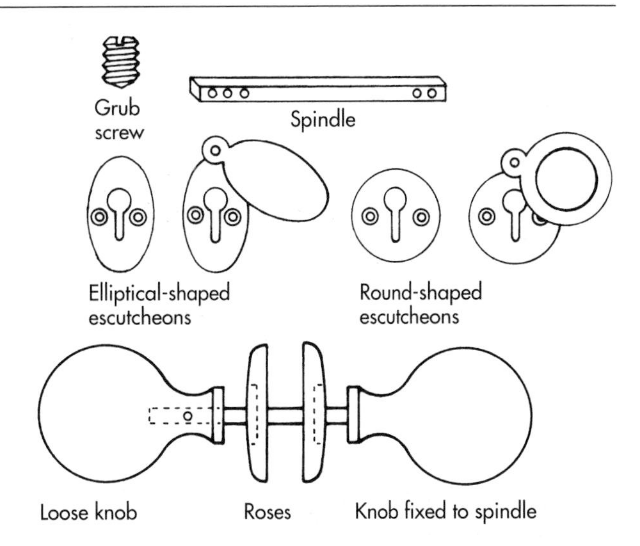

Figure 18.2(e) Door knob furniture

furniture (Figure 18.2(d)). Both types of lock have a striking plate (Figure 18.2(c)) which is housed into the jamb to receive the projecting latch and the turned bolt. Shallow-depth locks should only be fitted with lever furniture, as door knobs come too close to the face of the jamb or lining on the door-stop side, resulting in possible scraped knuckles. If door knobs are to be used, the lock should be at least 112 mm ($4\frac{1}{2}$ in) deep.

18.3 MORTICE LATCHES

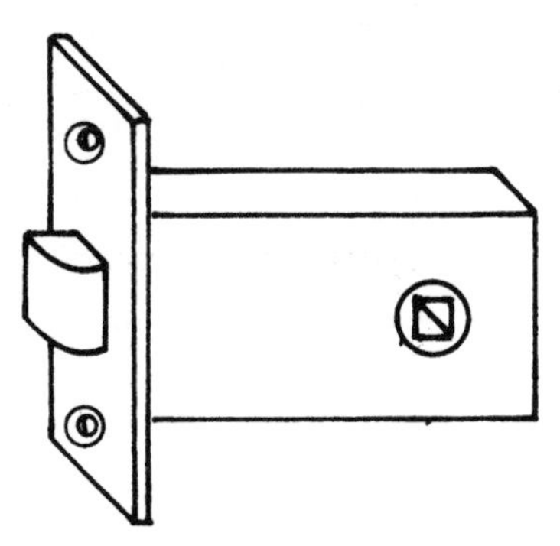

Figure 18.3(a) Mortice latch

Figure 18.3: Mortice latches are used on internal doors not requiring to be locked. Two types are available: one, as in Figure 18.3(a), is oblong-shaped for morticing in to a 16 × 38 mm × 64 or 75 mm deep mortice hole; the other, as in Figure 18.3(b), is tubular-shaped for inserting into a drilled hole of 22 mm diameter × 64 or 75 mm depth. A mortice latch is always supplied with a striking plate and fixing screws, but requires separately a set of latch lever-furniture (without the keyhole), as shown in Figure 18.3(c).

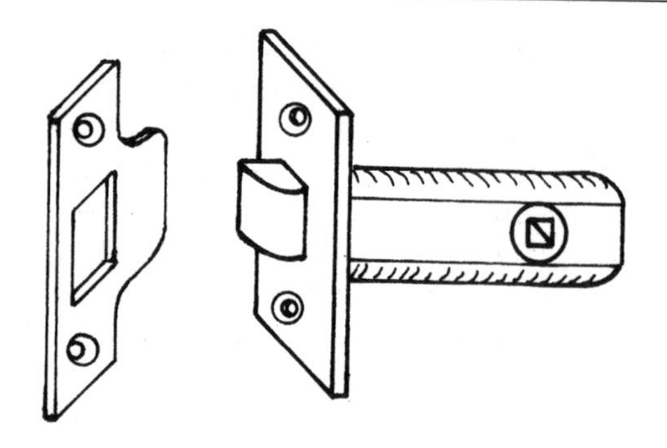

Figure 18.3(b) Tubular mortice latch

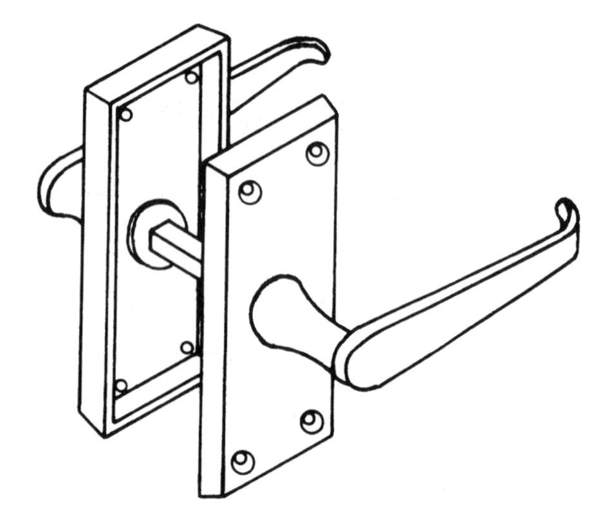

Figure 18.3(c) Latch lever-furniture

18.4 MORTICE DEAD LOCKS

Figure 18.4: Mortice dead locks are for extra security and contain only a locking bolt – no latch. They are fitted to external doors in addition to a latch-type cylinder lock and are recommended by insurance companies to have five levers. The more expensive locks of this kind have a box-recessed striking plate or *keep* – and the brass bolt contains two hardened steel rollers to resist being cut with a hacksaw blade. Ironically, these locks can be more of a deterrent on the inside of a property than on the outside. The reason for this is that if burglars have gained access through a window – which is quite common – they like to leave by a door, which is less suspicious, easier and quicker than carrying stolen goods through a window. My preference with these locks is to make a keyhole only on the outside of the door, which is done to stop anyone deadlocking themselves in the property at night, in case there is a fire. The only door furniture required is supplied in the form of two escutcheons, one being a *drop escutcheon* with a pivoting cover plate that drops down to cover the keyhole on the inside of the door, as shown in Figure 18.2(e).

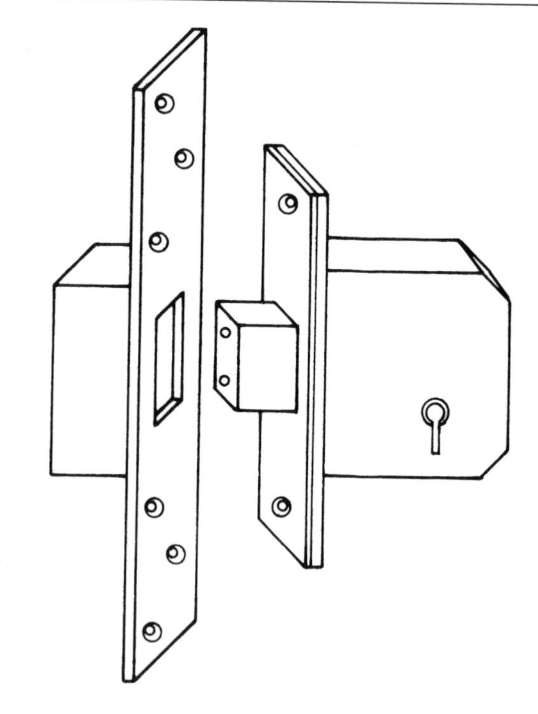

Figure 18.4 Mortice dead-lock and box-recessed striking plate

18.5 CYLINDER NIGHT LATCHES

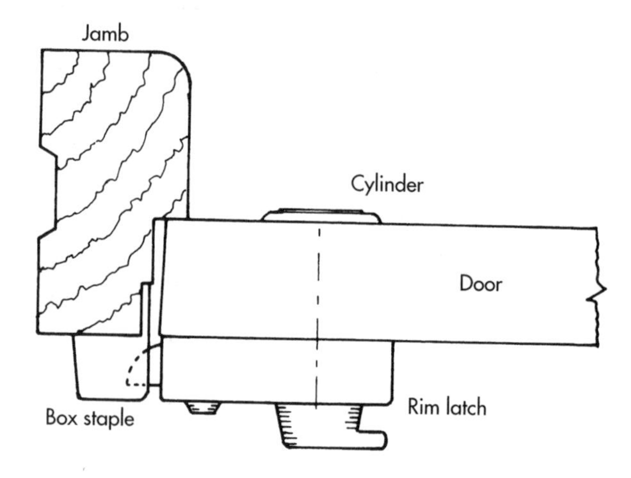

Figure 18.5(a) Part horizontal section through normal inward-opening door

Figure 18.5(a) and (b): This type of lock/latch is commonly used on front entrance doors and consists of the rim latch itself, as illustrated, with a turn-knob and a small sliding latch-button for holding the latch in an open or closed (locked) position, a cylinder with a bar that connects into the latch, a loose rose plate that provides a rim for the cylinder (or alternatively, a cylinder door pull), a back plate with connecting screws for

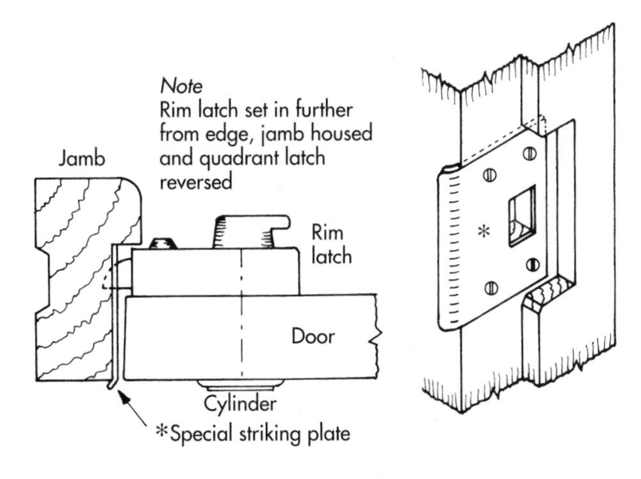

Figure 18.5(b) Uncommon outward-opening door

securing the cylinder to the door and a box staple (also referred to as a *keep*) for receiving the striking quadrant-shaped latch. These locks are obtainable in standard sizes or narrow sizes. The latter is sometimes required on glazed entrance doors with narrow stiles. A standard cylinder night latch requires a 32 mm ($1\frac{1}{4}$in) diameter hole to be drilled through the door at 61 mm in from the edge to the centre, to receive the cylinder. The narrow latch type requires the same size hole to be drilled, but at 40 mm in from the edge to the centre.

18.6 FITTING A LETTER PLATE

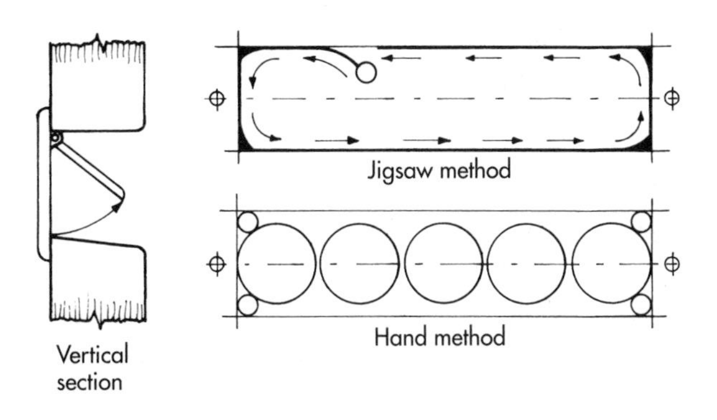

Figure 18.6 Fitting a letter plate

Figure 18.6: Measurements for the oblong aperture must be carefully taken from the letter plate and plotted on the outside *and* the inside face of the door. This can best be done by marking a level centre line, as illustrated, on each face, to use as a datum line for the other measurements. This is also the line on which to mark the critical position of the 6 mm diameter holes to be drilled for the connecting bolts. If confident of the accuracy of your marking out, these holes are best drilled halfway through from each side. Slight misalignments midway can be overcome by the reaming effect gained by using a Sandvik combination auger bit.

The aperture can be cut out easily with a good jigsaw. Note that when reaching each corner, the saw is worked back and forth a few times to create space for the blade to turn through 90°. When cut, the hole is then cleaned-up with a wide bevelled-edge chisel and/or a Surform file.

Alternatively, a line of large-diameter holes can be drilled, with smaller holes drilled at each end to make it easier to cut the end grain with a sharp bevelled-edge chisel. By using this hand method, once the ends are chopped through from each side, the remaining timber above and below the large holes pares out quite easily with a wide bevelled-edge chisel. The exposed arrises of the aperture should be removed with a chisel and/or glasspaper.

18.7 FITTING A MORTICE LOCK

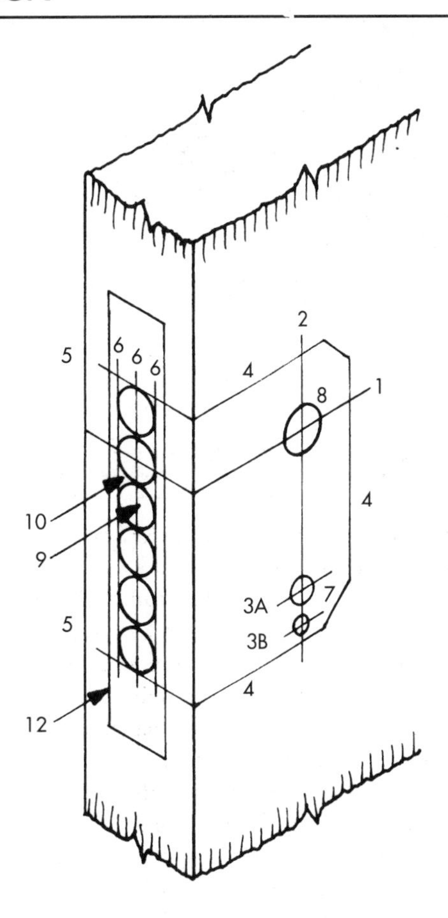

Figure 18.7 Fitting a mortice lock

Figure 18.7: The following technique for fitting a standard mortice lock can – with the omission of the spindle hole – be modified for fitting a mortice dead lock. The auger bits used can be power-driven or fitted into a ratchet brace. If power-driven, the drill must have a reversing facility and, ideally, a variable speed.

1. If no predetermined height exists, measure down half the door's height to mark the spindle level and square this around the door.
2. Measure the lock from the outer edge of the face to the centre of the spindle hole and mark this on the door on each face with a pencil line gauged from the blade-end of the combination mitre square. Make a slight allowance for the undershot edge of the door.
3. Measure the lock vertically from the centre of the spindle hole to the top-centre and bottom-centre of the keyhole and mark cross-lines A and B on both faces of the door.
4. Hold the lock against the door, sight through the spindle hole to line it up with the spindle cross-mark, then mark lightly around the outer edge of the lock.
5. Square this outline across the edge of the door.
6. Set up a marking gauge and mark the centre of the door thickness, then reset the gauge to be 8 mm from the centre and mark a line on each side of the first gauge line.
7. Drill a 10 mm diameter hole at 3A and a 6 mm diameter hole at 3B, preferably from each side of the door to avoid spelching out.
8. Drill a 16 mm diameter hole for the spindle, again from each side of the door.
9. As indicated, drill a series of 16 mm diameter holes close to each other for the lock-mortice. Use a depth gauge on the drill or bind masking tape around the auger bit to achieve a slightly oversize depth. Hold a flat rule or the blade of the combination square on the face of the door when drilling, to check on drill-alignment occasionally.
10. Clean out the mortice hole with, say, a 25 mm bevelled-edge chisel and a 16 mm firmer chisel. On mortices for deep locks, a so-called swan-neck chisel is sometimes necessary.
11. Complete the keyhole shape with a small chisel or a pad saw.
12. Try the lock in the mortice until, after easing, you achieve a slightly loose fit, then adjust it for central position edgewise and mark around the face plate with a sharp pencil. The edge-lines can be scored heavily with a marking gauge to reduce the risk of splitting the edges with a chisel, then chisel-chopping across the grain is carried out before hand-routering the face-plate depth, by judgement with the chisel held in a position similar to when it is being sharpened. Keep trying the lock until it fits slightly under the surface. Then screw into position.
13. Now close the door a few times until the latch marks the jamb's edge, or mark this position with a pencil, and square these marks across to relate to the striking plate's mortice hole. Position the striking plate onto the protruding latch and mark the plate's face where it protrudes past the face of the door. Now hold the striking plate against the marks on the jamb, with the edge-mark on the plate in line with the jamb's edge and mark around the plate with a sharp pencil.
14. Carefully chop out a shallow housing for the striking plate and try in position. Mark the outline for the latch and bolt mortice holes, remove the plate and drill/chop out for these. Fix the striking plate and try closing the door for easy latching and locking.

18.8 FITTING DOOR FURNITURE

When fitting the door furniture, which is usually done after the doors have been painted or sealed, or – better still – has been done previously and then removed and replaced after painting or sealing, care must be taken with the vertical and lateral positioning and screwing of the furniture. This is because the 6.35 × 6.35 mm square-sectioned spindle sits quite loosely in the lock's spindle bush and if strained over to one extreme or the other by ill-positioned screws in the door furniture, binding can occur which may cause the latch and the lever handles or knobs to stick in the levered or turned position, without the springs being able to effect a self-return action. To avoid this, always feel for the correct position by gently moving the furniture from left to right and up and down – and by settling in the middle of these extremes. To help with this, on lock lever furniture, the keyholes can be sighted through for alignment. After fixing, always check lever handles or knobs for a smooth, easy movement and a self-returning action.

19

Fixing Pipe Casings and Framed Ducts

19.1 INTRODUCTION

Occasionally, according to the design of a property, there is a need to conceal vertical and/or horizontal pipes to improve the appearance of the room(s) that they pass through. These rooms are usually the bathroom or kitchen. The pipes may be 110 mm diameter soil pipes, 40 mm diameter waste pipes, or various small-diameter copper supply pipes. If the supply pipes are fitted with stopcocks, provision must be made for their access when the pipes are to be concealed. Basically, the two arrangements for concealing pipes are pipe casings and framed ducts.

19.2 PIPE CASINGS

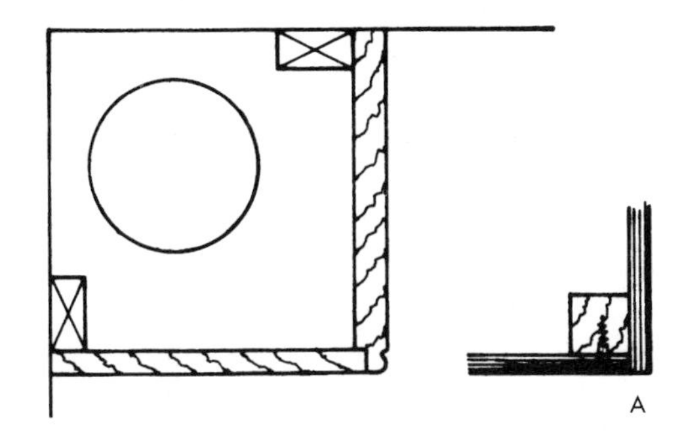

Figure 19.1 Horizontal section through traditional pipe casing

Figure 19.1: Traditionally, solid timber of about 225 mm prepared width was used in the construction of pipe casings. The side casing had a beaded and rebated edge to receive a thinner timber casing. Vertical battens of about 50 × 25 mm section were fixed to the finished wall surfaces, carefully positioned laterally to ensure that the completed casing fitted squarely in the corner and was square in itself. Any under-achievement in this respect showed up badly at ceiling and floor levels. When working out the lateral position of the battens in relation to the width of each casing, allowances had to be made if it was decided that the casings would need to be scribed to the walls. Modern pipe casings are of similar construction, but the timber casings have been replaced by plywood of usually 12 mm thickness, with the rebated edge in the form of a planted, vertical corner batten, as at detail A.

19.3 FRAMED DUCTS

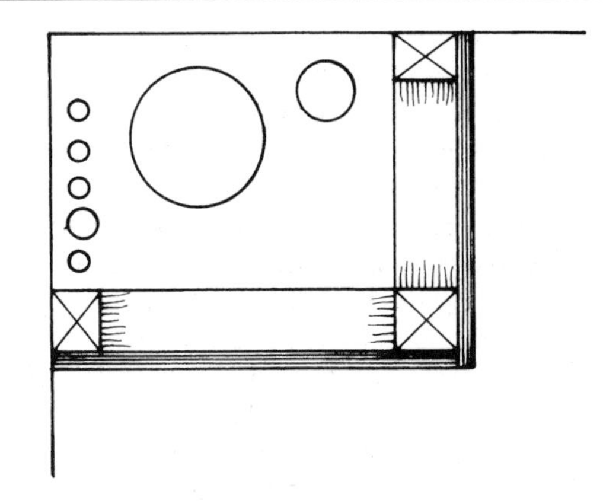

Figure 19.2 Horizontal section through modern framed duct

Figure 19.2: Framed ducts take over from ordinary plywood casings when there are a greater number of pipes to conceal, or the concealment involves a more complex arrangement of vertical and/or horizontal pipework. The sawn or prepared timber framing may be of 50 × 50 mm, 50 × 38 mm or 38 × 38 mm section. As illustrated, vertical or horizontal battens of 50 × 25 mm section are still used as before to establish the ducting's position and anchorage to the walls or – in cases of horizontal ducting – to the wall and the floor or the wall and the ceiling. The other framing consists of a longitudinal corner-member and cross-noggings spaced at 600 mm centres. The framing is built up *in situ*, like stud partitioning. The butt-jointed noggings are skew-nailed to the longitudinal battens and nailed through the corner member at the other end. When the skeletal framework is completed, it is then covered with 6 mm or 12 mm plywood. If future access to the pipes or stopcocks is required, the face panels should be neatly screwed.

Appendix: Glossary of Terms

The terms and other technical names listed here for explanation, are relevant to those used in this book only – not to the industry as a whole. For continuity, some terms are explained in the appropriate chapter and may or may not be repeated here.

Aggregate: Stone, flint and finer particles used in concrete.
Apron lining: A horizontal board, covering the rough-sawn vertical face of a trimmer or trimming joist in a stairwell.
Architrave: A plain or fancy moulding, mitred and fixed around the face-edges of door openings, to add a visual finish and to cover the joint between plaster(board) and door frame or lining.
Arris, arrises: The sharp corner edges on timber or other material.
Ashlaring or ashlars: Vertical timber studs fixed in an attic room from floor to rafters, to partition off the lower, acute angle of the roof slope.

Balusters: Lathe-turned wooden posts, fixed between the handrail and string capping or handrail and landing nosing, as part of the balustrade of a staircase.
Baluster sticks: As above, but only square posts with no turning.
Balustrade: The barrier or guarding at the open side of a staircase or landing, comprising newel posts, handrail and balusters (spindles).
Bare-faced tenon: A tenon with only one shoulder.
Bearer: A timber batten, usually ex. 50 × 25 mm, that supports a shelf.
Bearing: The point of support for a beam, lintel or joist.
Bed, bedding or bed joint: A controlled thickness of mortar – usually 10 mm – beneath timber plates, bricks and blocks.
Birdsmouth: A vee-shaped notch in timber, that is thought to represent a bird's mouth in appearance.
Bits: Parallel-shank and square-taper shank tools that fit a power drill and/or a carpenter's ratchet brace for drilling and countersinking.
Block partitions: Partition walls built of aerated insulation blocks, usually measuring 440 mm long × 215 mm high × 100 mm thick.
Bow: A segmental-shaped warp in the length of a board, springing from the wide face of the material.
Boxwood: A yellow-coloured hardwood with close, dense grain – still used in the manufacture of four-fold rules and chisel handles, to a limited extent.
Brace: 1. A diagonal support. 2. A tool for holding and revolving a variety of drill bits.
Brad head: The head of a nail (oval brad) or awl (bradawl) whose shape is scolloped from the round or oval to a flat point.
Bullnose step: A step at the bottom of a flight of steps, projecting past the newel post, with a quadrant-shaped (quarter of a circle) shaped end.
Burr: A sharp metal edge in the form of a lip, projecting from the true arris of the metal.
Butt-joint: A square side-to-side, end-to-end or end-to-side abutment in timber, without any overlapping.

Casement: Hinged or fixed sash windows in a casement frame.
Centimetre: One hundredth of a metre, i.e. ten millimetres (10 mm).
Chamfer: An equal bevel (45° × 45°) removed from the arrises of bearers or slatted shelves.
Chase, chased, chasing: Rough channels or grooves cut in walls or concrete floors to accommodate pipes, conduits or cables; or cut in the face of mortar beds to take the top-turned edge of apron flashings.
Chord: A reference to trussed-rafter rafters (top chord) and trussed-rafter ceiling-joist ties (bottom chord).
Chuck: The jaws of a drill or brace.
Cladding: The clothing of a structure in the form of a relatively thin outer skin, such as horizontal weather-boarding or tiles.
Cleats: Short boards or battens, usually fixed across the grain of other boards to give laminated support to the join.
Clench-nailed: Two pieces of cross-grained timber held together by nails with about 6–10 mm of projecting point bent over and flattened on the timber, in the direction of the grain for a visual finish, or across the grain for extra strength.
Coach bolt: This has a thread, nut and washer at one

end and a dome-shaped head and partly square shank at the other. The square portion of shank is hammered into the round hole to stop the bolt turning while being tightened.
Common brickwork: Rough brickwork or brickwork built with cheaper non-face bricks, to be plastered or covered.
Concave: Shaped like the inside of a sphere.
Concentric: Sharing the same centre point.
Conduit: A metal pipe for housing electrical cables; although nowadays plastic and fibre tubes are mostly used.
Convex: Shaped like the outside of a sphere.
Corbel or corbelling: A structural projection from the face of a wall in the form of stone, concrete or stepped brickwork, to act as a bearing for wall plates and purlins, etc.; straight or hooked metal corbel plates, being the forerunner of modern joist hangers, were used at about 1 m centres, projecting from the face of a wall, to support suspended wall plates.
Course: One rise of bricks or blocks laid in a row.
Cramp: 1. Sash cramp – a tightening device for holding framed timber components together under pressure, usually while being glued. 2. Frame cramp or tie – a galvanized steel bracket holding-device, fixed to the sides of frames and bedded in the mortar joint.
Cross-halving: A half-lap joint between crossed timbers.
Cup or cupping: Concave or convex distortion across the face of a board, usually caused by the board's face being tangential to the growth rings.

Datum: A fixed and reliable reference point from which all levels or measurements are taken, to avoid cumulative errors.
Decimetre: One tenth of a metre, i.e. 100 millimetres (100 mm).
Dihedral angle: The angle produced between two surfaces, or geometric planes, at the point where they meet. For example, two vertical surfaces meeting at right-angles to each other, produce a dihedral angle of 90°, but incline the surfaces from their vertical state, to represent a hip or valley formation, and the dihedral angle thus produced is different, according to the degree of inclination.
Door joint: The necessary gap of 2–3 mm around the edges of a door for opening clearances.
Dovetail key: The locking effect of a dovetail, or nails driven in to form a dovetail shape.
Dowel: A round wooden (usually hardwood) or metal pin.
Draw-bore pin: A front-tapered wooden dowel, driven into an offset hole drilled (separately) through a mortice and tenon joint, to pull up the shoulder-fit and permanently reinforce the joint.

Easing: 1. Removing shavings from an edge to achieve a better fit. 2. Concave and convex shaping of the top edges of stair strings (especially on the two right-angled wall strings containing tapered steps in a quarter-space turn).
Eaves: The lowest edge of a roof, which usually overhangs the structure from as little as the fascia-board thickness up to about 450 mm, where rainwater drainage is effected via a system of guttering and downpipes.
Eccentric point: The bent portion of a trammel-head pin, which causes the axis (centre) of the pin to move, when rotated, in an eccentric orbit, even though the pin-pointed position remains concentric. This allows fine adjustments to be made to the trammel distance without altering the trammel head itself.

Facework or face brickwork: Good quality face bricks, well-laid to give a finished appearance to the face of walls.
Fair-faced brickwork: Common brickwork, roughly pointed and bagged (rubbed) over with an old sack.
Fillet: A narrow strip of wood, rectangular or triangular in section, usually fixed between the angle of two surfaces, as a covermould.
Firring: Building up the edges of joists, with timber strips which may be parallel or wedge-shaped, to achieve a level, a sloping or a higher surface when boarded or covered in sheet material.
Flange: The bottom or top surface of a steel I beam or channel section.
Flashing: A lead or felt apron that covers various roof junctions.
Fletton: An extensively used common brick, named after a village near Peterborough and made from the clay of that neighbourhood.
Floating: see Rendering.
Flush: A flat surface, such as a flush door, or in the form of two or more components or pieces of timber being level with each other.

Gablet: The triangular end of a roof, known as a gablet when separated from the gable wall below, as in a gambrel roof.
Glue blocks: Short, triangular-shaped blocks, glued – and sometimes pinned – to the inside angles of steps in a wooden staircase and other joinery constructions.
Going: The horizontal distance, in the direction of flight, of one step or of all the steps (total going) of a staircase.
Grain: The cellular structure and arrangement of fibres, running lengthwise through the timber.
Green brickwork: Freshly laid or recently laid brickwork, not fully set.
Groove: A channel shape sunk into the face or edge of timber or other material.
Grounds: Sawn or planed (prepared) battens, which may be preservative-treated, used to create a true and receptive fixing surface.

Gullet: The lower area of the space between saw teeth.
Gusset plate: A triangular-shaped metal (or timber; usually plywood) joint connector.

Half-brick-thick wall: A stretcher-bond wall, where bricks are laid end-to-end only, in one thickness of brick.
Hardcore: Broken brick and hard rubble used as a substrata for concrete oversites.
Hardwood: A commercial description for the timber used in industry, which has been converted from broad-leaved, usually deciduous trees, belonging to a botanical group known as angiosperms. Occasionally, the term hardwood is contradictory to the actual density and weight of a particular species. For example, balsa wood is a hardwood which is of a lighter weight and density than most softwoods.
Heel: The back, lower portion of a saw or plane.
Hone: Sharpen.
Housing: A trench or groove usually cut across the timber.

Inner skin: The wall built on the dwelling-side of a cavity wall, usually constructed in blockwork.
Inner string or wall string: One of the two long, deep boards that house the steps at the side of a staircase, being on the side against the wall.
***In situ* concrete**: Concrete units or structures cast in their actual and final location, controlled by *in situ* formwork (timber shuttering).

Jamb: The name given to the side of a door frame or window frame.
Joists: Structural timbers that make up the skeleton framework in timber floors, ceilings and flat roofs.

Kerf: The cut made by a saw during its progress across the material.
Knots: Roots of a tree's branches, sliced through during timber conversion. Healthy-looking knots are known as *live* knots, and those with a black ring around them are likely to fall out and are known as *dead* knots.
Knotting: Shellac used for sealing knots (to stop them *bleeding* or exuding resin) prior to priming. Shellac is derived from an incrustation formed by lac insects on the trees in India and nearby regions.

Lag or lagged: Wrapped or covered with insulation material.
Landing nosing: The narrow, projecting board, equal in thickness and shape to the front-edge of a tread board, that is fixed on all top edges of the landing stairwell. It is often rebated on the underside to meet a reduced-thickness of floor material.
Lignum vitae: Dark brown, black-streaked hardwood with extremely close grain. It is very hard and dense, about twice the weight of British elm, grown in the West Indies and tropical America.
Lintel: A concrete or metal beam over door or window openings.

Mitre: Usually a 45° bisection of a right-angled formation of timber (or other material) members – but the bisection of angles other than 90° is still referred to as mitring.
Muzzle velocity: The speed of a nail in the barrel of a cartridge tool.

Newel posts: Plain or lathe-turned posts in a staircase, usually morticed and jointed to the outer string and the handrail tenons and attached to the floor at the lower end and to the landing trimmer at the other. The newel posts assist in creating good anchorage of the staircase at both ends, as well as providing stability and strength to the remainder of the balustrade.
Noggings: Short timber struts, usually between studs or joists and rafters.
Normal: The geometrical reference to a line or plane at right-angles to another, especially in the case of a line radiating from the centre of a circle, in relation to a right-angled tangent on the outside.
Nosing: The projecting front-edge of a tread board past the face of the riser, reckoned to be not more than the tread's thickness.

Open-riser stairs: Stairs without riser boards.
Outer skin: The wall built on the external side of a cavity wall.
Outer string: One of the two long, deep boards that house the steps at the open side of a staircase, away from the wall.
Oversite: An *in situ* concrete slab of 100 mm minimum thickness, laid over hardcore on the ground as part of the structure of this type of ground floor.

Paring: Chiselling – usually across the grain.
Pellets: Cork-shaped plugs for patching counterbored holes when involved in screwing and pelleting.
Perpends: Perpendicular cross-joints in brickwork or masonry.
Pilot hole: A small hole made with a twist drill or bradawl to take the wormed thread of a screw.
Pin or pinned: Fixed with wooden-dowel pins, but more commonly the reference is to fixing with nails or panel pins.
Pinch rod: A gauge batten for checking internal distances for parallel.
Pitch: 1. The angle of inclination to the horizontal of a roof or staircase. 2. Repetitive, equal spacing of the tips or points of saw teeth or other equally spaced objects.
Plant: Equipment.
Planted mould or stop: Separate mouldings or door stops, pinned or fixed by nails to the base material.
Plate: 1. A horizontal timber that holds the ends of vertical or inclined timbers in a state of alignment and framed

spacing, as in the case of roof wall-plates and stud-partition sill plates and head plates. 2. Metal components such as striking plates for latches and letter plates.
Plumb or plumbing: Checking or setting up work in a true, vertical position.
Plumb cut: The vertical face of a cut angle.
Pocket screwing: Screws which are angled or skew-screwed into shallow niches and shank holes drilled at an angle through the (usually) hidden face of the timber being fixed – an example being the top riser and nosing piece of a stair.
Precast concrete: Concrete units cast in special mould boxes in a factory or on site, but not cast in their actual and final location.
Primed: Painted with the first coat of paint (priming) after being knotted.
Profile: 1. A horizontal board attached to stakes or pegs driven in the ground, across the line of an intended foundation strip – one at each end, set well clear of the digging area, has saw cuts or nails in the top edge of the board to mark the foundation and wall positions. When initially digging or building, ranging lines are set up across the boards to establish the required positions. In the case of mechanical digging, a thin line of sand is trickled vertically beneath the lines to mark the trench position, then the lines are removed and reinstated later when the building work is to be started. 2. Any object or structure acting as a template in guiding the shape of something being made or built.

Quadrant: 1. A right-angled sector shape, equalling a quarter of a circle. 2. A small, wooden bead of this shape.

Rebate: A return or inverted right-angle removed from the edge of a piece of wood or other material.
Rendering and/or floating coats: Successive coats of coarse plaster built up to an even and true surface for skimming.
Resin bonded: This is usually a reference to the cross-laminates of plywood having been bonded (glued) with synthetic resins. According to the type of resin used, the plywood may be referred to as moisture resistant (MR), boil resistant (BR), or – better still – weather and boil proof (WBP).
Retaining wall: A wall built to retain high-level ground on a split-level site.
Reveals: The narrow, return edges or sides of an opening in a wall.
Rise: The vertical distance of one step or of all the steps (total rise) of a staircase.
Riser: The vertical face or board of a step.
Runners: Sawn-timber beams, used in formwork.

Sarking: Roof boarding or sheeting material and/or roofing felt.
Scribe; scribing: Techniques used in joinery and second-fixing carpentry for marking and fitting mouldings against mouldings, or straight timbers against irregular shapes or surfaces.
Seat cut: The horizontal face of a cut angle.
Set, setting: 1. The alternate side-bending of the tips of saw teeth. 2. The chemical setting action that brings initial hardening of glue, concrete, mortar or plaster. 3. A coat of finishing plaster (see Skimming).
Shank: The stem or shaft of a tool or screw.
Shank hole: A small hole made with a twist drill to take the stem or shank of a screw.
Sheathing: Close-boarding or sheet material such as plywood or Sterling OSB, fixed to vertical framing (studs) as a strengthening-skin and a base for cladding with weather-boarding or tiles.
Sherardized: Ironmongery (hardware), such as nails and screws, coated with zinc dust in a heated, revolving drum and achieving a penetrated coating, claimed to be more durable than galvanizing.
Shuttering: Temporary structures formed on site to contain fluid concrete until set to the required shape; shuttering is also known as *formwork*.
Skew-nailing: Nailing at an angle of about 30–45° to the nailed surface, through the sides of the timber, instead of squarely through the edge or face.
Skimming or setting coat: The fine finishing plaster, traditionally applied to ceilings and walls in a 3–5 mm thickness and trowelled to a smooth finish.
Soffit: The underside of a lintel, beam, ceiling, staircase or roof eaves' projection.
Softwood: A commercial description for the timber used in industry, which has been converted from needle-leaved, usually coniferous evergreen trees, belonging to a botanical group known as gymnosperms. Occasionally, the term softwood is contradictory to the actual density and weight of a particular species. For example, parana pine is a softwood that is quite heavy and dense, like most hardwoods.
Spall, spalling: A breaking or flaking away of the face material of concrete, brick or stone.
Span: 1. Clear span – the horizontal distance measured between the faces of two opposite supports. 2. Structural span – for design calculations, is measured between half the bearing-seating on one side to half the bearing on the other. 3. Roof span - measured in the direction of the ceiling joists, from the outer-edge of one wall plate to the outer-edge of the other.
Spindles: In carpentry and/or joinery terms, this is an alternative name for balusters and therefore refers to lathe-turned wooden posts, fixed between the handrail and string capping or handrail and landing nosing, as part of the balustrade of a staircase.
Spotting: Marking a trowel-line through a slither of trowelled mortar when setting out walls and partitions on concrete foundations or oversites.
Spring or sprung: Warping that can occur in timber after conversion and seasoning, producing a *sprung*,

cambered or segmental-shaped edge adjacent to the wide face of the material. Joists and rafters should be placed with the sprung edge uppermost.
Stretcher: 1. The temporary timber batten at the base of a door frame or lining that stretches the legs apart to the correct dimension until the fixing operation takes place. 2. The long face of a brick.
Strut: A timber prop, supporting a load vertically, horizontally or obliquely.
Stub tenon: A shortened tenon, usually morticed into its opposite member by only a half to two-thirds its potential size.
Stuck mould or rebate: Moulded shapes or rebates cut into the face of solid timber members.
Studs, studding, stud partitioning: Vertical timber posts.

Tamp, tamped, tamping: A term used in concreting, referring to the level surface being zig-zagged and tamped (compacted) with a levelling board. The tamping is effected by bumping the board up and down as it is moved across the surface.
Tang: A square-taper shape at the end of a round-shanked tool.
Tangent: This is a line that lays at right-angles to another line – known geometrically as a *normal* – that radiates from the centre of a circle.
Toe: The reference in carpentry to the front of a saw or plane.
TRADA: Timber Research and Development Association.
Tread: The horizontal face or board of a step.
Twist: 1. Warping that can occur in timber after seasoning and conversion, producing distortion in length to a spiral-like propeller-shape. 2. Distortion in a framed-up unit caused by one or more of the members being twisted, or by ill-formed corner joints.

Voussoir: A tapered brick in a gauged brick arch.

Warp: Distortion of converted timber, caused by changing moisture content (see Bow, Spring, Twist, Cup and Wind).
Web: The connecting membrane between the flanges of a steel I beam or channel section.
Wind, winding: These terms are the equivalent of Twist and Twisting. The expression *in wind* means twisted and *out of wind* means not twisted.

Index